Applications of Instrumental Methods for Food and Food By-Products Analysis

Applications of Instrumental Methods for Food and Food By-Products Analysis

Editor

Agata Górska

MDPI • Basel • Beijing • Wuhan • Barcelona • Belgrade • Manchester • Tokyo • Cluj • Tianjin

Editor
Agata Górska
Warsaw University of Life Sciences (WULS-SGGW)
Poland

Editorial Office
MDPI
St. Alban-Anlage 66
4052 Basel, Switzerland

This is a reprint of articles from the Special Issue published online in the open access journal *Applied Sciences* (ISSN 2076-3417) (available at: https://www.mdpi.com/journal/applsci/special_issues/Instrumental_Methods).

For citation purposes, cite each article independently as indicated on the article page online and as indicated below:

LastName, A.A.; LastName, B.B.; LastName, C.C. Article Title. *Journal Name* **Year**, *Volume Number*, Page Range.

ISBN 978-3-0365-4521-9 (Hbk)
ISBN 978-3-0365-4522-6 (PDF)

Contents

About the Editor

Agata Górska

Prof. Dr. Agata Górska became a Doctor of pharmaceutical sciences in 2004. She has worked as a professor of agricultural sciences in the discipline of food and nutrition technology at the Institute of Food Sciences in the Department of Chemistry since 2016. Her current research interests are mainly focused on the development of methods for the efficient extraction of bioactive compounds from waste products of the fruit industry and the characterization of the obtained fractions, with particular emphasis on lipid fractions. Her research aims to obtain valuable substances from waste products in order to reuse them in the production of food and cosmetics, which is in agreement with the principles of the circular economy policy.

applied sciences

MDPI

Editorial

Special Issue on Application of Instrumental Methods for Food and Food By-Products Analysis

Agata Górska

Department of Chemistry, Institute of Food Sciences, Warsaw University of Life Sciences, 159c Nowoursynowska Street, 02-776 Warsaw, Poland; agata_gorska@sggw.edu.pl; Tel.: +48-(22)-593-7613

Citation: Górska, A. Special Issue on Application of Instrumental Methods for Food and Food By-Products Analysis. *Appl. Sci.* **2022**, *12*, 3888. https://doi.org/10.3390/app12083888

Received: 30 March 2022
Accepted: 8 April 2022
Published: 12 April 2022

Publisher's Note: MDPI stays neutral with regard to jurisdictional claims in published maps and institutional affiliations.

1. Introduction

The application of various analytical procedures and methods determining the properties and safety of food and food constituents is a particularly important topic when dealing with food and food by-product analyses. Advanced analytical tools are of great importance for food quality determination, including the chemical and physicochemical characteristics, thermal properties and stability of food products, and recently also the by-products of the food industry. Thanks to the use of instrumental methods it is also possible to investigate innovative, newly formulated food products and technologies.

Taking the aforementioned reasons into account, this Special Issue aims to collect information and results dealing with procedures, instrumental analytical techniques and methods that are applied to study food and food processing by-products. A discussion of the application of advanced methods in food research to detect and characterize specific food components of significance to food science and technology, such as lipids, proteins, and carbohydrates seems to be of significant importance.

In total, there were 14 manuscripts submitted to this Special Issue, 13 of which were published. The articles belong to two main groups: (i) instrumental methods used in food analysis; (ii) innovative procedures enabling bioactive compounds and oil extraction from by-products.

2. Instrumental Methods Used in Food Analysis

Three of the published manuscripts are concerned with an analysis of oil and fat fraction by using advanced techniques.

The article titled "Phytochemical Profile of Eight Categories of Functional Edible Oils: A Metabolomic Approach Based on Chromatography Coupled with Mass Spectrometry" published by Carmen Socaciu and coauthors [1] aimed to define the influence of botanical origins on the phytochemical components of 30 edible oils from eight categories and to define their specific identity biomarkers. Advanced techniques based on gas and liquid chromatography coupled with diode array, mass spectrometry, or fluorescence detection were very helpful in the study of fatty acids, volatiles, carotenoids, tocopherols, and phenolic components. Taking into account the obtained results, the specific recognition biomarkers were defined and a so called "identity card" was proposed for each category of oil. The results of investigations demonstrated that the application of instrumental methods combined with chemometric models can be helpful to determine the oils' botanical origin based on the studies of qualitative and quantitative differences between samples. In the study of Wirkowska M. and coauthors [2], chromatographic and thermal methods were applied to characterize oils isolated from quinoa and amaranth seeds. The oxidative stability was determined by using pressure differential scanning calorimetry and the Rancimat test and the kinetic parameters of the oxidation process were determined using differential scanning calorimetry. Additionally, differential scanning calorimetry was applied to define the melting characteristics of oils and gas chromatography was employed for an analysis of the fatty acid composition and their distribution in triacyglycerols. Based on the obtained results, the authors stated that both types of seeds can be a good source of unsaturated fatty acids. Taking into account the values of the induction time of oxidation, it can be suggested that quinoa oil presents worse resistance to oxidation than amaranth oil. The melting characteristics of the oils confirmed the presence of low-melting triacylglycerol fractions with unsaturated fatty acids. The authors concluded that the obtained results of the kinetics of the oxidation reaction can be useful to predict the oxidation process under various conditions and may be helpful when assessing the oxidation rate of oils from other pseudocereals. In the manuscript titled "Lipid Fraction Properties of Homemade Raw Cat Foods and Selected Commercial Cat Foods" published by Górska A. and coauthors [3], lipid fractions from five self-prepared and seven commercial cat foods were characterized using gas chromatography in order to define the composition of fatty acids and the pressurized differential scanning calorimetry technique was used to determine oxidative stability. Based on the obtained results, the authors stated that self-prepared cat foods contained a high level of essential fatty acids but showed low oxidative stability, especially for those with significant amounts of polyunsaturated fatty acids. It was found that the omega-6 to omega-3 fatty acids ratio was beneficial, despite the low amount of essential fatty acids. Furthemore, in the case of fats extracted from commercial cat foods a longer induction time was determined in comparison to fats isolated from self-prepared samples, indicating higher oxidative stability of fats. The authors also pointed out that the distribution of fatty acids in the triacylglycerol structure for cat food was determined in their article for the first time.

Four of published articles were focused on an investigation of new products and the impact of applied processing technologies on food-sample properties by using advanced analytical tools.

In the article titled "The Influence of a Chocolate Coating on the State Diagrams and Thermal Behaviour of Freeze-Dried Strawberries" authored by Ostrowska-Ligęza E., Szulc K., Jakubczyk E., Dolatowska-Żebrowska K., Wirkowska-Wojdyła M., Bryś J. and Górska A. [4], DSC curves, sorption isotherms, glass transition temperature and state diagrams of freeze-dried strawberries and dark and milk chocolate coated freeze-dried strawberries were determined. The obtained results allowed them to conclude that the shape and course of the sorption isotherms of the freeze-dried strawberries and dark and milk chocolate-coated strawberries were influenced by the method of snack preparation. It is worth mentioning that after coating with milk or dark chocolate changes in the hygroscopicity tendency of the freeze-dried strawberries, a significant reduction was observed. Such behaviour can consequently result in improving the shelf life of a final product. In the study of Jakubczyk E. and coauthors [5], the influence of the addition of apple puree and maltodextrin to agar sol on the sorption properties and structure of the dried gel was determined. Additionally, different drying methods were used to determine their effect on the sorption behaviour of apple puree gels. Among the analyzed samples, the air-dried apple puree gels were considered as producing the lowest hygroscopicity and the highest stability at room conditions. In addition, changes in sample composition, namely a decrease in the amount of apple puree from 40 to 25% in dried gels, improved the sorption properties. Based on the obtained results, the authors stated that it is possible to obtain new products with a tailored structure and sorption properties by designing a composition of products and via the application of certain drying methods. In the article titled "Sous Vide Cooking Effects on Physicochemical, Microbiological and Sensory Characteristics of Pork Loin", Kurp L. and coauthors [6] aimed to study the characteristics (cooking loss, instrumentally measured colour and texture, microbiological quality and sensory properties) of pork loin and the influence of different temperature-time combinations applied in sous vide pork cooking. Based on the obtained results, it was concluded that in the case of cooking at 60 or 65 °C for 4 h, pork loin showed the most attractive and acceptable sensory traits. Texture attributes were thought to be of most importance for pork perception. Sous vide processing of meat at 60 °C/4 h caused less cooking loss than at 65 °C/4 h, however, this was not confirmed for moisture content and sensorially assessed juiciness. Authors showed that the applied parameters of sous vide cooking were sufficient to reduce the microflora in the pork loin to the level safe for consumption. In the next article published by Bingman M. and coauthors [7], the profile of volatile organic compounds from pressed apple juice, processed through fermentation and dry hopping was monitored by Headspace-Solid Phase Microextraction-Gas Chromatography-Mass Spectrometry analysis in order to define the process of the occurrence of aroma compounds in cider production. It is worth mentioning that 89 volatile organic compounds, such as higher alcohols, acetate esters, ethyl esters, other esters, aldehydes, ketones, volatile acids, and terpenes were detected throughout processing. The authors reported that both the variety of hops and the time taken to complete key processing steps significantly influence the profile and concentration of aroma-important volatile organic compounds in dry-hopped cider.

In this Special Issue, a group of articles are provided that aim to study certain food ingredients, components and additives that can affect the quality of the final products.

In the article titled "The Effect of Essential Oils on the Survival of Bifidobacterium in In Vitro Conditions and in Fermented Cream", Kozłowska and coauthors [8] performed an in vitro study investigating how the viability of selected Bifidobacterium strains is influenced by the presence of essential oils extracted from clove buds, lemon peels, and juniper berries. The authors also studied, using cream samples during fermentation and after storage, the effect of the addition of selected essential oils on the viability of the Bifidobacterium strain. The solid-phase microextraction (SPME)-GC–time-of-flight (TOF)-MS method was used to determine the volatile aroma compound profiles of sour cream samples with essential oils. In their in vitro study, the authors found that the studied strains of Bifidobacterium were sensitive to the influence of the selected essential oils. Among the tested strains of the genus Bifidobacterium, B. animalis subsp., lactis Bb-12 showed the most sensitive behavior when essential oils of clove and juniper were applied. Additionally, there was no influence of the concentration and type of essential oils on the number of cells of this strain in the cream samples after fermentation and after 21-day storage. The authors pointed out that the tested essential oils could be added to dairy products as natural substances, thereby enriching sensory attributes and health-promoting properties. In the study of Kasapidou E. and coauthors [9], the possibility of near infrared reflectance spectroscopy application for the estimation of the chemical composition of traditional sausages was tested. The obtained results of the calibration model revealed that near infrared reflectance spectroscopy can be useful in the traditional village-style sausages analysis including in terms of fat content, with very good precision. Moisture and protein content can be assessed with the application of this method with good accuracy. Additionally, the authors concluded that the external validation confirmed the ability of near infrared reflectance spectroscopy to predict the chemical composition of sausages and that the method can be applied as a screening technique when a high but not an absolute level of accuracy is required. Methods and systems for determining the lycopene content in fruits were presented by Villaseñor-Aguilar M.J. and coauthors [10]. The authors pointed out that among the optical systems focused on the estimation and identification of lycopene are high-performance liquid chromatography, the colorimeter, infrared near NIR spectroscopy, UV-VIS spectroscopy, Raman spectroscopy, and the systems of multispectral imaging and hyperspectral imaging. Based on the literature analysis, it was summarized that high-performance liquid chromatography and spectrophotometry methods can provide more efficient results, but they present some limitations due to long and complicated measurement procedures. On the other hand, multispectral, hyperspectral, and colorimeter imaging techniques are characterized as fast, non-contact and suitable for online applications but still require further study regarding their accuracy. Other methods such as NIR spectroscopy, UV-Vis spectroscopy, and Raman spectroscopy have proven to be moderately reliable with respect to high-performance liquid chromatography. According to authors, it is worth mentioning that the introduction of artificial intelligence algorithms, the internet of things, parallel processing hardware, and the reduction of equipment costs are areas of future study that can lead to the early translation of laboratory results to field applications.

The development of effective methods that can be used in the effective separation of active compound from certain mediums is an important topic in food technology. This problem was discussed by Małajowicz J. and coauthors [11] in their study verifying the possibility of secreting gamma-decalactone separation from biotransformation media, in which a lactone was synthesized from castor oil via Yarrowia lipolytica KKP379 yeast. The effectiveness of the following three techniques: liquid–liquid extraction, hydrodistillation and adsorption was compared with regard to fragrance recovery. Based on the obtained results, and by taking into account the selectivity of the process, its efficiency, and the speed of execution, adsorption on Amberlite XAD-4 can be recommended. In the context of the purity of the released gamma-decalactone, the hydrodistillation process seems to be more advantageous. According to the authors, it is worth mentioning that the effect of many factors should be considered before choosing a proper separation method. It should be concluded that the development of an effective method of gamma-decalactone separation from biotransformation media seems to be important in the context of the attempts to apply a biotechnological synthesis of peach lactone in industry.

In the article published by Wierzchowska K. and coauthors [12] they attempted to study the efficiency of microbial oil production, fatty acid composition and the growth of Y. lipolytica yeast, influenced by inorganic phosphorus and nitrogen sources limitations. The authors highlighted the importance of nitrogen limitation in culture media. It is worth mentioning that further phosphorus limitation may consequently lead to the higher efficiency of microbial lipid biosynthesis. According to the authors, the phosphorus and nitrogen level as well as the ratio should be taken into account as one of the important factors that influences microbial lipid production and satisfactory biomass yield. Interestingly, the authors pointed out that such approaches seem to be practically essential when the possibility of the use of oily waste as a substrate in microbial culture is discussed.

3. Alternative Methods of Bioactive Compounds and Oil Extraction from By-Products

By-products of the food industry can be considered as a source of valuable bioactive substances and oils with the potential of reuse, so it is of urgent need to search for novel extraction methods that may help to obtain extracts or oils with improved properties. Particularly, Piasecka and coauthors [13], in the article titled "Alternative Methods of Bioactive Compounds and Oils Extraction from Berry Fruit By-Products—A Review" summarized possible extraction methods, including alternative, innovative techniques and their impact on the composition of extracts and oils obtained from berry fruit by-products. In the article, it was pointed out that application of conventional solvent extraction techniques may present the disadvantage of being insufficient in the achievement of high polyphenol or lipid fraction yields and in the selective isolation of characteristic compounds. The authors further indicated the significance of alternative extraction methods such as ultrasound-assisted extraction, pulsed electric field-assisted extraction, microwave-assisted extraction and supercritical fluid extraction as possible ways by which to improve the efficiency of the isolation of bioactive compounds and oils from berry fruit by-products. It is of great importance that these non-conventional techniques are classified as green extraction methods due to the lower energy, solvent volume and time consumption. In the article, the authors pointed out major differences between the applied methods with respect to their usefulness. Additionally, important factors such as the selection of solvents and solid–liquid ratios used in the extraction process were discussed. The time required for the extraction process is another important factor that was different across the studied methods. By applying alternative extraction methods such as ultrasound-assisted extraction or microwave-assisted extraction, the time required for extraction processes can be reduced in comparison to traditional solid–liquid extraction.

Funding: This research received no external funding.

Acknowledgments: I would like to thank all the authors and reviewers for their valuable contributions to this Special Issue.

Conflicts of Interest: The authors declare no conflict of interest.

Short Biography of Authors

Prof. Dr. Agata Górska Doctor in 2004 in pharmaceutical sciences. Full professor of agricultural sciences, in the discipline of food and nutrition technology 2016, at the Institute of Food Sciences, at the Department of Chemistry. Current research interests are mainly focused on the development of methods for the efficient extraction of bioactive compounds from the waste products of the fruit industry and the characterization of the obtained compounds, with particular emphasis on the lipid fraction. The research aims to obtain valuable substances from waste products in order to reuse them in the production of food and cosmetics, which is in agreement with the principles of the circular economy policy.

References

1. Socaciu, C.; Dulf, F.; Socaci, S.; Ranga, F.; Bunea, A.; Fetea, F.; Pintea, A. Phytochemical profile of eight categories of functional edible oils: A metabolomic approach based on chromatography coupled with mass spectrometry. *Appl. Sci.* **2022**, *12*, 1933. [CrossRef]
2. Wirkowska-Wojdyła, M.; Ostrowska-Ligęza, E.; Górska, A.; Bryś, J. Application of Chromatographic and Thermal Methods to Study Fatty Acids Composition and Positional Distribution, Oxidation Kinetic Parameters and Melting Profile as Important Factors Characterizing Amaranth and Quinoa Oils. *Appl. Sci.* **2022**, *12*, 2166. [CrossRef]
3. Górska, A.; Mańko-Jurkowska, D.; Bryś, J.; Górska, A. Lipid Fraction Properties of Homemade Raw Cat Foods and Selected Commercial Cat Foods. *Appl. Sci.* **2021**, *11*, 10905. [CrossRef]
4. Ostrowska-Ligęza, E.; Szulc, K.; Jakubczyk, E.; Dolatowska-Zebrowska, K.; Wirkowska-Wojdyła, M.; Bryś, J.; Górska, A. The Influence of a Chocolate Coating on the State Diagrams and Thermal Behaviour of Freeze-Dried Strawberries. *Appl. Sci.* **2022**, *12*, 1342. [CrossRef]
5. Jakubczyk, E.; Kamińska-Dwórznicka, A.; Ostrowska-Ligęza, E.; Górska, A.; Wirkowska-Wojdyła, M.; Mańko-Jurkowska, D.; Górska, A.; Bryś, J. Application of Different Compositions of Apple Puree Gels and Drying Methods to Fabricate Snacks of Modified Structure, Storage Stability and Hygroscopicity. *Appl. Sci.* **2021**, *11*, 10286. [CrossRef]
6. Kurp, L.; Danowska-Oziewicz, M.; Kłębukowska, L. Sous Vide Cooking Effects on Physicochemical, Microbiological and Sensory Characteristics of Pork Loin. *Appl. Sci.* **2022**, *12*, 2365. [CrossRef]
7. Bingman, M.T.; Hinkley, J.L.; Bradley, C.P., II; Cole, C.A. Aroma Profiles of Dry-Hopped Ciders Produced with Citra, Galaxy, and Mosaic Hops. *Appl. Sci.* **2022**, *12*, 310. [CrossRef]
8. Kozłowska, M.; Ziarno, M.; Rudzińska, M.; Majcher, M.; Małajowicz, J.; Michewicz, K. The Effect of Essential Oils on the Survival of Bifidobacterium in In Vitro Conditions and in Fermented Cream. *Appl. Sci.* **2022**, *12*, 1067. [CrossRef]
9. Kasapidou, E.; Papadopoulos, V.; Mitlianga, P. Feasibility of Application of Near Infrared Reflectance (NIR) Spectroscopy for the Prediction Chemical of the Composition of Traditional Sausages. *Appl. Sci.* **2021**, *11*, 11282. [CrossRef]
10. Villaseñor-Aguilar, M.-J.; Padilla-Medina, J.-A.; Botello-Álvarez, J.-E.; Bravo-Sánchez, M.-G.; Prado-Olivares, J.; Espinosa-Calderon, A.; Barranco-Gutiérrez, A.-I. Current Status of Optical Systems for Measuring Lycopene Content in Fruits: Review. *Appl. Sci.* **2021**, *11*, 9332. [CrossRef]
11. Małajowicz, J.; Górska, A.; Bryś, J.; Ostrowska-Ligęza, E.; Wirkowska-Wojdyła, M. Attempt to Develop an Effective Method for the Separation of Gamma-Decalactone from Biotransformation Medium. *Appl. Sci.* **2022**, *12*, 2084. [CrossRef]
12. Wierzchowska, K.; Zieniuk, B.; Nowak, D.; Fabiszewska, A. Phosphorus and Nitrogen Limitation as a Part of the Strategy to Stimulate Microbial Lipid Biosynthesis. *Appl. Sci.* **2021**, *11*, 11819. [CrossRef]
13. Piasecka, I.; Wiktor, A.; Górska, A. Alternative Methods of Bioactive Compounds and Oils Extraction from Berry Fruit By-Products—A Review. *Appl. Sci.* **2022**, *12*, 1734. [CrossRef]

applied sciences

Review

Alternative Methods of Bioactive Compounds and Oils Extraction from Berry Fruit By-Products—A Review

Iga Piasecka [1,*], Artur Wiktor [2] and Agata Górska [1]

[1] Department of Chemistry, Institute of Food Sciences, Warsaw University of Life Sciences, 159c Nowoursynowska Street, 02-776 Warsaw, Poland; agata_gorska@sggw.edu.pl
[2] Department of Food Engineering and Process Management, Institute of Food Sciences, Warsaw University of Life Sciences, 159c Nowoursynowska Street, 02-776 Warsaw, Poland; artur_wiktor@sggw.edu.pl
* Correspondence: iga_piasecka@sggw.edu.pl; Tel.: +48-22-5937607

Abstract: Berry fruit by-products are a source of polyphenol compounds and highly nutritious oils and can be reused to fulfill the requirements of the circular economy model. One of the methods of obtaining polyphenol-rich extracts or oils is extraction. Applying conventional solvent extraction techniques may be insufficient to reach high polyphenol or lipid fraction yields and selectivity of specific compounds. Alternative extraction methods, mainly ultrasound-assisted extraction, pulsed electric field-assisted extraction, microwave-assisted extraction and supercritical fluid extraction, are ways to improve the efficiency of the isolation of bioactive compounds or oils from berry fruit by-products. Additionally, non-conventional techniques are considered as green extraction methods, as they consume less energy, solvent volume and time. The aim of this review is to summarize the studies on alternative extraction methods and their relationship to the composition of extracts or oils obtained from berry waste products.

Keywords: berry fruit by-products; alternative extraction methods; waste management; green extraction; PEF-assisted extraction; ultrasound-assisted extraction

Citation: Piasecka, I.; Wiktor, A.; Górska, A. Alternative Methods of Bioactive Compounds and Oils Extraction from Berry Fruit By-Products—A Review. *Appl. Sci.* **2022**, *12*, 1734. https://doi.org/10.3390/app12031734

Academic Editor: Dino Musmarra

Received: 15 January 2022
Accepted: 6 February 2022
Published: 8 February 2022

Publisher's Note: MDPI stays neutral with regard to jurisdictional claims in published maps and institutional affiliations.

1. Introduction

From a botanical point of view, berry fruit is an artificial fruit classification. However, it is the term that is commonly used to refer to the the group of fruits from *Rubus* (raspberry, blackberry), *Ribes* (currants, gooseberry), *Aronia* (chokeberry), *Vaccinium* (cranberry, blueberry) and *Fragaria* (strawberry) genera. According to FAO statistics, berry fruit production reaches an amount of over 12.2 million tons worldwide [1]. The latest data, including area and quantity of production, is presented in Figure 1.

Berries, as well as other fruits, may be consumed raw or can be processed to such products as, e.g., frozen, dried or canned fruits, juices. Those products may be further processed as well [2]. However, every step of processing and transport may generate losses reaching, according to the FAO, even 45% of fruits and vegetables produced [3]. Such significant percentages of loss and waste not only have economic consequences, but also affect the natural environment, especially water use [4]. In order to obtain economic and environmental benefits it is widely recommended to apply circular economy model concepts in the food production chain. The model includes the further use of by-products as a way of managing and minimizing production of wastes when they are still a source of bioactive, highly nutritive compounds [5]. Berries are mainly processed to juices and concentrates. The technological scheme of juice production leads to pomace generation—a major fruit processing by-product, which contains stem cells, skins and seeds of fruits. The aim of this review is to present and systematize possible methods of bioactive compounds and oils extraction from chosen berry fruit pomaces and the impact of certain extraction methods on the quality and composition of those extracts.

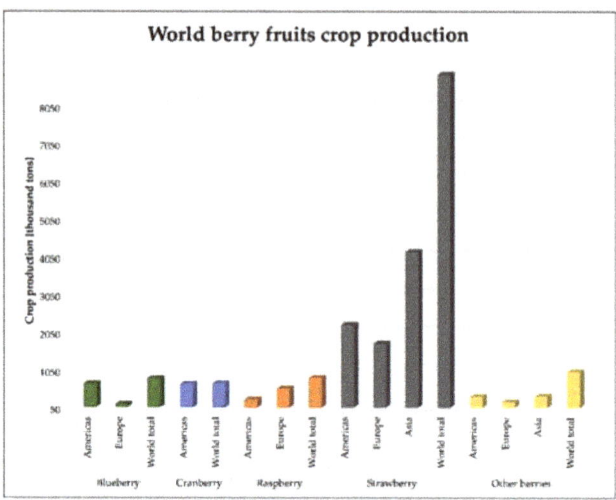

Figure 1. World berry fruits crop production (2019).

2. Composition of Berry Pomaces

2.1. General Information

Contents of pomaces depend on the berry species, but as they are rich in the seeds and skins of fruits, dietary fiber is a dominant component of pomaces. For instance, chokeberry pomace powder consists of ca. 3.61% of fat, 5.97% of protein, 28.8% of carbohydrates and 57.8% [6] to 59.5% [7] of fiber (mostly insoluble fraction). McDougall and Beames [8] studied the composition of raspberry pomace and the composition of the studied by-product was described as follows: 11.1% of fat, 10.0% of protein and 59.5% dietary fiber. In a different study conducted by Górnaś et al. [9], the concentrations of constituents in raspberry pomace were: 9.1%, 8.7% and 54.2% for fat, protein and dietary fiber, respectively. In the same study, the composition of strawberry and blackcurrant pomaces was determined. In strawberry pomace, concentrations of nutrients were at levels 3.4%, 9.2% and 33.9% and in blackcurrant pomace at levels 0.7%, 6.9% and 38.5% for fat, protein and dietary fiber, respectively. Based on the studies conducted by Reißner et al. [7], it can be stated that the physicochemical properties of currant pomace depend on the color group of the berry. Blackcurrant (*Ribes nigrum*) pomace powder was found to contain about 61.0% of seeds. The percentage contents of the nutritional components of blackcurrant pomace were as follows: 20.21% fat, 15.71% protein, 2.20% carbohydrates and 59.13% fiber, with a predominance of insoluble fiber. Redcurrant (*Ribes rubrum*) pomace consisted of 40.4% seeds and, based on the obtained results, it can be treated as a source of 14.23% fat, 11.76% protein, 12.65% carbohydrates and 58.1% fiber, mostly insoluble fiber. Gooseberry pomace powder in turn consisted of 34.2% seeds, 10.93% fat, 12.40% protein and 56.6% dietary fiber, over 87% of which was an insoluble fraction.

2.2. Polyphenols

2.2.1. Role of Polyphenols

Polyphenols belong to a group of secondary metabolites present in plant-derived food. They are the most common antioxidants in the human diet and consist of different compounds; however, their classification is not strict. Generally, the term 'polyphenols' refers to flavonoids (with the subgroups: anthocyanins, flavanols, flavanones, flavones, flavonols and isoflavonoids), tannins, stilbenes and phenolic acids and their derivatives [10].

Although polyphenols are considered as non-nutritive compounds, they play a role in disease prevention and help to improve health due to their ability to neutralize free radicals [11]. There are a number of results of meta-analyses available concerning the influence of polyphenols on human functions. They may have potential in improving performance in groups of healthy humans [12], stimulating the growth of health-promoting species but inhibiting development of pathogenic organisms in gut microbiota [13]. Supplementation of polyphenolic compounds can be helpful in inflammatory bowel disease therapy [14], brain functions [15], lipid profile and inflammation status [16] improvement. In addition, an association between anthocyanins intake and lowered risk of hypertension [17] and cancer [18] was found.

2.2.2. Bioavailability of Polyphenols

However, there are limitations related to the bioavailability of some polyphenols and their stability during processing or storage. There are several factors that may affect the stability of phenolic compounds, for instance, pH, temperature, interactions with other food components, access to light and oxygen and metal ions presence and abundance [19]. Polyphenols tend to be more stable in acidic than in alkaline conditions; also, storing or processing foods at high temperatures leads to decreases in polyphenol content [20–22]. There are some studies which indicate that increased temperature may result in the appearance of other polyphenols in heated material as compared to unheated samples. This might be an effect of polymerization or polyphenol release from certain components [23].

2.2.3. Polyphenol Content in Berry Pomaces and Their Antioxidant Activity

Industrial chokeberry pomace consists of a solid number of bioactive compounds. Total polyphenol content (TPC) of chokeberry pomace reaches 5.5 g/100 g dm (dry mass), expressed as catechin monohydrate, determined using the Folin–Ciocalteu method. The main groups of polyphenols were found to be anthocyanins at 1.80 g/100 g dm, followed by phenolic acids at 0.31 g/100 g dm and flavonols at 0.184 g/100 g dm. AA (antioxidant activity) was measured as 1111 μmol FE/g in the ferric reducing antioxidant power (FRAP) assay [6]. Consistent results were obtained in a composition analysis of a store-bought product which contained 100% chokeberry pomace. TPC measured using the Folin–Ciocalteu method reached 4233 mg GAE/100 g dm and anthocyanins content determined by the pH differential method was 1165 mg CGE/100 g dm. AA, measured in a FRAP assay amounted to 47.38 mmol FE/100 g dm and was measured in a DPPH assay as 131.06 mmol TE/100 g dm [24]. The main phenolic compounds in chokeberry pomace, detected using HPLC, were polymeric procyanidins with a concentration of 9586 mg/100 g dm, which was two-fold higher than the result for fresh berries and almost seven-fold higher than that for juice [25].

The TPC of industrially obtained seedless blackcurrant pomace, measured using the Folin–Ciocalteu method, was in a range of 1855.5–2241.6 mg EE/100 g pomace. Specific polyphenol composition determined using HPLC indicated that anthocyanins are the dominant components, reaching values of 344.6–1046.1 mg/100 g pomace and depending on the year of fruit harvest. AA values determined in a DPPH assay ranged from 93.3–126.5 μmol TE/g pomace [26]. A comprehensive study of redcurrant, raspberry and blackberry pomaces conducted by Jara-Palacios et al. [27] showed that redcurrant pomace was characterized by the highest values of TPC, as determined by the Folin–Ciocolteau method, which were equal to 3446.59 mg GAE/100 g dm, followed by 2014.66 mg GAE/100 g dm for raspberry pomace and 1699.62 mg GAE/100 g dm for blackberry pomace. Anthocyanin concentrations, determined in a HPLC/MS analysis, were similar for all pomaces and ranged from 149.91 mg/100 g dm for redcurrant pomace to 188.05 mg/100 g dm for raspberry pomace. As AA was correlated with TPC, redcurrant was characterized by the highest AA (tested using ABTS), 60.83 mmol TE/100 g dm, while lower AA values for raspberry and blackberry were observed, these being, respectively, 29.75 and 22.54 mmol TE/100 g dm.

TPC, as determined by the Folin–Ciocalteu method, in blackberry pomace extract obtained from wild fruits ranged from 48.28–50.16 mg GAE/g dm and from cultivated fruits ranged from 26.30–35.40 mg GAE/g dm, which indicates that phenolic compound concentrations are higher among wild blackberries [28]. The phenolic composition of raspberry pomace was described. TPC, measured using HPLC, was determined as 238.36 mg/100 g dm.

Anthocyanins were the dominant phenolic compounds, reaching nearly 83% of the TPC, followed by ellagic acid and flavanols [29].

2.2.4. Applications of Polyphenolic Extracts

As berry pomaces are rich in polyphenols, they may yield polyphenol-rich extracts. The use of extracts obtained from berry pomaces is gaining the interest of researchers. There are papers reporting applying chokeberry pomace extract to enrich the composition of apple juice. Fortified products represent increased acidity, levels of vitamin C, TPC, total flavonoids, total anthocyanins and higher AA, as determined by the ABTS method [30]. Raspberry pomace extract can be recognized as an antioxidative but also as an antibacterial ingredient [31]. Extracts from chokeberry pomace were considered for use as an ingredient of chitosan-based packaging films. Adding the extract in film formulation resulted in its decreased solubility and has a possible application as a pH-indicating film due to the high stability in acidic conditions of anthocyanins [32]. Berry pomace extracts may also be applied as natural, antioxidant colorants [27].

2.3. Lipid Fraction

As pomaces contain seeds, they are a source of lipophilic components. Oils obtained from berry by-products using traditional methods (solid–liquid extraction, maceration, cold-pressing) vary in composition and concentration of fatty acids, phospholipids, tocopherols, sterols and other bioactive compounds, e.g., carotenoids. They may also have different oxidative and thermal stabilities or shelf lives. This variation is caused by species differences in fruits [33], fruit growing conditions [34] and conditions of storage of material and oil [35].

2.3.1. Fatty Acid Composition

Fatty acid profiles may differ significantly even in the same genus of a berry plant. The results of the research conducted by Šavikin et al. [36] on *Ribes* sp. show variation of specific FAs depending on the color of fruit, with blackcurrant reaching the highest values of LA and GLA, but the lowest for ALA and OA. Red- and white currant presented similar concentrations of LA, GLA and ALA, whereas the white type was characterized by the highest values of OA. Results showing SFA content were not diverse. Table 1 presents the fatty acid profiles of berry oils. According to this summary, it can be concluded that the considered berry seed oils are rich in unsaturated, mostly polyunsaturated (PUFA), fatty acids. However, they differ in terms of specific fatty acid profiles. The dominant fatty acid in berry seed oils is linoleic acid (C18:2, *n*6); its content ranges from 33.86% for gooseberry [37] to 71.1% for chokeberry oil [38]. Additionally, the content of linolenic acids is high, except for chokeberry oil. The composition of fatty acids results in oil properties. High PUFA contents (especially linoleic and α-linolenic acids) lower the stability of plant oils, so they are more susceptible to oxidation and are characterized by shorter shelf lives [39–41]. Moreover, a high amount of MUFAs results in reduced stability values, although to a lesser degree than with PUFAs [42]. However, the nutritional value of unsaturated fatty acids is significant. FAO/WHO recommends replacing intake of saturated fatty acids with unsaturated fatty acids, especially PUFAs [43]. Numerous meta-analyses and review papers describe the positive impact on human health of marine-derived PUFAs, EPA and DHA [44–47], although plant-derived PUFAs can be elongated and desaturated into AA, EPA or DHA in the human system [48]. In addition, some reports claim that all-source-derived PUFA intake reduced all-cause mortality [49], while *n*3 PUFA intake reduced the risk of metabolic syndrome [50]. Plant-derived PUFAs' ability to regulate serum insulin levels has been described [51]. The most common FAs in berry oils, linoleic acid and α-linolenic acid, are classified as essential fatty acids and have to be delivered by food consumption due to the human disability for their endogenic production [52]. MUFAs can be produced in the human organism [53], but food-derived MUFAs were also found to have a role in disease prevention, especially in glucose–insulin management and reduced risk of co-existing diseases, in a group of diabetic patients [54–56].

Table 1. Fatty acid profiles (%) of oils extracted from berry pomaces. (C16:0—palmitic acid, C16:1—palmitoleic acid, C18:0—stearic acid, C18:1—oleic acid, C18:2—linoleic acid, C18:3—linolenic acid, C18:4—stearidonic acid, C20:0—arachidic acid, C20:1—paullinic acid, C20:2—eicosadienoic acid, C22:0—behenic acid.)

Source of Oil	C16:0	C16:1	C18:0	C18:1	C18:2	C18:3	C18:4	C20:0	C20:1	C20:2	C22:0	Reference
Chokeberry	5.1–7.22	0.15–0.53	1.1–1.39	*n9 cis*: 23.47, *n9 trans*: 0.93 or total: 17.48–21.4	64.67–71.1	*n3*: 0.34–0.92 or total: 0.5	-	0.6–0.81	0.25	5.26	0.38–0.8	[38,57,58]
Raspberry	2.43–2.92	0.08–0.12	0.87–1.45	*n9*: 11.74–11.76 *n7*: 0.80 or total: 10.87–11.99	51.44–54.52	*n3*: 6.68–31.68 *n6*: 0.07 or total: 29.11	-	0.37–0.62	0.13–0.14	0.03–0.33	0.10–0.34	[37,38,58–60]
Raspberry (wild)	2.61	0.06	1.19	*n9 cis*: 26.22 *n9 trans*: 0.23	51.07	*n3*: 17.93	-	0.43	0.05	-	0.14	[57]
Blackberry	3.47–4.52	0.03–0.13	2.10–2.87	*n9*: 7.50–12.17 *n7*: 0.56 or total: 14.72	61.22–67.96	*n3*: 15.60–17.60 *n6*: 0.07	-	0.47–1.06	0.31–0.38	0.15	0.12–0.16	[37,58,59]
Blackcurrant	4.49–6.5	0.03–0.1	1.4–1.93	*n9*: 10.2–13.79 *n7*: 0.35–0.7 or total: 16.1	41.41–57.8	*n3*: 12.91–14.9 *n6*: 13.9–15.6 or total: 13.2	2.7–3.89	0.04–0.2	0.16–1.0	0.06–0.3	0.1–4.7	[38,59,61,62]
Redcurrant	4.8–6.88	0.09	1.29–3.0	*n9*: 9.61–17.8 *n7*: 0.6–0.72	40.7–44.0	*n3*: 23.34–24.5 *n6*: 5.6–9.16	3.0–4.48	0.13	0.71	0.27	0.09	[59,63]
Blueberry	4.98–7.64	0.08	1.93–3.31	*n9 cis*: 50.74, *n9 trans*: 0.38 *n9 total*: 18.00 *n7*: 0.56	30.0–35.84	*n3*: 7.06–36.08 *n6*: 0.14	-	0.19–0.49	0.14	0.05	0.11	[57,59]
Gooseberry	8.12	-	1.83	*n9*: 14.32	33.86	*n3*: 20.54 *n6*: 8.48	5.45	-	-	-	-	[64]
Strawberry	4.32	-	1.68	14.55	42.22	*n3*: 36.48	-	0.71	-	-	-	[37]

2.3.2. Tocopherols and Tocotrienols

Tocopherols and tocotrienols (also commonly called tocols or, generally, vitamin E) are phenolic compounds, lipid-soluble antioxidants. Due to their oxidation-preventive ability, they protect PUFAs from oxidation and are widely present in edible, unsaturated fatty acid-rich oils [65]. However, the loss of their antioxidant efficacy and even an adverse oxidation-promoting effect of tocopherols have been observed at high temperatures [66] or in oils enriched with high levels of tocopherols, especially in highly unsaturated oils [67]. In studies describing the decomposition of tocols during storage, some differences were found: α-tocopherol is described as the least stable, with a rapid decrease in stability index observed, and δ-tocopherol as the most stable section of tocopherols [68]. In addition, there are findings confirming that higher levels of α-tocopherol in oils lead to a decreased oxidative stability index of oils [69]. The composition of tocols fractions may differ depending on oil processing conditions and berry cultivar. [65]. Tocopherol composition in chosen berry oils is presented in Table 2. The values vary in relation to the species of the berries. However, certain common characteristics exist. Tocopherols reach higher concentrations than tocotrienols in all of the described studies [37,38,59,60,70]. For blackberry, raspberry, redcurrant and blackcurrant oils, in almost any case, γ-tocopherol is their main tocopherol. For chokeberry oil, the major tocols fraction was α-tocopherol [38]. From a nutritional point of view, vitamin E proper uptake is essential, however, α-tocopherol is the most active antioxidant in the human system and the only tocol that is able to cover human vitamin E demand [71].

Table 2. Tocopherol composition of berry oils (mg/100 g oil). (TP- tocopherols; T3- tocotrienols.)

Source of Oil	α-TP	β-TP	γ-TP	δ-TP	α-T3	β-T3	γ-T3	δ-T3	Reference
Chokeberry	70.6	28.2	0.2	0.2	-	-	0.8	-	[38]
Raspberry	27.74–46.1	0.65	58.19–164.0	5.83–22.59	-	2.71	7.2	-	[37,59,60]
Blackberry	0.89–2.54	0.18	42.41–131.1	3.17–6.97	-	0.44	2.0	-	[37,59]
Blackcurrant	28.85–36.9	0.2–0.55	23.01–55.4	4.09–6.9	0.09–0.1	0.3–0.65	0.2–0.26	-	[38,59]
Redcurrant	3.04–5.75	0.56–0.79	33.64–156.39	19.38–41.13	0.10	0.31	0.13	0.03	[59,70]
Gooseberry	5.26	0.20	60.35	3.32	-	-	-	-	[70]
Blueberry	0.44	-	3.44	-	-	-	33.04	0.6	[37]
Strawberry	-	-	26.03	2.0	-	-	-	-	[37]

2.3.3. Sterols

Plant-derived sterols, also called phytosterols, are amphiphilic steroid alcohols. They play a role as a plant cell membrane compound. Most common are sitosterol, stigmasterol and campesterol [72]. Table 3 presents the sterol contents of analyzed berries oils. In all of them, sterols composition corresponds with common values and the main sterol occurring is β-sitosterol. Specific sterol contents may differ depending on variety, year, fruit maturity and processing conditions, e.g., temperature [73,74]. Additionally, sterols' thermal stability depends on their structure, mainly on the number and location of double bonds. In the structure of β-sitosterol there is one double bond that results in its thermosensitivity at a medium range [75]. Afinisha-Deepam et al. [76] and T. Wang [77] reported that sterols do not affect the stability of oils, so their concentration in products does not influence oxidative reactions or the length of the shelf life of oil. However, there are also findings describing the sterol fraction stigmasterol as prooxidative at temperatures around 60 °C but antioxidative at frying or baking temperatures around 180 °C [78]. Furthermore, sterol esters added to oils or to different fats, e.g., margarine, can decrease their oxidation stability [79,80]. In human nutrition, sterols are believed to be competitors of cholesterol and as a result of that they reduce cholesterol absorption from dietary sources, which leads to reductions in concentrations of plasma cholesterol. This property may be useful in lipids-correlated disorders prevention and treatment, e.g., hypercholesterolemia [81,82].

Table 3. Sterol composition of lipid fraction of berry pomaces (%).

Source of Oil	Cholesterol	Campesterol	Stigmasterol	β-Sitosterol	D5-Avenasterol	D7-Stigmasterol	D7.25-Stigmasterol	Reference
Chokeberry	2.95	5.5	3.85	81.8	1.85	1.8	1.8	[38]
Blackberry	0.33	5.3–7.03	1.8–4.87	77.77–84.7	3.02–7.0	1.41	-	[37,59]
Raspberry	0.43	4.5–4.51	0.84–1.2	79.6–83.95	5.35–7.2	1.24	-	[37,59]
Blackcurrant	2.5	1.25	4.9	86.6	1.3	0.85	1.4	[38]
Blackcurrant	0.31	8.14	0.42	81.09	3.10	1.92	-	[59]
Redcurrant	0.36	10.01	0.24	87.58	0.36	-	-	[59]
Blueberry	0.24	3.4–4.63	0.3–0.37	66.5–82.85	2.14–13.8	3.97	-	[37,59]
Strawberry	-	5.4	2.3	71.1	8.7	-	-	[37]

3. Processing and Extraction

Bioactive compounds may be isolated from pomaces by physical techniques, such as cold pressing, or chemical techniques, i.e., extraction, and extracts as well as oils may be products of these processes. Conventional methods of extraction may, however, require extended energy intake and use of organic solvents in large quantities. To help to reduce the environmental and financial impact of extraction processes caused by both high energy and organic solvent consumption, non-conventional extraction methods have been proposed. Particular novel extraction methods may lead to the obtention of extracts or oils with improved properties. A scheme for the procedures of extraction of bioactive compounds and oils from berry fruit by-products is presented in Figure 2.

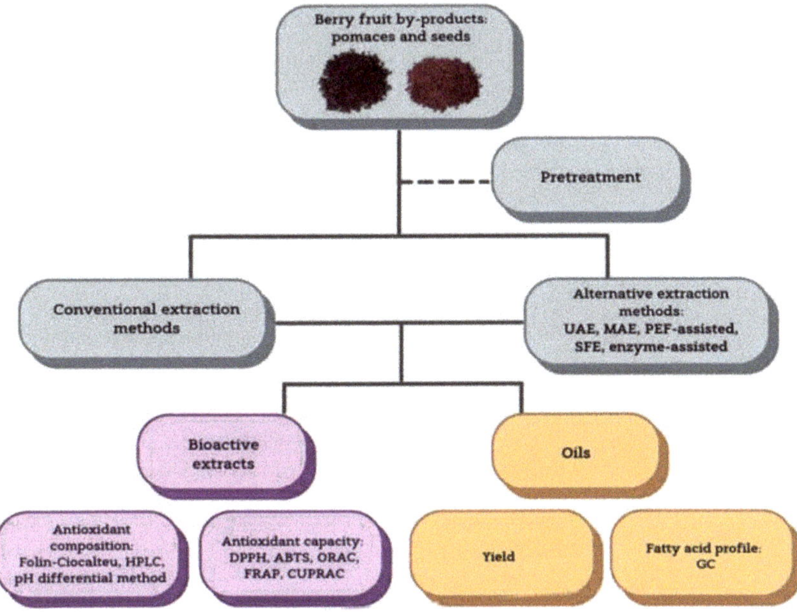

Figure 2. A scheme for the extraction procedures for bioactive compounds and oils from berry fruit by-products.

3.1. Conventional Extraction Methods

Conventional, liquid–liquid and solid–liquid extraction methods are the most commonly used at a laboratory scale. In these processes, usually organic solvents, like methanol, ethanol, hexane and acetone, are used, but aqueous solvents might also be employed. The mechanism of the process involves removing a soluble fraction from an insoluble solid. The concentration of compounds released from plant tissues to the solvent reaches equilibrium with the concentrations of unreleased substances as described by the equilibrium distribution constant. Fick's second law of diffusion describes how fast the compounds are able to dissolve and reach equilibrium [83]. Factors that may improve extraction efficiency are increased solvent concentration and reduced particles of solids. Increasing solvent concentration enhances the gradient of concentrations in two phases and reducing the size of particles decreases the diffusion distance of a solution within the solid and, additionally, increases the concentration gradient, too. Temperature also helps to increase diffusivity, according to the Einstein equation. However, due to solvents' toxicity with respect to the environment or human health, conventional methods are perceived as inappropriate in the food processing industry. What is more, not all of the phenolic or lipid compounds are possible to extract using these methods. Increased temperature may also lead to damage of some thermolabile structures [83]. The efficiency of solvent extraction is determined mainly by the type of solvent and the material/solvent ratio [84]. Tables 4 and 5 compare the conditions and effectiveness (in the case of oils, yield and fatty acid profiles; in the case of phenolic extracts, antioxidant composition and antioxidant capacity of extracts) of conventional extractions of antioxidants and oils from berry by-products.

3.1.1. Lipid Fraction

The composition of fatty acids in extracted oils is mainly determined by the species of fruit as the source of seeds. However, the yield appears to be dependent on extraction conditions, i.e., solvent type, time of extraction and pomace pretreatment method. In the case of the solvent type, organic solvents used in conventional extraction methods are characterized by different polarities. They are defined by different dielectric constant values—a measure of solvent polarity which determines the solute–solvent correlation. The optimum dielectric constant values range from 6–8 and result in higher oil yields. However, the higher polarity of solvents results in limitations in the solubility of extracted lipids and can lead to their hydrolysis. What is more, TAGs are amphipathic compounds and some solvents may cause hydrogen bonding of TAG ester groups. So, in conclusion, higher polarity may be a reason of lower oil yield, despite the fact that increasing the polarity of the solvent causes the opening of cell walls and improves compound release. What is more, in the case of plant oils, they also contain more polar constituents, e.g., phospholipids and tocopherols, which are the source of components that may present greater affinity to more polar solvents [85,86]. Hexane is a widely employed solvent in fat extraction processes due to its low polarity and it was used in most of the reviewed studies. However, it is a chemical substance with proven toxicity and water-polluting ability, so it should not be used in the food industry [87]. Considering berry seed oils extraction, ethanol is used as a solvent in the extraction of fat from blackberry seeds, resulting in higher yields compared to hexane—11.8% and 14.2%, respectively [88]. The higher oil yield value, when a more polar solvent was used, may be connected to the phenomena of partly polar compounds in the fat fraction. The time of the process is also a factor influencing efficiency. Comparing the extraction of oil from dried raspberry seeds using hexane, a higher yield was connected with a longer extraction time: the process lasting 8 h gave a 14.33% yield [89] and a 2 h extraction resulted in a 10.7% yield [60]. The analyzed studies also showed the significant impact of seed pretreatment methods on extraction efficiency. Considering raspberry by-products, dried pomaces [59] or seeds [60] were more susceptible to oil extraction processes than wet material [57]. In addition, replacing drying at room temperature with oven-drying resulted in higher fat yield values [89].

3.1.2. Antioxidant Fraction

Conventional extraction methods for polyphenolic compounds from berries are also conducted usually using organic solvents; however, the polarity of the solvent has to be adjusted, as polyphenols are more polar components. On a laboratory scale, alcohols are the most commonly used solvents, but these substances may still be harmful to organisms or the environment, e.g., methanol. The specific antioxidant component composition and the antioxidant capacity of obtained extracts are influenced, as in the case of oils, by the species of fruit, pretreatment methods and sample preparation. Other factors that have an influence on the properties of extracts are solvent type, solid–liquid ratio, extraction time and solvent pH.

Table 4. Characteristics of lipid fractions extracted from berry by-products using conventional methods.

Source of Waste	Pretreatment Method	Extraction Conditions (S—Solvent, S–L—Solid–Liquid Ratio, M—Mass of Solid, t—Time, T—Temperature)	Oil Yield	Fatty Acids Profile (%) (PUFA/MUFA/SFA, Dominant FA)	Reference
Blackberry (*Rubus fruticosus* cv. Tenac) Raspberry (*Rubus idaeus* cv. Meeker) Blueberry (*Vaccinium myrtillus* cv. Ivanhoe) Blackcurrant (*Ribes nigrum* cv. Junifer) Redcurrant (*Ribes rubrum* cv. Smoothstem) pomaces	Seeds separated from pomace dried at room temperature using sieves, then ground	S: hexane Soxhlet apparatus	*Blackberry pomace:* 15.68% *Raspberry pomace:* 10.55%, Blueberry pomace: 13.27% *Blackcurrant pomace:* 26.15% *Redcurrant pomace:* 9.11%	*Blackberry pomace* **PUFA:** 83.78 **MUFA:** 8.40 **SFA:** 6.49 Dominant: C18:2 *n6* *Raspberry pomace* **PUFA:** 81.05 **MUFA:** 12.81 **SFA:** 4.13 Dominant: C18:2 *n6* *Blueberry pomace* **PUFA:** 72.11 **MUFA:** 18.78 **SFA:** 6.75 Dominant: C18:3 *n3* *Blackcurrant pomace* **PUFA:** 73.16 **MUFA:** 14.33 **SFA:** 6.46 Dominant: C18:2 *n6* *Redcurrant pomace* **PUFA:** 81.25 **MUFA:** 11.13 **SFA:** 8.39 Dominant 18:2 *n6*	[59]
Blackberry (*Rubus fruticosus* cv. Ćačanska beztrna) Raspberry (*Rubus idaeus* cv. Willamette) seeds	Seeds obtained from the pomace dried at room temperature or in oven	Standard laboratory method S: hexane t: 8 h	*Blackberry seeds:* 13.97–14.34% *Raspberry seeds:* 13.44–14.33% (Higher values for pomaces dried in oven)	-	[89]
Raspberry (*Rubus idaeus*) different cv. seeds	Seeds air-dried in fluid bed dryer for 2 h/25 °C, then ground	S: hexane 1 L M: 100 g t: 2 h T: 4 °C Extraction performed 3 times	10.7%	Crude oil: **PUFA:** 83.63 **MUFA:** 11.99 **SFA:** 3.66 Dominant: C18:2 *n6*	[60]

Table 4. *Cont.*

Source of Waste	Pretreatment Method	Extraction Conditions (S—Solvent, S–L—Solid–Liquid Ratio, M—Mass of Solid, t—Time, T—Temperature)	Oil Yield	Fatty Acids Profile (%) (PUFA/MUFA/SFA, Dominant FA)	Reference
Wild: Blueberry (*Vaccinium myrtillus*) Cowberry (*Vaccinium vitis-idaea*) Raspberry (*Rubus idaeus*) Cultivated: Blueberry (*Vaccinium myrtillus*) Chokeberry (*Aronia melanocarpa*) pomaces	Wet pomaces obtained after juice pressing	Methanol/chloroform homogenization procedure S: methanol (50 mL) and chloroform (100 mL) M: 5 g	*Wild* *Blueberry pomace:* 3.93% *Cowberry pomace:* 3.75% *Raspberry pomace:* 7.00% *Cultivated* *Blueberry pomace:* 2.80% *Chokeberry pomace:* 5.50%	*Wild* *Blueberry pomace* **PUFA:** 41.78 **MUFA:** 48.35 **SFA:** 9.80 Dominant: C18:1 *n9* *Cowberry pomace* **PUFA:** 44.47 **MUFA:** 50.87 **SFA:** 4.65 Dominant: C18:1 *n9* *Raspberry pomace* **PUFA:** 69.00 **MUFA:** 26.56 **SFA:** 4.44 Dominant: C18:2 *n6* *Cultivated* *Blueberry pomace* **PUFA:** 37.00 **MUFA:** 51.20 **SFA:** 11.70 Dominant C18:1 *n9* *Chokeberry pomace* **PUFA:** 65.01 **MUFA:** 24.93, **SFA:** 10.06 Dominant: C18:2 *n6*	[57]
Blackcurrant (*Ribes nigrum* cv. Ben Lomond and cv. Ben Tirran) pomaces	Pomaces were air-dried in oven or freeze-dried and (A) ground or (B) seeds separated from unground pomace using 500 µm sieve	Soxhlet extraction S: isohexane M: 10 g t: 30 min Residue left after 5 cycles was reground and extracted in another 5 cycles	From seeds: 14.5% from ground pomace: 7.8% (w/dry)	*Seeds* **PUFA:** 79.4 **MUFA:** 12.0 **SFA:** 8.7 Dominant: C18:2 *n6* *Pomace* **PUFA:** 72.0 **MUFA:** 11.7 **SFA:** 16.3 Dominant: 18:2 *n6*	[62]
Raspberry (*Rubus idaeus*) dust as a by-product from fruit lyophilization	Seeds separated from the dust by sieving and then ground	Soxhlet apparatus S: petroleum ether M: 100 g t: 6 h	14.5%	**PUFA:** 78.9 **MUFA:** 16.9 **SFA:** 4.2 Dominant: C18:2 *n6*	[90]

Table 4. *Cont.*

Source of Waste	Pretreatment Method	Extraction Conditions (S—Solvent, S–L—Solid–Liquid Ratio, M—Mass of Solid, t—Time, T—Temperature)	Oil Yield	Fatty Acids Profile (%) (PUFA/MUFA/SFA, Dominant FA)	Reference
Blackberry (*Rubus fruticosus* cv. Cacanska beztrna) and Raspberry (*Rubus idaeus* cv. Wllamette) pomaces obtained from pressing long-term frozen fruits	Pomaces dried at room temperature or in oven	Standard laboratory method S: hexane t: 8 h	–	*Blackberry pomace (room temperature-dried):* **PUFA:** 74.94 **MUFA:** 17.87 **SFA:** 7.13 *Blackberry pomace (oven-dried):* **PUFA:** 75.66 **MUFA:** 19.03 **SFA:** 7.48 Dominant: C18:2 *n6* *Raspberry pomace (room temperature-dried)* **PUFA:** 82.52 **MUFA:** 13.21 **SFA:** 4.23 *Raspberry pomace (oven-dried):* **PUFA:** 87.30 **MUFA:** 12.57 **SFA:** 4.26 Dominant: C18:2 *n6*	[91]
Blackberry (*Rubus fruticosus*) pomace	Pomace dried in the sun and milled	Soxhlet apparatus S: hexane 250 mL/ethanol 250 mL t: 8 h	*Hexane* 11.8% *Ethanol* 14.2%	*Hexane extracted* **PUFA:** 71.4 **MUFA:** 17.5 **SFA:** 11.1 Dominant: 18:2 *n6* *Ethanol extracted* **PUFA:** 69.4 **MUFA:** 17.4 **SFA:** 13.2 Dominant: 18:2 *n6*	[88]
Chokeberry (*Aronia melanocarpa*), strawberry (*Fragaria vesca*), blackcurrant (*Ribes nigrum*) pomaces	–	Sample homogenized S: chloroform and methanol (*v/v*, 2:1) M: 5.0 g	–	*Chokeberry pomace* **PUFA:** 73.58 **MUFA:** 16.91 **SFA:** 9.51 Dominant: C18:2 *n6* *Strawberry pomace* **PUFA:** 55.77 **MUFA:** 18.16 **SFA:** 26.07 Dominant: C18:2 *n6* *Blackcurrant pomace* **PUFA:** 69.11 **MUFA:** 11.56 **SFA:** 19.33 Dominant: C18:2 *n6*	[92]

Table 5. Antioxidant composition and capacity of berry by-product extracts obtained with the use of conventional methods.

Source of Waste	Pretreatment Method	Extraction Conditions (S—Solvent, S–L—Solid–Liquid Ratio, M—Mass of solid, t—Time, T—Temperature)	Antioxidant Composition *	Antioxidant Capacity *	Reference
Blackcurrant (*Ribes nigrum* cv. Mortti), Green currant (*Ribes nigrum* cv. Verti), Redcurrant (*Ribes rubrum* cv. Red Dutch), White currant (*Ribes rubrum* cv. White Dutch) pomaces	None (thawed pomace)	S: 92% ethanol S–L: 1:2 (*w/v*)	*Blackcurrant pomace* **TPC:** 55.3 µmol GAE/g (fw) *White currant pomace* **TPC:** 24.7 µmol GAE/g *Redcurrant pomace* **TPC:** 20.5 µmol GAE/g *Green currant pomace* **TPC:** 17.1 µµmol GAE/g	*Blackcurrant pomace* **TRAP:** 25,7 µmol TE/g (fw) **ORAC:** 88.8 µmol TE/g *Redcurrant pomace* **TRAP:** 11.6 µmol TE/g **ORAC:** 23.0 µmol TE/g *Green currant pomace* **TRAP:** 8.7 µmol TE/g **ORAC:** 32.9 µmol TE/g *White currant pomace* **TRAP:** 8.4 µmol TE/g **ORAC:** 16.8 µmol TE/g	[93]
Cranberry (*Vaccinium macrocarpon*), Blueberry (*Vaccinium angustifolium*) pomace	Pomace freeze-dried, then ground	S: 80% ethanol S–L: 1:5 (*w/v*) t: 1 h Mixed, obtained extracts were freeze-dried	*Cranberry pomace* **TPC (Folin–Ciocalteu):** 54.35 mg GAE/g (dm) **TPC (Glories):** 36.25 mg GAE/g **Tartaric esters:** 10.29 mg CAE/g **Flavanols:** 11.74 mg QE/g **TAC:** 11.14 mg C3 GE/g **Tannins:** 48.09 mg GAE/g *Blueberry pomace* **TPC (Folin–Ciocalteu):** 72.01 mg GAE/g **TPC (Glories):** 55.67 mg GAE/g **Tartaric esters:** 15.03 mg CAE/g **Flavanols:** 18.34 mg QE/g **TAC:** 38.53 mg C3GE/g **Tannins:** 58.87 mg GAE/g	*Cranberry pomace* **ABTS:** 306.77 µmol TE/g (dm) **FRAP:** 243.61 µmol TE/g *Blueberry pomace* **ABTS:** 468.79 µmol TE/g **FRAP:** 372.22 µmol TE/g	[94]
Bilberry (*Vaccinium myrtillus*), Blackberry (*Rubus fruticosus*), Raspberry (*Rubus idaeus*), Strawberry (*Fragaria vesca*) pomaces	None	S: 80% methanol with 0.05% acetic acid M: 20 g 3 times extracted: 60 min (160 mL of S), 30 min (80 mL of S) and 30 min (80 mL of S)	*Bilberry pomace* **TPC:** 11.16 mg GAE/g (fw) **TFC:** 10.47 mg RE/g **TAC:** 12.79 mg C3GE/g *Blackberry pomace* **TPC:** 8.05 mg GAE/g **TFC:** 2.45 mg RE/g **TAC:** 1.49 mg C3GE/g *Raspberry pomace* **TPC:** 6.38 mg GAE/g **TFC:** 5.92 mg RE/g **TAC:** 0.65 mg C3GE/g *Strawberry pomace* **TPC:** 4.88 mg GAE/g **TFC:** 2.96 mg RE/g **TAC:** 0.19 mg C3GE/g	*Bilberry pomace* **DPPH IC$_{50}$:** 0.040 mg/mL (pomace extract) *Blackberry pomace* **DPPH IC$_{50}$:** 0.017 mg/mL *Raspberry pomace* **DPPH IC$_{50}$:** 0.040 mg/mL *Strawberry pomace* **DPPH IC$_{50}$:** 0.038 mg/ml	[95]

Table 5. *Cont.*

Source of Waste	Pretreatment Method	Extraction Conditions (S—Solvent, S–L—Solid–Liquid Ratio, M—Mass of solid, t—Time, T—Temperature)	Antioxidant Composition *	Antioxidant Capacity *	Reference
Blackberry (*Rubus fruticosus*) wild and cultivated (cv. Cacanska Bestrna and cv. Chester Thornless) pomaces	–	Soxhlet apparatus S: 80% ethanol t: 6 h extracts dried in a vacuum desiccator	*Blackberry cv. Chester Thornless pomace* **TPC:** 35.40 mg GAE/g fresh pomace (dm) **TFC:** 5.66 mg QE/g **Flavanols:** 6.63 mg QE/g **Monomeric anthocyanins:** 17.31 mg C3GE/g *Blackberry cv. Cacanska Bestrna pomace* **TPC:** 26.30 mg GAE/g **TFC:** 3.32 mg QE/g **Flavanols:** 2.55 mg QE/g **Monomeric anthocyanins:** 8.43 mg C3GE/g *Blackberry wild pomace* **TPC:** 48.28–50.16 mg GAE/g **TFC:** 7.45–7.73 mg QE/g **Flavanols:** 6.13–6.39 mg QE/g **Monomeric anthocyanins:** 13.05–13.40 mg C3GE/g	*Blackberry cv. Chester Thornless pomace* **DPPH IC$_{50}$:** 0.178 mg/mL (pomace extract) **ABTS IC$_{50}$:** 0.035 mg/mL *Blackberry cv. Cacanska Bestrna pomace* **DPPH IC$_{50}$:** 0.206 mg/mL **ABTS IC$_{50}$:** 0.047 mg/mL *Blackberry wild pomace* **DPPH IC$_{50}$:** 0.106–0.127 mg/mL **ABTS IC$_{50}$:** 0.024–0.027 mg/mL	[28]
Raspberry (*Rubus idaeus* cv. Meeker and cv. Willamette) pomace	None	S: 80% methanol with 0.05% acetic acid M: 20 g T: room temperature Two extractions in: 160 mL, 60 min and 80 mL, 30 min	*cv. Meeker pomace* **TPC (HPLC):** 338.80 mg/100 g (pomace) *cv. Willamette pomace* **TPC:** 410.66 mg/100 g	*cv. Meeker pomace* **DPPH IC$_{50}$:** 0.67 mg/mL (pomace extract) *cv. Willamette pomace* **DPPH IC$_{50}$:** 0.54 mg/mL	[96]
Cranberry (*Vaccinium macrocarpon*), blueberry (*Vaccinium myrtillus*) and raspberry (*Rubus idaeus*) pomace	Dehydrated, ground, separated into 2 groups depending on particle size (smaller particle size, 0.15 mm, and larger, 1 mm)	S: methanol S–L: 1:20 t: 1–24 h T: 40 °C using orbital shaker	Obtained after the most effective extraction parameters *Cranberry pomace (1 h, larger particles)* **TPC:** 138 mg GAE/g (fresh extract) *Blueberry pomace (6 h, smaller particles)* **TPC:** 172 mg GAE/g *Raspberry pomace (18 h, smaller particles)* **TPC:** 270 mg GAE/g	Obtained after the most effective extraction parameters *Cranberry pomace (1 h, larger particles)* **DPPH EC$_{50}$:** 3.73 mg/mL (pomace extract) *Blueberry pomace (6 h, smaller particles)* **DPPH EC$_{50}$:** - *Raspberry pomace (18 h, smaller particles)* **DPPH EC$_{50}$:** 0.30 mg/mL	[97]

Table 5. *Cont.*

Source of Waste	Pretreatment Method	Extraction Conditions (S—Solvent, S–L—Solid–Liquid Ratio, M—Mass of solid, t—Time, T—Temperature)	Antioxidant Composition *	Antioxidant Capacity *	Reference
Strawberry (*Fragaria vesca*) pomace	Pomace dried in convection dryer (temp. 65–70 C, 8 h), sieved, particles over 5 mm were ground to obtain final material, with particle size between 2 and 5 mm	*Water extraction* S: water S–L: 4:1 M: 1500 g t: 1 h T: 65–70 °C 3 times extracted *Ethanol extraction* S: 60% ethanol (4 L) S–L: 4:1 M: 3.5 kg of aqueous extract t: 24 h T: 20 °C Repeated once using 3.5 L of ethanol	*Water extract* **TPC** (HPLC): 5.8 g/100 g (dm) *Ethanolic extract* **TPC:** 29.71 g/100 g	-	[98]
Blueberry (*Vaccinium myrtillus*) Raspberry (*Rubus idaeus*) Redcurrant (*Ribes rubrum*) and Blackberry (*Rubus fruticosus*) pomaces	Lyophilized and ground pomaces	S: 75% methanol with 1% HCl (5 mL) M: 1 g t: 12 h T: 25 °C	*Blueberry pomace* **TPC:** 19.55 mg GAE/g (dm) **TAC** (HPLC): 11.88 mg/g *Raspberry pomace* **TPC:** 20.15 mg GAE/g **TAC:** 1.88 mg/g *Redcurrant pomace* **TPC:** 34.47 mg GAE/g **TAC:** 1.50 mg/g *Blackberry pomace* **TPC:** 17.00 mg GAE/g **TAC:** 1.92 mg/g	*Blueberry pomace* **ABTS:** 269.8 µmol TE/g (dm) *Raspberry pomace* **ABTS:** 297.5 µmol TE/g *Redcurrant pomace* **ABTS:** 608.3 µmol TE/g *Blackberry pomace* **ABTS:** 225.4 µmol TE/g	[27]
Raspberry (*Rubus idaeus* cv. Meeker and cv. Willamette)	None	S: 80% methanol with 0.05% acetic acid M: 70 g T: room temperature 2 times extracted (A) t: 60 min, 560 mL of solvent (B) t: 30 min, 280 mL of solvent	*cv. Meeker pomace* **TPC:** 26.3 mg GAE/g (dm) **TFC:** 25.2 mg RE/g **TAC:** 4.28 mg C3GE/g *cv. Willamette pomace* **TPC:** 43.7 mg GAE/g **TFC:** 22.0 mg RE/g **TAC:** 2.32 mg C3GE/g	*cv. Meeker pomace* **DPPH EC$_{50}$:** 0.072 mg/mL (pomace extract) *cv. Willamette pomace* **DPPH EC$_{50}$:** 0.042 mg/mL	[31]
Blackberry (*Rubus fruticosus*) residues after pulp processing and blueberry (*Vaccinium myrtillus*) residues after juice processing	None	S: ethanol 200 mL M: 5 g t: 5 h T: 80 °C Soxhlet apparatus	*Blackberry pomace* **TPC:** 7.84 mg GAE/g (dm) **Monomeric anthocyanins:** 2.82 mg C3GE/g *Blueberry pomace* **TPC:** 6.83 mg GAE/g **Monomeric anthocyanins:** 2.58 mg C3GE/g	*Blackberry pomace* **DPPH:** 66.92 µmol TE/g (dm) **ABTS:** 124.14 µmol TE/g **FRAP:** 120.90 µmol TE/g *Blueberry pomace* **DPPH:** 40.38 µmol TE/g **ABTS:** 100.66 µmol TE/g **FRAP:** 63.90 µmol TE/g	[99]

* Results are expressed as written in the bracket after first given result.

3.2. Ultrasound Assisted Extraction

Ultrasound-assisted extraction (UAE) involves using ultrasound generating devices with the proper solvent to extract bioactive compounds. In a review by Medina-Torres et al. [100], the mechanism and effects on plant tissue of UAE is summed up. UAE utilizes the phenomenon of acoustic cavitation which results in damage to the cell walls of plant material, as shown in Figure 3. This leads to increased release of bioactive compounds. The operating principles of ultrasound are mechanical waves characterized by length, amplitude, frequency, speed, power and intensity. Ultrasonic wave frequency ranges from 20 kHz to 10 MHz. The sustainability of UAE is due to decreased solvent and energy consumption according to lower time and temperature requirements as compared to conventional extraction methods.

Table 6 presents a summary of pretreatment and extraction parameters used in ultrasound assisted extraction. UAE was successfully used as a method for the isolation of bioactive compounds from fresh berry fruits. It led to the improved yield of polyphenols (and therefore antioxidant activity) of extract obtained from chokeberry fruits. In addition, temperature increase and addition of ethanol to the solvent enhanced the efficiency of the process [101]. UAE prior to the separation and analysis of polyphenol compounds in strawberry fruits results in decreased extraction time [102]. It was also found that UAE allowed the use of a lower temperature and lower solvent concentrations in anthocyanin extraction from blueberry fruits and that it results in monomeric anthocyanin-rich extracts [103]. UAE has been applied to extract the lipid fractions from seeds other than berry fruit seeds. The crucial parameter for oil extraction yield from apricot kernels was temperature [104]. However, in UAE from papaya seeds, the most significant factors for oil extraction yield, AA and oil stability were time and temperature [105]. Fatty acid composition and the TAG profile of oil from papaya seed extracted by UAE does not vary significantly from oil extracted conventionally [106].

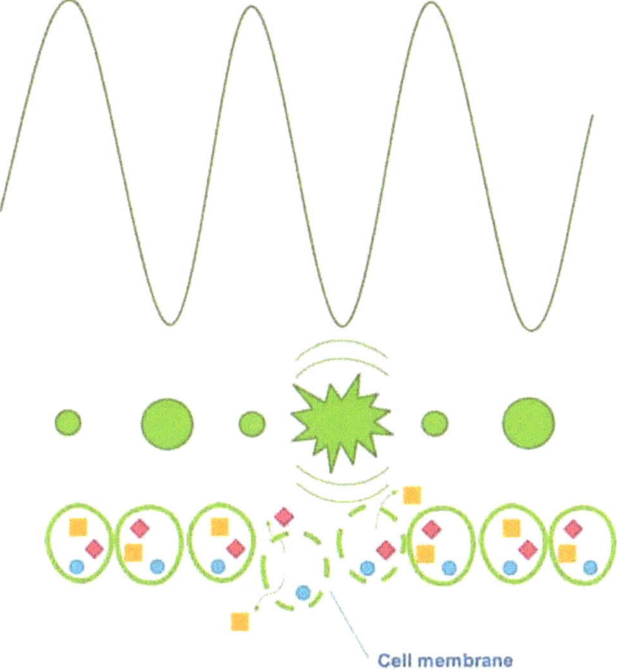

Figure 3. A scheme of ultrasound impact on the plant material.

3.2.1. Polyphenolic Compounds

Krivokapić et al. [107] examined raspberry pomace extraction using the UAE method and compared it to conventional maceration. UAE was conducted in an ultrasonic cleaner for 20 min at 50 °C and 50 kHz with acidulated methanol as a solvent. The obtained results showed that measured TPC, TFC and TAC were all significantly increased when UAE was used. It led to levels of 27.79 mg GAE/L of extract, 8.02 mg QE/g of pomace and 7.13 mg CGE/L of extract for TPC, flavonoid content and anthocyanin content, respectively. In addition, AA measured in FRAP and DPPH assays was higher in extracts treated with ultrasound. The values obtained for FRAP and DPPH were 1002.72 µmol FE/L of extract and 567.00 µmol TE/100 g of pomace, respectively. Ellagic acid was detected by HPLC as the most abundant polyphenol. Bamba et al. [108] conducted UAE of polyphenols from blueberry wine pomace. Before extraction, the pomace was stored as freeze-dried material. In this paper, it was summarized how individual parameters of UAE—sonication time, solid–liquid ratio, solvent type, temperature and pH—affected the quality of extracts. Qualitative parameters consisted of TPC, TFC, TAC and AA. The time of extraction significantly affected only TAC, which values were significantly higher in the case of a 90 min run. Solid–liquid ratio was a factor determining changes in all measured parameters. With water used as a solvent, TPC, TFC and TAC were highest for the lowest value of the solid–liquid ratio—1/40. A solid–liquid ratio of 50% ethanol at the lowest level of 1/20 resulted in the highest AA, TPC and TAC in extracts. However, TFC did not follow that trend and the highest values were reached with a 1/15 ratio. In addition, concentrations for all of the compounds, as well as the AA value, were significantly higher when 50% ethanol was used in UAE. Ethanol concentration was also a parameter affecting bioactive compound yield and AA of extracts. The most effective concentration was 50% (v/v) ethanol in water. The lowest TPC, TFC, TAE and AA values were obtained using the highest concentration of ethanol—90% (v/v). pH determined significantly TPC, TAC and AA, and for TPC and AA, pH 8.3 was the most effective; however, it resulted in decreased TAC, which was highest with pH 3.3. The temperature set at 60 °C resulted in significantly increased TFC and AA but decreased TPC, for which a temperature set at 20 °C was the most preferable. TAC was not significantly affected by temperature. The anthocyanidin profile was specified by HPLC. In both water and ethanolic extracts, the most abundant compound was malvidin and subsequently delphinidin, petunidin and cyanidin. Zafra-Rojas et al. [109] described UAE optimization using RSM. The experiment was held under constant conditions of frequency 20 kHz, S–L ratio 1:24 and a temperature of 4 °C at the beginning of the process and measured as 25 °C at the end. Variable parameters were the amplitude of ultrasounds, in a range of 80–90%, and the time of extraction, in a range of 10–15 min. Mathematical analysis revealed an amplitude and time of 91% and 15 min, respectively, as the most beneficial parameters for TPC, TAC and AA extracts at dm basis. Predicted values were 1200 mg GAE/100 g TPC, 380 mg/100 g TAC and 6300 µmol TE/100 g AA in an ABTS assay and 9600 µmol TE/100 g in a DPPH assay. The experimental values obtained at optimum UAE conditions were: 1201.23 mg GAE/100 g, 379 mg/100 g, 6318 µmol TE/100 g and 9617.22 µmol TE/100 g for TPC, TAC, AA (determined by ABTS) and AA (determined by DPPH), respectively. The UAE method was compared to conventional SLE with use of both water and ethanol. UAE yielded the highest values for TPC and AA in both assays. However, TAC in the extract obtained by UAE was not significantly different from the TAC of the extract obtained by SLE with water and was significantly lower compared to the result for SLE with ethanol. The impact of ultrasound treatment on other unconventional methods of extraction was also investigated. Xue et al. [110] combined UAE with enzymatic treatment of pectinase. The mechanism of enzyme impact on the extraction process was provided in the following text. Raspberry pomace obtained from the wine industry was freeze-dried until the moisture content was <5%. After 12 h at −18 °C, it was milled to 0.45 mm as the maximum size of particles. Conditions of UAE were determined using RSM. The most efficient parameters of anthocyanin extraction were chosen as follows: temperature 43.94 °C, ultrasound power 290.9 W, time 30 min and pectinase dosage 0.16%. S–L ratio and type of solvent were constant and ranged from 1/30 and 60% (v/v) acidulated ethanol, respectively. To evaluate the mathematical method the following conditions were set: temperature 44 °C, power of ultrasound 290 W, time 30 min and pectinase dosage 0.16%. The obtained value of anthocyanin yield was 0.853 mg/g of pomace (dm) vs. the predicted 0.888 mg/100 g. Enzymatic extraction with US treatment was compared with conventional extraction methods: SLE with hot water, SLE with acidulated ethanol and EAE (without US). US resulted in the highest extraction efficiency and TAC values in obtained extracts. AA measured in DPPH and RP (reducing power) assays reached the highest values when US was used. The AA value of the product extracted using US, as determined in an ABTS assay, was not significantly different from conventional EAE but was significantly increased compared to the AA values of extracts obtained in both SLEs. Ramić et al. [111] used UAE to extract polyphenolic compounds from chokeberry by-products from a filter-tea factory, produced from pomace remaining after juice pressing. The most efficient parameters were chosen using RSM for every quantitative factor individually (TPC, TFC, monomeric anthocyanin content and proanthocyanins content). The common optimal parameter for all the measured properties of extracts was the temperature set at 70 °C. Other values of conditions predicted as the most efficient were as follows: for TPC—206.64 W and 80.1 min; for TFC—210.24 W and 75 min; for monomeric anthocyanins content—216 W and 70 min; and for proanthocyanins content—199.44 W and 70 min for power and time, respectively. The observed values of yields for individual compounds were 15.058 mg GAE/mL extract (vs. predicted 15.41 mg GAE/mL), 10.436 mg CE/mL (vs. predicted 9.86 mg CE/mL), 2.09 mg C3GE/mL (vs. predicted 2.26 mg C3GE/mL) and 19.82 mg CE/mL (vs. predicted 20.67 mg CE/mL) for TPC, TFC, monomeric anthocyanins content and proanthocyanins content, respectively. He et al. [112] used UAE to extract polyphenolic compounds from blueberry wine pomace. The experiment was designed using RSM and optimal parameters were predicted as follows: temperature 61 °C, S–L 1:22, time 24 min. It resulted in extraction yields of 16.03 mg GAE/g pomace and 4.19 mg C3GE/g pomace for TPC and TAC, respectively. (Predicted values were 15.81 mg GAE/g pomace for TPC and 4.12 mg C3GE/g pomace for TAC.) Compared to the conventional method, the SLE method with acidulated ethanol (70%) used as a solvent running for 35 min and with applied optimal conditions for UAE (temperature—61 °C; S–L ratio—1/22), UAE appeared to be significantly more efficient, despite a shorter extraction time. There were seven anthocyanins also identified, which were present in both extracts: delphinidin-3-O-glucoside, delphindin-3-O-arabinoside, petunidin-3-O-glucoside, cyanidin-3-O-arabinoside, cyanidin-3-O-glucoside, malvidin-3-O-glucoside and malvidin-3-O-arabinoside. Loncarić et al. [113] designed an experiment with different extraction methods (PEF-assisted—mentioned in the following text and high voltage electrical discharge-assisted). UAE with variable conditions of time, temperature and solvent type was conducted. The results showed that TPC, AA, TAC, flavanol and flavonol contents reached the highest levels when UAE was performed at 80 °C and lasted 15 min. The type of used solvent did not affect the obtained values. Only phenolic acid yield was associated with methanol used as a solvent, an applied temperature of 40 °C and an extraction time of 15 min. Compared to PEF-assisted and HVED-assisted extraction methods, UAE resulted in the lowest yields of polyphenolic compounds and AA.

The UAE method was also used to obtain blackberry seed extracts from three cultivars: Dircksen, Thornfree and Black Satin. Extract yields were dependent on the cultivar and, in the case of Black Satin, the highest yield of 7% was reached. Polyphenolic compounds of obtained extracts were also determined. LC/UV/MS analysis enabled the identification of 64 polyphenols in extracts: 47 ellagitannins, 10 ellagic acid derivatives, 4 gallic acid derivatives and traces of protocatechuic, chlorogenic and salicylic acid. The most abundant compound for both Thornfree and Dircksen seed extracts was free ellagic acid and the main constituent of Black Satin seed extract was lambertianin C. Additionally, three polyphenolic compounds were successfully isolated from extracts using a semipreparative HPLC method [114].

3.2.2. Lipid Fraction

Teng et al. [115] extracted oil from raspberry seeds using UAE. In the experimental design, optimal values of time and temperature were predicted in order to obtain the highest yield, vitamin E content and AA. The parameters for time and temperature were reported as 37 min and 54 °C, respectively. These conditions resulted in a 22.78% extraction yield (23% predicted), 15.21 mg/g dm vitamin E content (15 mg/g dm predicted) and 80.94 μmol TE/g dm AA (81.65 μmol TE/g dm predicted). Compared to conventional SLE using the Soxhlet apparatus method, all the examined determinants were increased when US was applicated. The extract obtained using US presented an improved FA profile, containing significantly less SFA than conventionally obtained extract. However, in both extracts, the dominant FA was linoleic acid, followed by γ-linolenic acid.

3.2.3. Impact of Processing Conditions

The efficiency of the UAE process expressed as quantity of antioxidants and antioxidative capacity depends mostly on the species of fruit. There are also conditions of extraction that modulate the effectiveness of UAE, listed below.

Time: Researchers claim that time of sonication influenced polyphenol or lipid fraction yields. Only in the case of one analyzed study, sonication time did not influence the polyphenol content in extracts (besides TAC). With increasing sonication time, yields were higher; however, after reaching specific cutoff values, which may vary among the studies and among the examined compounds, yields decreased [109–113,115].

Temperature: The impact of temperature on extract yield and composition is reported. Generally, the relationship between increasing temperature and polyphenol yield is similar to the impact of time on the yield of extraction. With increasing temperature, polyphenol content increases but beyond a specified point starts to decrease [109,112]. Optimal temperature conditions may vary when its impact on specific fractions of polyphenols is being measured. For example, extracts with high TAC and phenolic acids are obtained at lower temperatures [110,113]. By contrast, higher temperatures result in increased quality and yield of oil obtained by UAE [115].

Solvent type: When different solvent types were compared, the use of ethanol resulted in the highest polyphenol yields. Reports, however, differ with respect to the data provided for the specific concentrations that are most effective in UAE. Machado et al. [99] found 70% ethanol to be optimal for polyphenol extraction yields from blackberry and blueberry pomaces, in contrast to Bamba et al.'s [108] research, which found 50% ethanol to be optimal. However, Loncarić et al. [113] reports that solvent type did not affect specific polyphenol extraction yields except for phenolic acids, whose content was highest when methanol was used.

Table 6. Conditions and results for ultrasound-assisted extraction of oils and bioactive compounds from berry by-products.

Source of Waste	Pretreatment Method	Extraction Conditions (E—Equipment, P—Power, T—Temperature, t—Time, f—Frequency, S—Solvent, S–L—Solid–Liquid Ratio, A—Amplitude)	Antioxidant Composition *	Antioxidant Capacity *	Reference
Raspberry (*Rubus idaeus*) pomace	None	E: ultrasonic cleaner t: 120 min T: 50 °C f: 50 kHz S: acidulated methanol (80%)	**TPC:** 27.79 mg GAE/L (extract) **TFC:** 8.02 mg QE/g (pomace) **TAC:** 7.13 mg C3GE/L (extract)	**FRAP:** 1007.72 µmol/L FRAP **DPPH:** 567.00 µmol/100 g TE **DPPH IC$_{50}$:** 20.00 µL/mL	[107]

Table 6. *Cont.*

Source of Waste	Pretreatment Method	Extraction Conditions (E—Equipment, P—Power, T—Temperature, t—Time, f—Frequency, S—Solvent, S–L—Solid–Liquid Ratio, A—Amplitude)	Antioxidant Composition *	Antioxidant Capacity *	Reference
Blueberry (*Vaccinium angustifolium*) pomace	Freeze-dried pomace	E: ultrasonic cleaner bath S–L: 1/10; 1/15; 1/20; 1/40 S: water or ethanol (10/50 or 90% *v/v* in water) t: 30;40;60;90 min T: 20/40/60 °C pH: 3.3/5.0/6.3/8.3 f: 35 kHz Study was divided into parts where one of the factors was modulated while the others were constant	*Effect of time* (S: water, S–L: 1/20, T: 40 °C): **TPC, TFC, TAC** highest in 90 min *Effect of S–L* (S: 50% ethanol, T: 40 °C, t: 60 min) **TPC, TAC, DPPH** highest in 1/20; **TFC** in 1/15 *Effect of ethanol concentration* (S–L: 1/15, T: 40 °C, t: 40 min): **TPC, TFC, TAC, DPPH** highest in 50% ethanol *Effect of pH* (S: 50% ethanol, S–L: 1/15, T: 40 °C, t: 40 min) **TPC, TFC, DPPH** highest at pH 8.3, **TAC** in pH 3.3 *Effect of temperature* (S: 50% ethanol, S–L: 1/15, t: 40 min): **TPC, TAC** highest at 20 °C, **TFC, DPPH** in 60 °C	[108]
Blackberry (*Rubus fruticosus*) pomace	Lyophilized, milled and sieved (500 um particle size)	E: ultrasound processor f: 20 kHz S: water S–L: 1/24T: 4 °C (at the beginning), 25 °C (at the end of extraction) A: 80–90% t: 10–15 min	Optimum conditions A: 91% and t: 15 min **TPC:** 1201.23 mg GAE/100 g (dm) **TAC:** 379.12 mg/100 g	**ABTS:** 6318 µmol TE/100 g (dm) **DPPH:** 9617.22 µmol TE/100 g	[109]
Raspberry (*Rubus idaeus*) pomace	Freeze-dried, milled (0.45 mm)	UAE combined with enzymatic extraction E: not specified S–L: 1/30 S: acidulated 60% ethanol (*v/v*) t: 20/30/40 min T: 40/45/50 °C +enzyme (pectinase dosage): 0.10/0.15/0.20%	Optimum conditions T: 44 °C, P: 290 W, t: 30 min, pectinase dosage: 0.16% **Anthocyanin yield:** 0.853 mg/g (fw)	**DPPH:** 417.15 TE/g (extract) **ABTS:** 520.07 TE/g	[110]
Chokeberry (*Aronia melanocarpa*) by-products from filter-tea production (tea produced from pomace)	None	M: 10.0 g S: 50% ethanol E: sonication water bath S–L: 1/5 f: 40 kHz P: 72/144/216 W T: 30/50/70 °C t: 30/60/90 min	Optimum conditions for each property in brackets **TPC:** 15.058 mg GAE/mL (extract) (P: 206.64 W, T: 70 °C, t: 80.1 min) **TFC:** 10.436 mg CE/mL (P: 210.24 W, T:70 °C, t: 75 min) **Monomeric anthocyanins:** 2.09 mg C3GE/mL (P: 216 W, T:70 °C, t: 45.6 min) **Proanthocyanins:** 19.82 mg CE/mL (P: 199.44 W, T:70 °C, t: 89.7 min)	-	[111]

Table 6. *Cont.*

Source of Waste	Pretreatment Method	Extraction Conditions (E—Equipment, P—Power, T—Temperature, t—Time, f—Frequency, S—Solvent, S–L—Solid–Liquid Ratio, A—Amplitude)	Antioxidant Composition *	Antioxidant Capacity *	Reference
Blueberry (*Vaccinium ashei*) pomace	Pomace dried in air-circulating oven at 30 °C for 48 h, then milled	E: not specified S: acidulated ethanol (70% *v/v*) P: 400 W T: 50/60/70 °C t: 15/25/35 min S–L: 1/15, 1/20, 1/25	Optimal conditions T: 61 °C, S–L: 1/22 **TPC:** 16.03 mg GAE/g (pomace) **TAC:** 4.19 mg C3GE/g	-	[112]
Blueberry pomace	Freeze-dried, milled	E: ultrasonic bath F: 35 kHz S: acidulated ethanol (50%)/acidulated methanol (50%) t: 5/10/15 min T: 20/40/80 °C	Highest values for: **TPC:** 5.46 mg GAE/g (dm) (Ethanol, 15 min, 80 °C) **TAC:** 953.91 µg/g (Methanol, 15 min, 80 °C) **Phenolic acids:** 561.26 µg/g (Methanol, 15 min, 80 °C) **Flavanols:** 156.04 µg/g (Methanol, 15 min, 80 °C) **Flavonols:** 98.63 µg/g (Methanol, 15 min, 80 °C)	Highest values for: **DPPH:** 0.25 mmol TE/g (dm) (Methanol, 15 min, 80 °C)	[113]

Table 6. *Cont.*

Source of Waste	Pretreatment Method	Extraction Conditions (E—Equipment, P—Power, T—Temperature, t—Time, f—Frequency, S—Solvent, S–L—Solid–Liquid Ratio, A—Amplitude)	Antioxidant Composition *	Antioxidant Capacity *	Reference
Blueberry (*Vaccinium myrtillus*) pomace and blackberry (*Rubus fruticosus*) residues after pulp processing	None	E: ultrasonic bath F: 37 kHz P: 580 W S–L: 1/22,5 S: ethanol 50%/ethanol 70%/acidified water t: 90 min T: 80 °C	*Blackberry residues* *Ethanol 50%* **TPC:** 5.28 mg GAE/g (dm) **Monomeric anthocyanins:** 2.37 mg C3GE/g *Ethanol 70%* **TPC:** 5.86 mg GAE/g **Monomeric anthocyanins:** 2.38 mg C3GE/g *Acidified water* **TPC:** 2.08 mg GAE/g **Monomeric anthocyanins:** 1.26 mg C3GE/g *Blueberry residues* *Ethanol 50%* **TPC:** 4.40 mg GAE/g **Monomeric anthocyanins:** 2.07 mg C3GE/g *Ethanol 70%* **TPC:** 5.75 mg GAE/g **Monomeric anthocyanins:** 2.33 mg C3GE/g *Acidified water* **TPC:** 2.47 mg GAE/g **Monomeric anthocyanins:** 1.36 mg C3GE/g	*Blackberry residues* *Ethanol 50%* **DPPH:** 49.50 µmol TE/g (dm) **ABTS:** 67.35 µmol TE/g **FRAP:** 81.59 mg TE/g *Ethanol 70%* **DPPH:** 51.50 mol TE/g **ABTS:** 70.01 µmol TE/g **FRAP:** 85.09 mg TE/g *Acidified water* **DPPH:** 17.63 µmol TE/g **ABTS:** 24.34 µmol TE/g **FRAP:** 37.03 mg TE/g *Blueberry residues* *Ethanol 50%* **DPPH:** 33.90 µmol TE/g **ABTS:** 55.11 µmol TE/g **FRAP:** 49.94 mg TE/g *Ethanol 70%* **DPPH:** 42.51 µmol TE/g **ABTS:** 55.25 µmol TE/g **FRAP:** 54.82 mg TE/g *Acidified water* **DPPH:** 19.94 µmol TE/g **ABTS:** 19.36 µmol TE/g **FRAP:** 50.16 mg TE/g	[99]
Chokeberry (*Aronia melanocarpa* cv. Nero) pomace	Freeze-dried and ground pomace	E: ultrasonic processor t: 10/13/20/27/30 min S–L: 1/10 S: 60/65/78/90/96% ethanol T: 25 °C A: 50 µm	Highest values for: **TPC:** 188 mg GAE/g (dm) (60% ethanol, 20 min) **TAC:** 89.3 mg C3GE/g (65% ethanol, 13 min)	Highest values for **DPPH:** 49.2 mmol TE/100 g (dm) (60% ethanol, 20 min)	[116]
Chokeberry (*Aronia melanocarpa*) pomace	-	E: horn-type transducer with cooling bath S: ethanol-water (1:1)S–L: 1/10 f: 20 kHz A: 14 µm t: 600 s	**TPC:** 1046 mg/L GAE **Monomeric anthocyanins:** 631 mg/L C3GE	-	[32]
Chokeberry (*Aronia melanocarpa*) stems	Dried stems	E: ultrasonic water bath S: water t: 30 min S–L: 1/25	**TPC:** 5.22 mg GAE/g (dm) **TFC:** 3.94 mg RE/g	**ABTS IC$_{50}$:** 10.09 µg/mL (extract)	[117]

Table 6. *Cont.*

Source of Waste	Pretreatment Method	Extraction Conditions (E—Equipment, P—Power, T—Temperature, t—Time, f—Frequency, S—Solvent, S–L—Solid–Liquid Ratio, A—Amplitude)	Antioxidant Composition *	Antioxidant Capacity *	Reference
Chokeberry (*Aronia melanocarpa* cv. Galicjanka) pomaces obtained from juice pressing from crushed and uncrushed fruits	Freeze-dried, ground pomaces	E: ultrasonic bath S: methanol with 2% formic acid S–L: 1/25 t: 25 min	*Pomace from crushed fruits* **TPC** (UPLC): 15.61 g/100 g (dm) *Pomace from uncrushed fruits* **TPC:** 24.45 g/100 g	*Pomace from crushed fruits* **ABTS:** 59.94 mmol TE/100 g (dm) **FRAP:** 32.61 mmol TE/100 g *Pomace from uncrushed fruits* **ABTS:** 81.63 mmol TE/100 g **FRAP:** 52.22 mmol TE/100 g	[118]

Lipid fractions					
Source of waste	**Pretreatment method**	**Procedure**	**Oil yield**	**Fatty acid composition (%)**	
Raspberry (*Rubus coreanus*) seeds	Seeds dried in a convection oven at 60 °C for 24 h, then milled	E: sonication cleaning bath f: 40 kHz P: 250 W S–L: 1/40 S: ethanol t: 10/20/30/40/50 min T: 30/40/50/60/70 °C	Optimal conditions: 54 °C, 37 min 22.78%	**SFA:** 2.45 **MUFA:** 0.55 **PUFA:** 92.25	[115]

* Results are expressed as written in the bracket after first given result.

3.3. Pulsed Electric Field-Assisted Extraction

A review by Kumari et al. [119] described a pulsed electric field (PEF) applied in short duration pulses of moderate voltage to a material placed between two electrodes. The effect of PEF includes electroporation caused by damage to cell membranes. The formation of pores leads to mechanical breakdown of cell membranes and the material is defined as disintegrated, as presented at Figure 4. Factors involved in PEF-assisted extraction are the intensity of the electric field, the duration of treatment, the waveform of the pulse, conductivity, the porosity of the material, pH and the ionic strength of the solvent. PEF technology results in improved extraction of intracellular compounds due to increased diffusivity of intracellular substances and increased mass transfer rates. The conditions and effects of PEF applied in the pretreatment of berry fruits before juice pressing on antioxidant content and antioxidant capacity in pomace extracts are summarized in Table 7.

PEF was applied to improve the extraction of juice from blueberries. Anthocyanins from blueberry pomace were investigated by Pataro et al. [120]. Blueberries, cut in half, were pretreated by PEF with different input energy values—1 kJ/kg, 5 kJ/kg or 10 kJ/kg—before juice pressing. In all the samples of pomace obtained after pressing blueberries pretreated with PEF, TAC and AA were increased compared to the control (a sample of pomace remaining after pressing the juice from blueberries untreated with PEF). A correlation between increased energy input and increased values of both TAC and AA was noticed. PEF was also applied to crushed blueberries before juice-pressing in a study conducted by Bobinaitė et al. [121]. An influence of field strength value on the SLE of pomace efficiency was studied and it was found that the highest used field strength—5 kV/cm—resulted in the highest antioxidant content and capacity of blueberry pomace extracts. The values of antioxidant content and antioxidant activity were increased compared to extracts obtained from blueberry pomace untreated with PEF before juice pressing. Loncarić et al. [113]

applied PEF to freeze-dried and milled pomace, which was then extracted (SLE) with acidulated ethanol or with acidulated methanol used as a solvent. The correlation between specific phenolic yields and variable numbers of pulses, energy inputs and solvent types was studied. For TPC, phenolic acids and flavonols, the most suitable conditions were: 100 pulses, a field strength of 20 kV/cm and acidulated ethanol used as a solvent. For TAC and flavanols, the number of pulses and the field strength defined as most efficient did not change, but acidulated methanol was preferable as a solvent. The DPPH of an extract was highest (830 µmol TE/g dm) when the number of pulses was 100, the field strength was 20 kV/cm and acidulated ethanol was used as a solvent. Another example of PEF application in antioxidant extraction from blueberry pomace was described by Zhou et al. [122]. The researchers treated with PEF pomace thawed to a liquid state and ground in a colloid mill. The influence of variable conditions of treatment on anthocyanin yield was studied: the number of pulses, field strength and liquid–liquid ratio. The highest extraction yields were obtained after 10 pulses of PEF with a field strength of 20 kV/cm and using acidic ethanol in a ratio of one-to-six. Compared with UAE, PEF extraction resulted in higher anthocyanin extract yields despite a shorter time and lower temperature for the process.

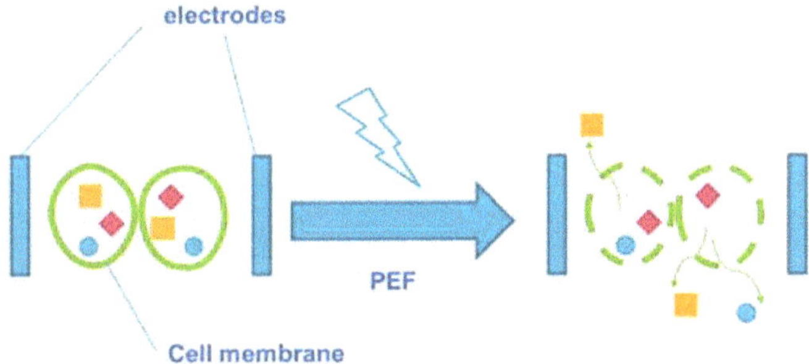

Figure 4. A scheme of PEF impact on plant material.

As the above results for PEF-assisted extraction from by-products only concern blueberry pomace treatment, an example of extraction of polyphenolic compounds from blackcurrant juice preceded by PEF application could be mentioned (not included in the following table). The PEF procedure was conducted before juice production under optimized conditions, as determined by RSM modelling. The chosen conditions for the electric field—1318 V/cm and 315 pulses (pulse width 100 ms)—followed by methanol (for TPC and AA determination) or ethanol (for monomeric anthocyanin content determination) liquid extraction of blackcurrant juice resulted in extracts with significantly lowered pH but increased TPC, AA and monomeric anthocyanin content compared to the control (blackcurrant juice untreated with PEF) [123]. Conditions of extraction also influence the effectiveness of PEF, as listed below.

Impact of Processing Conditions

Field strength: Most studies show that a higher field strength results in increased antioxidant compound yields. When field strength is greater, the potential difference outside and inside the cell membrane is higher than the critical membrane potential and this improves the dissolution rate of the cell membrane. However, applying too strong a field may promote antioxidant degradation [119].

Number of pulses: Study results are inconclusive regarding the influence of the number of PEF pulses on the antioxidant content and capacity of extracts. Loncarić et al. [113] claim that the highest number of pulses used (100) resulted in the highest antioxidant content and capacity, while Zhou et al. [122] claim that 12 pulses was less efficient than 10, this being the optimal number of PEF pulses for the anthocyanin extraction process.

Solvent type: The properties of solvents which may determine extraction efficiency are described in the paragraph focused on SLE.

Table 7. Conditions and results of pulsed electric field-assisted extraction of bioactive compounds from berry by-products.

Material	PEF Conditions (E—Equipment, P—Power, FS—Field Strength, f—Frequency, W—Pulse Width, I—Energy Input, S—Solvent, L–L—Liquid–Liquid Ratio)	Procedure	Antioxidant Composition *	Antioxidant Activity *	Reference
Blueberry (*Vaccinium myrtillus*)	E: generator of monopolar square wave pulses FS: 3 kV/cm f: 10 Hz W: 20 μs I: 1/5/10 kJ/kg	PEF was applied to blueberry fruits cut in half, before pressing, obtained pomace examined	Optimum conditions I: 10 kJ/kg **TAC** (HPLC): 1574.1 mg/100 g (pomace)	Optimum conditions I: 10 kJ/kg **DPPH:** 34.2 μmol TE/g (pomace) **FRAP:** 68.0 μmol TE/g	[120]
Blueberry (*Vaccinium myrtillus*)	E: cylindrical PEF treatment chamber (monopolar square pulses) FS: 1/3/5 kV/cm f: 10 Hz W: 20 μs P: 20 kW (average) I: 10 kJ/kg	PEF applied to crushed fresh berries before juice pressing. By-product was extracted in SLE and examined.	Optimum conditions FS: 5 kV/cm **TPC:** 1782.64 mg GAE/100 g (fw) **TAC:** 1698.55 mg/100 g	Optimum conditions FS: 5 kV/cm **FRAP:** ca. 72 μmol TE/g (fw)	[121]
Blueberry pomace	E: laboratory PEF treatment chamber Pulse duration: 2 μs No. of pulses: 10/50/100 FS: 10/15/20 kV/cm	PEF applied to lyophilized and milled pomace, followed by SLE extraction with acidulated ethanol or acidulated methanol	Optimum conditions in brackets: **TPC:** 10.52 mg GAE/g (dm) (Ethanol, 20 kV/cm, 100 pulses) **TAC:** 1757.32 μg/g (Methanol, 20 kV/cm, 100 pulses) **Phenolic acids:** 625.47 μg/g (Ethanol, 20 kV/cm, 100 pulses) **Flavanols:** 297.86 μg/g (Methanol, 20 kV/cm, 100 pulses) **Flavonols:** 157.54 μg/g (Ethanol, 20 kV/cm, 100 pulses)	Optimum conditions in brackets: **DPPH:** 830 μmol TE/g (dm) (Ethanol, 20 kV/cm, 100 pulses)	[113]
Blueberry pomace	E: PEF system Pulse duration: 2 μs S: acidic ethanol L/L: 1:5/1:6/1:7 FS: 15/20/25 kV/cm No of pulses: 8/10/12 Flow rate: 7 mL/min	Thawed to liquid, grinded in colloid mill pomace was treated with PEF in liquid material chamber	Optimum conditions No of pulses: 10, FS: 20 kV/cm, L/L: 1:6 **TAC:** 223.13 mg C3GE/L (extract)	-	[122]

* Results are expressed as written in the bracket after first given result.

3.4. Microwave-Assisted Extraction (MAE)

Microwave-assisted extraction employs microwaves—non-ionizing electromagnetic waves—which cause changes in plant cell structure. The phenomena of heat and mass transfer which proceed in one direction occur in this type of extraction. The microwave energy is applied directly to material due to molecular interactions with the electromagnetic field through the conversion of electromagnetic energy to thermal energy. The heat must be then dissipated volumetrically inside the sample. These phenomena improve cell penetration and the internal and external diffusion of compounds is what finally leads to improved extraction yields [124]. The examples of microwaves applied in bioactive compound extraction processes for berry fruit wastes are presented in Table 8.

Pap et al. [125] studied the optimum conditions for MAE of anthocyanins from blackcurrant pomace. Variable values for power, time, solid–liquid ratio and the pH of the solvent were applied. Results have shown that the highest power (700 W) and solid–liquid ratio (1:20) but the shortest time (10 min) and the lowest pH (2) values were the most efficient conditions for anthocyanin extraction. In a HPLC study, delphinidin-3-rutoside turned out to be the most abundant anthocyanin. Davis et al. [126] applied various types of solvents and levels of power in MAE of pectin and polyphenolic rich extracts from cranberry pomace. The highest yields of polyphenols were obtained with an alkaline extraction process with a power value of 36 W/g. When SLE and MAE used with different solvents of cranberry press residues were compared, MAE resulted in a higher yield of extraction in every variant of the experiment. Values of quercetin equivalents of powdered cranberry residues were highest for MAE with 100% acetone used as a solvent and were significantly increased compared to water and ethanol extraction processes [127]. Klavins et al. [128] compared different methods of extraction of phenolic compounds from cranberry pomace: SLE, UAE and MAE, using ethanol and trifluoroacetic acid as a solvent mixture. The extract obtained in MAE featured the lowest anthocyanin and polyphenol contents across all the studied samples.

Impact of Processing Conditions

Power: Inconclusive results of applied microwave power impact on efficiency were observed. A higher power value applied in MAE improved the extraction of anthocyanins from blackcurrant pomace [125] and also led to lower phenolic content in extracts from cranberry pomace [126]. These results are associated with the temperature increase when extended power is applied. The efficiency of MAE increases with increasing microwave power till the optimum temperature point is reached, after which it starts to decrease while the power (and temperature) is still rising [124].

Time: The shorter the time of microwave input, the better MAE efficiency was observed [125]. This may be related to the destructive impact of microwaves and the increased temperature on the structure of bioactive compounds. A prolonged duration of microwave power applied to a sample can promote the degradation of antioxidants [124].

Solvent type: The solvent influence on the MAE process is similar to that described in SLE. However, the capacity of the solvent to absorb microwave energy should be taken into consideration while analysing MAE experiments. Considering the analysed studies, alkaline solvents result in higher TPCs of extracts [126], and replacing ethanol with acetone was found to improve the TPCs of extracts [127]. Acetone is a solvent polar enough to be heated by microwave energy, which results in better cell heating and improved diffusion of extracted compounds [124].

Table 8. Conditions and results of microwave-assisted extraction of bioactive compounds from berry by-products.

Source of Waste	Pretreatment	Extraction Procedure (E—Equipment, f—Frequency, T—Temperature, P—Power, M—Sample Weight, S–L—Solid–Liquid Ratio, t—Time, S—Solvent)	Antioxidant Content *	Reference
Blackcurrant (*Ribes nigrum*) pomace	Pomace was obtained from enzymatically treated fruits	E: single-mode cavity resonator f: 2.45 GHz T: 69.7 °C P: 140/420/700 W M: 28 g S–L: 1:10, 1:13.3, 1:20 t: 10/20/30 min solvent pH: 2/4.5/7	In optimum conditions P: 700 W, t: 10 min, S–L: 1:20, pH 2 **TAC** (HPLC): 20.4 mg/g (fw)	[125]
Cranberry (*Vaccinium macrocarpon*) pomace	Freeze-dried and ground pomace	E: microwave reactor S–L: 1:30 M: 1 g P: 36/72 W/g pomace t: 4 min *Acidic extraction* S: 0.1 M HCl *Alkaline extraction* S: 0.15 M NaOH *Sequential acidic and alkaline extraction* S: 0.1 M HCl + 0.15 M NaOH	*Acidic extraction* P: 36 W/g **TPC:** 3.01 mg GAE/g (fw) P: 72 W/g **TPC:** 0.92 mg GAE/g *Alkaline extraction* P: 36 W/g **TPC:** 22.78 mg GAE/g P: 72 W/g **TPC:** 11.79 mg GAE/g *Sequential* P: 36 W/g TPC: 11.90 mg GAE/g P: 72 W/g **TPC:** 11.63 mg GAE/g	[126]
Cranberry (*Vaccinium macrocarpon*) pomace	Oven-dried (1 h/100 °C) press cake, then ground	E: microwave press S: water/ethanol/acetone M: 3.5 g T: 125 °C t: 10 min	*Water* **TPC:** 0.02 mmol QE/g (extract) *50% ethanol* **TPC:** 0.12 mmol QE/g *100% ethanol* **TPC:** 0.30 mmol QE/g *50% acetone* **TPC:** 0.27 mmol QE/g *100% acetone* **TPC:** 0.53 mmol QE/g	[127]
Cranberry (*Vaccinium macrocarpon*) pomace	Freeze-dried and homogenized pomace	E: microwave extraction unit S: 96% ethanol and 0.5% trifluoroacetic acid 50 mL M: 0.5 g P: 600 W T: 80 °C t: 20 min	**TAC** (pH differential method): 0.054 g/100 g (berry powder) **TPC** (Folin–Ciocalteu): 1.09 g/100 g	[128]

* Results are expressed as written in the bracket after first given result.

3.5. Supercritical Fluid Extraction

Supercritical fluid extraction (SFE) includes the use of solvents at temperatures and pressures above the critical values for temperature and pressure. These conditions exhibit both the gaseous and liquid properties of solvents. SFE is commonly performed using carbon dioxide (CO_2), as it has a low critical pressure and temperature and is considered non-toxic, non-flammable and not expensive. It is also a non-polar and hydrophobic solvent. This is the reason why SFE is mostly used to extract lipid fractions from plant material. SFE also provides high selectivity of extraction which may be modulated by changing the conditions of extraction [129].

Examples of supercritical fluid extraction application are given in Table 9. Campalani et al. [130] compared SFE with the SLE of lipid fractions from raspberry, blueberry, blackcurrant, blackberry and strawberry pomaces. Oil yield obtained with SFE was increased compared to SLE only in the case of raspberry and blackberry oils. The study also showed particular fatty acid yields, and despite the general value of the oil yield, applying SFE resulted in higher percentage of FAs for each extract. This proves the better selectivity of SFE towards FAs. Marić et al. [131] studied differences between cold pressing and

SFE of raspberry seeds. A higher oil yield was reached when cold pressing was applied, and a lower tocopherol content was observed in oil obtained with SFE. Applying SFE, Milala et al. [132] extracted oils from raspberry, chokeberry and strawberry by-products. Oil yields ranged from 12% for raspberry to 18% for strawberry pomaces. The lipid fraction was collected at particular times of the extraction process. The properties of oil—tocopherol content and fatty acid and pigment composition—were dependent on the oil collection time. Correa et al. [133] studied different conditions of SFE of oil from blackberry seeds and a comparison between SFE and SLE methods. The difference between oil yields under optimum conditions for SFE and SLE was significant. Using propane as a solvent at a temperature of 70 °C and a pressure of 20 MPa during SFE resulted in a 2.32% oil yield. By comparison, SLE conducted using hexane resulted in a 10.51% oil yield. Antioxidant content was also measured and extracts obtained in SFE with CO_2 as a solvent under optimum conditions were characterized by higher TPC values than oils extracted traditionally. Wajs-Bonikowska et al. [88] also studied properties of blackberry seed oil extracted using the SFE method. Lipid fraction yield in the case of SFE was similar to the yield when SLE with hexane was applied. SFE resulted in oil with a lower tocopherol content compared to SLE. Despite the results summarized in the Table 9, Pavlić et al. [134], on the basis of their studies applying SFE in the process of oil isolation from raspberry seeds, concluded that higher oil yields were determined by higher pressure and CO_2 flow rate.

Basegmez et al. [135] obtained polyphenol-rich extracts using SFE in their research. The optimum conditions of the process were 45 MPa, 60 °C and 120 min, and resulted in a TPC of 24.34 mg GAE/g extract.

Impact of Processing Conditions

Pressure of solvent: In studies which aim was to optimize the SFE method it may be noticed that higher oil and polyphenol rich extract yields were noted when higher (but not the highest possible) pressures were applied [133,135]. This is connected to the increasing solubility of compounds when pressure is increased [136].

Temperature: The higher the temperature, the greater the oil and bioactive extract yields obtained [133,135]. Increased temperature improves the diffusion and solubility of substances [136].

Flow rate: In the reviewed studies, oils were extracted more effectively when the flow rate was high enough to be diffusion-limited [88,130–132,135]. However, this is connected with the increased amounts of solvents used in the process [136].

Table 9. Conditions and results of supercritical fluid extraction of oils and bioactive compounds from berry by-products.

Source of Waste	Pretreatment Method	Extraction Conditions (E—Equipment, M—Sample Weight, S—Solvent, p—Pressure, V—Volume, T—Temperature, t—Time, FR—Flow Rate)	Oil Yield	Fatty Acids Profile (PUFA/MUFA/SFA, Dominant FA)	Reference
Raspberry, blueberry, blackcurrant, blackberry, strawberry frozen pomaces	Air-dried with water rinsing (15 min water rinsing, 24 h drying) pomace	S: CO_2 M: 3 g FR: 5.0/2.5 cm^3/min p: 300 bar T: 70 °C t: 5 h Collected by venting into hexane	*Raspberry pomace* 7.5% *Blueberry pomace* 9.7% *Blackcurrant pomace* 4.6% *Blackberry pomace* 6.6% *Strawberry pomace* 13.5%	*Fatty acid yields in mg/g* *Raspberry pomace* **PUFA:** 191.0 **MUFA:** 61.2 **SFA:** 50.7 Dominant: C18:2 *Blueberry pomace* **PUFA:** 134.9 **MUFA:** 74.2 **SFA:** 25.6 Dominant: C18:2 *Blackcurrant pomace* **PUFA:** 60.3 **MUFA:** 0.0 **SFA:** 104.3 Dominant: C18:2 *Blackberry pomace* **PUFA:** 197.0 **MUFA:** 66.3 **SFA:** 31.4 Dominant: C18:2 *Wild strawberry pomace* **PUFA:** 145.8 **MUFA:** 64.0 **SFA:** 46.8 Dominant: C18:2	[130]
Raspberry (*Rubus idaeus*) cv. Willamette seeds	Milled seeds	E: high pressure extraction plant S: CO_2 M: 70 g P: 300 bar T: 40 °C FR: 0.194 kg/h t: 3 h	8.82%	**PUFA:** 77.90 **MUFA:** 14.47 **SFA:** 6.20 Dominant: C18:2 *n*6	[131]

Table 9. *Cont.*

Source of Waste	Pretreatment Method	Extraction Conditions (E—Equipment, M—Sample Weight, S—Solvent, p—Pressure, V—Volume, T—Temperature, t—Time, FR—Flow Rate)	Oil Yield	Fatty Acids Profile (PUFA/MUFA/SFA, Dominant FA)	Reference
Raspberry (*Rubus idaeus* cv. Polka and cv. Polana) Chokeberry (*Aronia melanocarpa* cv. Nero), Strawberry (*Fragaria vesca* cv. Honeoye, cv. Senga Sengana and cv. Polka) pomaces	Pomace dried convectively in industrial vacuum dryers at 70 °C for 8 h. Seeds separated in industrial sieving machines, then crushed in mill crusher and sieved again under CO_2 or nitrogen atmosphere	E: plant for extraction S: CO_2 M: 14.2 kg strawberry, 14.5 kg chokeberry, 13.3 kg raspberry p: 250 bar, one step separation at 53 bar T: 40 °C FR: 200 kg/h, t: 180–225 min with fractionation (particular collection times)	*Strawberry pomace* 18% *Chokeberry pomace* 15% *Raspberry pomace* 12%	Values for first collection (after 15 min) *Strawberry pomace* **PUFA:** 78.9 **MUFA:** 15.1 **SFA:** 5.5 Dominant: C18:2 *Chokeberry pomace* **PUFA:** 76.6 **MUFA:** 16.4 **SFA:** 5.4 Dominant: C18:2 *Raspberry pomace* **PUFA:** 84.1 **MUFA:** 10.8 **SFA:** 4.8 Dominant: C18:2	[132]
Blackberry (*Rubus* ssp. cv. Xavante) seeds	Seeds dried in air circulation oven (40 °C/48 h), milled, classified according to particle size using vibratory sieve shaker	E: special apparatus M: 30 g S: CO_2 and propane p: ranging 15.0–25.0 MPa when using CO_2 and 10.0–20.0 MPa when using propane T: 40.0–70.0 °C for CO_2 and 30.0–70.0 °C for propane t: for CO_2: 150 min, propane: 60 min	Optimum conditions in brackets S: CO_2 (70 °C, 25 MPa): 1.89% S: propane (70 °C, 20 MPa): 2.32%	-	[133]
Blackberry (*Rubus fruticosus*) pomace	Pomace dried in the sun and crushed four times in a cylinder mill	E: plant scale fluid extractor S: CO_2 M: 3500 g P: 300 bar T: 50 °C FR: 80 kg/h t: 150 min	11.4%	FA expressed as g/100 g **PUFA:** 58.2 **MUFA:** 12.6 **SFA:** 7.2	[88]

Polyphenol-rich extracts					
Source of waste	**Pretreatment**	**Extraction procedure**	**Antioxidant composition**	**Antioxidant activity ***	**Reference**
Blackcurrant (*Ribes nigrum*) pomace	Lyophilized, ground pomace	E: SFE system M: 15 g p: 30–55 MPa T: 30–60 °C t: 60–150 min FR: 3.6 g/min	Optimum conditions: 45 MPa, 60 °C, 120 min **TPC:** 24.34 mg GAE/g extract	**DPPH:** 1.59 mg TE/g (extract) **ORAC:** 11.35 mg TE/g **FRAP:** 25.00 mg TE/g	[135]

* Results are expressed as written in the bracket after first given result.

3.6. Other Alternative Methods of Extraction

3.6.1. Pressurized Liquid Extraction

Pressurized liquid extraction (PLE) is one of the pressurized approaches, next to SFE. PLE may, however, be used to extract high- and medium-polarity substances, whereas SFE results in extracts rich in non-polar compounds, as supercritical CO_2 is used as a solvent. Combining those two methods has the benefit of obtaining extracts containing compounds characterized by different polarities [137]. There are studies reporting the extraction of compounds from residues after SFE-CO_2. Brazdauskas et al. [138] studied PLE of compounds from black chokeberry pomace remaining after SFE-CO_2. Water, ethanol and formic acid mixture was used as a solvent, a constant pressure was set at 10.3 MPa and variable conditions of temperature, ethanol and formic acid concentrations were examined. The optimum values were 165 °C, 46% ethanol and 1.8% formic acid, and resulted in response variables of 72.53% yield, 236.64 mg GAE/g extract TPC (Folin–Ciocalteu method), 4.346 mmol TE/g extract TEAC and 5.92 µg/mL as the EC50 value in a DPPH assay. Grunovaite et al. [139] also described PLE of chokeberry pomace residues after SFE-CO_2. The procedure was performed at 70 °C, 10.3 MPa, using 96% ethanol or 100% acetone as a solvent and resulted in a 22.70% yield when ethanol was used and a 17.90% yield when acetone was used. Ethanol extract was characterized by a TPC of 89.41 mg GAE/g extract in a Folin–Ciocalteu assay (all values expressed per dm basis) and antioxidant activities: 2.52 mmol TE/g extract, 1.99 mmol TE/g extract and 3.03 mmol TE/g extract in ABTS, DPPH and ORAC assays, respectively. Pressurized acetone extraction resulted in a TPC value of 35.35 mg GAE/g extract and antioxidant activities for ABTS, DPPH and ORAC assays amounting to 0.91 mmol TE/g extract, 0.24 mmol TE/g extract and 3.84 mmol TE/g extract, respectively. In this study, also PLE alone, conducted with different conditions of temperature and using different solvents—hexane, methanol, water, acetone with water mixture and a methanol with water mixture—was examined. The optimum parameters of the process were: 130 °C and methanol used as a solvent. The conditions mentioned above allowed the obtention of the highest extraction yields (48.13%) and values for TPC (410.20 mg GAE/g extract) and antioxidant activity of the extract, as measured in an ABTS (2.17 mmol TE/g extract) assay. Antioxidant activity, as determined by ORAC and DPPH, was highest when an acetone and water mixture under 130 °C was used in extraction. It can be concluded that PLE extraction without a previous SFE-CO_2 procedure results in higher extraction yields. However, antioxidant activity determined in ABTS and DPPH assays assumes higher values when the residue after SFE-CO_2 is extracted again using PLE with ethanol as a solvent.

Kryževičūtė et al. [140] used pressurized liquid extraction with 50% ethanol of raspberry pomace remaining after SFE-CO_2 extraction to obtain antioxidative compounds which were used to prevent beef burger spoilage during prolonged storage. The extraction yield was 19.3% and the obtained extracts were characterized by the following antioxidant capacities: 936 µmol TE/g (extract) and 123 µmol TE/g in ORAC and ABTS assays, respectively. The TPC of the extract was 208.3 mg GAE/g. It was concluded that 1% PLE extract application helped to prevent oxidation in the burgers.

3.6.2. Enzyme-Assisted Extraction

Enzymes, such as cellulases, hemicellulases and pectinases, are used in the pretreatment of plant material prior to extraction processes to disintegrate the cell walls of material, resulting in enhanced penetration of solvents and increased extraction yields. Other benefits include shorter extraction times, reduced quantities of solvents and improved quality. However, the cost of enzymes and time and the limited ability of enzymes to complete cell wall disintegration along with the strict conditions for enzymes application are significant disadvantages of this extraction method [141].

For instance, enzymatic-assisted extraction (EAE) was used to obtain bioactive compounds from bilberry pomace, which was firstly defatted in an SFE-CO_2 procedure. After removing the lipophilic fraction, enzymes were applied in particular conditions (pH, volume of solution, temperature) and water-soluble fractions obtained under optimal chosen conditions (pH 4.5, 46 °C, 1 h, enzyme concentration 2 active units/g of pomace) were then examined. A water-soluble fraction obtained in SLE was used as a control sample. Values for yield, TPC, ABTS, ORAC and CUPRAC assays were higher when EAE was applied [142]. Kitryte et al. [143] optimized the EAE of bioactive compounds from chokeberry pomace using cellulolytic and xylanolytic enzymes. Variable values of E–S (enzyme–solid) ratio, temperature, pH and extraction time were applied. The optimal parameters were defined as: E/S 6%, temperature 40 °C, pH 3.5 and 7 h of extraction time. The influence of the type of enzyme on polyphenol and oil extraction yields from raspberry pomace was studied by Saad et al. [144]. The enzyme that improved extraction efficiency was alkaline protease. Using that enzyme, the optimization of EAE was conducted by the authors and optimal conditions were chosen: particle size of material 50–750 μm, pH 9, enzyme concentration of 1.2 units/100 g pomace, temperature of 60 °C, S–L of 9% and a hydrolysis time of 2 h, which resulted in a 5.87 g/100 g pomace (fw) extraction yield.

As the studies show, the conditions of EAE should be optimized to obtain the best possible extraction yield results. Enzymes are substances sensitive to different environmental factors and their efficiency is strictly connected with them. EAE may be considered a green extraction method which may be helpful in reducing solvent quantities or power levels used in the process; however, it is also a costly and demanding technique.

4. Conclusions

The study has shown a range of possible extraction methods which are applied to extract bioactive compounds or oils from berry fruit by-products: ultrasound-assisted, pulsed electric field-assisted, microwave-assisted, supercritical fluid, pressurized liquid and enzyme-assisted methods. There are some major differences between the presented methods with respect to their usefulness. The basic issue concerns the equipment needed for the extraction procedures. Some of the devices used in, e.g., SFE or PEF-assisted methods are advanced and costly units compared to the simple-to-operate SLE apparatus. Additionally, there are some preferable solvents and S–L ratios used in extraction, depending on the chosen method. Traditional SLE involves the use of high amounts of organic solvents, although in UAE, MAE or enzyme-assisted extraction these could be replaced with water or water–organic solvent mixtures, whereas SFE requires specific supercritical-state solvents. Time consumption is another factor that differs among the considered methods. SLE takes time, up to 24 h, and using alternative extraction methods can reduce the duration of extraction processes to a few hours or even minutes in the case of UAE or MAE. There are, also, specific, unique parameters of conditions for each extraction method which also influence the final results but which it is not possible to compare.

The conventional (solid–liquid) processing methods involve high energy and solvent consumption, which may be harmful to the environment and financially unfavorable. However, SLE is often the most effective extraction technique, considering extraction yield. Novel, alternative extraction methods are beneficial due to their high selectivity. They are also classified as 'green' extraction methods, which means they are environment-saving. The alternative extractions conditions are still being modified to obtain the most efficient model of extraction technique and further research in this area should be carried out.

Author Contributions: Conceptualization, I.P., A.W. and A.G.; methodology, I.P., A.W. and A.G.; investigation, I.P.; resources, I.P.; data curation, I.P.; writing—original draft preparation, I.P.; writing—review and editing, A.W. and A.G.; visualization, I.P. and A.W.; supervision, A.W. and A.G.; project administration, I.P., A.W. and A.G.; funding acquisition, A.G. All authors have read and agreed to the published version of the manuscript.

Funding: This research received no external funding.

Institutional Review Board Statement: Not applicable.

Informed Consent Statement: Not applicable.

Data Availability Statement: Not applicable.

Conflicts of Interest: The authors declare no conflict of interest.

Abbreviations

AA	Antioxidant activity
C3GE	Cyanidin-3-glucoside equivalent
CAE	Caffeic acid equivalent
CUPRAC	Cupric reducing antioxidant capacity
DF	Dietary fiber
dm	Dry mass
EAE	Enzyme-assisted extraction
EE	Epicatechin equivalent
FA	Fatty acid
FAO	Food and Agriculture Organization of the United Nations
FE	Iron (Fe^{2+}) equivalent
FRAP	Ferric reducing antioxidant power
fw	Fresh weight
GAE	Gallic acid equivalent
MAE	Microwave-assisted extraction
ME	Malvidin equivalent
MUFA	Monounsaturated fatty acid
ORAC	Oxygen radical absorbance capacity
PEF	Pulsed electric field
PLE	Pressurized liquid extraction
PUFA	Polyunsaturated fatty acid
QE	Quercetin equivalent
RE	Rutin equivalent
RSM	Response-surface methodology
SFA	Saturated fatty acid
SFE	Supercritical fluid extraction
SLE	Solid–liquid extraction
T3	Tocotrienol
TAC	Total anthocyanin content
TE	Trolox equivalent
TEAC	Trolox equivalent antioxidant capacity
TFC	Total flavonoid content
TP	Tocopherol
TPC	Total polyphenolic content
UAE	Ultrasound-assisted extraction
WHO	World Health Organization

References

1. FAO FAOSTAT Crop Statistics. Available online: http://www.fao.org/faostat/en/#data/QC (accessed on 1 March 2021).
2. Fruits and Derived Products. Available online: http://www.fao.org/es/faodef/fdef08e.htm (accessed on 6 May 2021).
3. Food Loss and Waste Facts. Available online: http://www.fao.org/resources/infographics/infographics-details/en/c/317265 (accessed on 6 May 2021).
4. FAO. *Moving Forward on Food Loss and Waste Reduction*; Food and Agriculture Organization of The United Nations: Rome, Italy, 2019.
5. Campos, D.A.; Gómez-García, R.; Vilas-Boas, A.A.; Madureira, A.R.; Pintado, M.M. Management of Fruit Industrial By-Products—A Case Study on Circular Economy Approach. *Molecules* **2020**, *25*, 320. [CrossRef]
6. Schmid, V.; Steck, J.; Mayer-Miebach, E.; Behsnilian, D.; Briviba, K.; Bunzel, M.; Karbstein, H.P.; Emin, M.A. Impact of defined thermomechanical treatment on the structure and content of dietary fiber and the stability and bioaccessibility of polyphenols of chokeberry (*Aronia melanocarpa*) pomace. *Food Res. Int.* **2020**, *134*, 109232. [CrossRef] [PubMed]
7. Reißner, A.-M.; Alhamimi, S.; Quiles, A.; Schmidt, C.; Struck, S.; Hernando, I.; Turner, C.; Rohm, H. Composition and physico-chemical properties of dried berry pomace. *J. Sci. Food Agric.* **2019**, *99*, 1284–1293. [CrossRef] [PubMed]

8. McDougall, N.; Beames, R. Composition of raspberry pomace and its nutritive value for monogastric animals. *Anim. Feed Sci. Technol.* **1994**, *45*, 139–148. [CrossRef]
9. Górnaś, P.; Juhņeviča-Radenkova, K.; Radenkovs, V.; Mišina, I.; Pugajeva, I.; Soliven, A.; Segliņa, D. The impact of different baking conditions on the stability of the extractable polyphenols in muffins enriched by strawberry, sour cherry, raspberry or black currant pomace. *LWT Food Sci. Technol.* **2016**, *65*, 946–953. [CrossRef]
10. Han, X.; Shen, T.; Lou, H. Dietary Polyphenols and Their Biological Significance. *Int. J. Mol. Sci.* **2007**, *8*, 950–988. [CrossRef]
11. Williamson, G. The role of polyphenols in modern nutrition. *Nutr. Bull.* **2017**, *42*, 226–235. [CrossRef]
12. Somerville, V.; Bringans, C.; Braakhuis, A. Polyphenols and Performance: A Systematic Review and Meta-Analysis. *Sports Med.* **2017**, *47*, 1589–1599. [CrossRef]
13. Ma, G.; Chen, Y. Polyphenol supplementation benefits human health via gut microbiota: A systematic review via meta-analysis. *J. Funct. Foods* **2020**, *66*, 103829. [CrossRef]
14. Liu, F.; Li, D.; Wang, X.; Cui, Y.; Li, X. Polyphenols intervention is an effective strategy to ameliorate inflammatory bowel disease: A systematic review and meta-analysis. *Int. J. Food Sci. Nutr.* **2020**, *72*, 14–25. [CrossRef] [PubMed]
15. Ammar, A.; Trabelsi, K.; Boukhris, O.; Bouaziz, B.; Müller, P.; Glenn, J.M.; Bott, N.T.; Müller, N.; Chtourou, H.; Driss, T.; et al. Effects of Polyphenol-Rich Interventions on Cognition and Brain Health in Healthy Young and Middle-Aged Adults: Systematic Review and Meta-Analysis. *J. Clin. Med.* **2020**, *9*, 1598. [CrossRef] [PubMed]
16. Shah, K.; Shah, P. Effect of Anthocyanin Supplementations on Lipid Profile and Inflammatory Markers: A Systematic Review and Meta-Analysis of Randomized Controlled Trials. *Cholest* **2018**, *2018*, 8450793. [CrossRef] [PubMed]
17. Godos, J.; Vitale, M.; Micek, A.; Ray, S.; Martini, D.; Del Rio, D.; Riccardi, G.; Galvano, F.; Grosso, G. Dietary Polyphenol Intake, Blood Pressure, and Hypertension: A Systematic Review and Meta-Analysis of Observational Studies. *Antioxidants* **2019**, *8*, 152. [CrossRef] [PubMed]
18. Fantini, M.; Benvenuto, M.; Masuelli, L.; Frajese, G.V.; Tresoldi, I.; Modesti, A.; Bei, R. In Vitro and in Vivo Antitumoral Effects of Combinations of Polyphenols, or Polyphenols and Anticancer Drugs: Perspectives on Cancer Treatment. *Int. J. Mol. Sci.* **2015**, *16*, 9236–9282. [CrossRef] [PubMed]
19. Cao, H.; Saroglu, O.; Karadag, A.; Diaconeasa, Z.; Zoccatelli, G.; Conte-Junior, C.A.; Gonzalez-Aguilar, G.A.; Ou, J.; Bai, W.; Zamarioli, C.M.; et al. Available technologies on improving the stability of polyphenols in food processing. *Food Front.* **2021**, *2*, 109–139. [CrossRef]
20. Zeng, L.; Ma, M.; Li, C.; Luo, L. Stability of tea polyphenols solution with different pH at different temperatures. *Int. J. Food Prop.* **2017**, *20*, 1–18. [CrossRef]
21. Chethan, S.; Malleshi, N. Finger millet polyphenols: Optimization of extraction and the effect of pH on their stability. *Food Chem.* **2007**, *105*, 862–870. [CrossRef]
22. Teleszko, M.; Nowicka, P.; Wojdyło, A. Effect of cultivar and storage temperature on identification and stability of polyphenols in strawberry cloudy juices. *J. Food Compos. Anal.* **2016**, *54*, 10–19. [CrossRef]
23. Chen, J.; Sun, H.; Wang, Y.; Wang, S.; Tao, X.; Sun, A. Stability of Apple Polyphenols as a Function of Temperature and pH. *Int. J. Food Prop.* **2014**, *17*, 1742–1749. [CrossRef]
24. Tolić, M.-T.; Jurčević, I.L.; Krbavčić, I.P.; Marković, K.; Vahčić, N. Phenolic Content, Antioxidant Capacity and Quality of Chokeberry (*Aronia melanocarpa*) Products. *Food Technol. Biotechnol.* **2015**, *53*, 171–179. [CrossRef]
25. Rodríguez-Werner, M.; Winterhalter, P.; Esatbeyoglu, T. Phenolic Composition, Radical Scavenging Activity and an Approach for Authentication of *Aronia melanocarpa* Berries, Juice, and Pomace. *J. Food Sci.* **2019**, *84*, 1791–1798. [CrossRef] [PubMed]
26. Sójka, M.; Król, B. Composition of industrial seedless black currant pomace. *Eur. Food Res. Technol.* **2009**, *228*, 597–605. [CrossRef]
27. Jara-Palacios, M.J.; Santisteban, A.; Gordillo, B.; Hernanz, D.; Heredia, F.J.; Escudero-Gilete, M.L. Comparative study of red berry pomaces (blueberry, red raspberry, red currant and blackberry) as source of antioxidants and pigments. *Eur. Food Res. Technol.* **2019**, *245*, 1–9. [CrossRef]
28. Jazić, M.; Kukrić, Z.; Vulić, J.; Četojević-Simin, D. Polyphenolic composition, antioxidant and antiproliferative effects of wild and cultivated blackberries(*Rubus fruticosus* L.)pomace. *Int. J. Food Sci. Technol.* **2019**, *54*, 194–201. [CrossRef]
29. Mildner-Szkudlarz, S.; Bajerska, J.; Górnaś, P.; Seglina, D.; Pilarska, A.; Jesionowski, T. Physical and Bioactive Properties of Muffins Enriched with Raspberry and Cranberry Pomace Powder: A Promising Application of Fruit By-Products Rich in Biocompounds. *Plant Foods Hum. Nutr.* **2016**, *71*, 165–173. [CrossRef]
30. Žlabur, J. Šic; Dobričević, N.; Pliestić, S.; Galić, A.; Bilić, D.P.; Voća, S. Antioxidant Potential of Fruit Juice with Added Chokeberry Powder (*Aronia melanocarpa*). *Molecules* **2017**, *22*, 2158. [CrossRef] [PubMed]
31. Četojević-Simin, D.D.; Velićanski, A.S.; Cvetković, D.D.; Markov, S.L.; Ćetković, G.S.; Šaponjac, V.T.T.; Vulić, J.J.; Čanadanović-Brunet, J.M.; Djilas, S.M. Bioactivity of Meeker and Willamette raspberry (*Rubus idaeus* L.) pomace extracts. *Food Chem.* **2015**, *166*, 407–413. [CrossRef]
32. Halász, K.; Csóka, L. Black chokeberry (*Aronia melanocarpa*) pomace extract immobilized in chitosan for colorimetric pH indicator film application. *Food Packag. Shelf Life* **2018**, *16*, 185–193. [CrossRef]
33. Raczyk, M.; Bryś, J.; Brzezińska, R.; Ostrowska-Ligęza, E.; Wirkowska-Wojdyła, M.; Górska, A. Quality assessment of cold-pressed strawberry, raspberry and blackberry seed oils intended for cosmetic purposes. *Acta Sci. Pol. Technol. Aliment.* **2021**, *20*, 127–133. [CrossRef]

34. Yang, H.Y.; Dong, S.S.; Zhang, C.H.; Wu, W.L.; Lyu, L.F.; Li, W.L. Investigation of Tocopherol Biosynthesis in Blackberry Seeds (*Rubus* spp.). *Russ. J. Plant Physiol.* **2020**, *67*, 76–84. [CrossRef]

35. Ying, Q.; Wojciechowska, P.; Siger, A.; Kaczmarek, A.; Rudzińska, M. Phytochemical Content, Oxidative Stability, and Nutritional Properties of Unconventional Cold-pressed Edible Oils. *J. Food Nutr. Res.* **2018**, *6*, 476–485. [CrossRef]

36. Šavikin, K.P.; Đorđević, B.S.; Ristić, M.S.; Krivokuća-Đokić, D.; Pljevljakušić, D.S.; Vulić, T. Variation in the Fatty-Acid Content in Seeds of Various Black, Red, and White Currant Varieties. *Chem. Biodivers.* **2013**, *10*, 157–165. [CrossRef]

37. Van Hoed, V.; De Clercq, N.; Echim, C.; Andjelkovic, M.; Leber, E.; Dewettinck, K.; Verhe, R. Berry Seeds: A Source of Specialty Oils with High Content of Bioactives and Nutritional Value. *J. Food Lipids* **2009**, *16*, 33–49. [CrossRef]

38. Zlatanov, M.D. Lipid composition of Bulgarian chokeberry, black currant and rose hip seed oils. *J. Sci. Food Agric.* **1999**, *79*, 1620–1624. [CrossRef]

39. Gao, F.; Birch, J. Oxidative stability, thermal decomposition, and oxidation onset prediction of carrot, flax, hemp, and canola seed oils in relation to oil composition and positional distribution of fatty acids. *Eur. J. Lipid Sci. Technol.* **2015**, *118*, 1042–1052. [CrossRef]

40. Li, H.; Fan, Y.-W.; Li, J.; Tang, L.; Hu, J.-N.; Deng, Z.-Y. Evaluating and Predicting the Oxidative Stability of Vegetable Oils with Different Fatty Acid Compositions. *J. Food Sci.* **2013**, *78*, H633–H641. [CrossRef]

41. Kochhar, S.P.; Henry, C.J.K. Oxidative stability and shelf-life evaluation of selected culinary oils. *Int. J. Food Sci. Nutr.* **2009**, *60*, 289–296. [CrossRef]

42. Nosratpour, M.; Farhoosh, R.; Sharif, A. Quantitative Indices of the Oxidizability of Fatty Acid Compositions. *Eur. J. Lipid Sci. Technol.* **2017**, *119*, 1700203. [CrossRef]

43. FAO/WHO Food and Agriculture Organization of the United Nations. *Fats and Fatty Acids in Human Nutrition*; Food and Agriculture Organization of The United Nations: Rome, Italy, 2010.

44. Dyall, S.C. Long-chain omega-3 fatty acids and the brain: A review of the independent and shared effects of EPA, DPA and DHA. *Front. Aging Neurosci.* **2015**, *7*, 52. [CrossRef]

45. Mozaffarian, D.; Wu, J.H.Y. Omega-3 Fatty Acids and Cardiovascular Disease: Effects on risk factors, molecular pathways, and clinical events. *J. Am. Coll. Cardiol.* **2011**, *58*, 2047–2067. [CrossRef]

46. Costantini, L.; Molinari, R.; Farinon, B.; Merendino, N. Impact of Omega-3 Fatty Acids on the Gut Microbiota. *Int. J. Mol. Sci.* **2017**, *18*, 2645. [CrossRef] [PubMed]

47. Hu, Y.; Hu, F.B.; Manson, J.E. Marine Omega-3 Supplementation and Cardiovascular Disease: An Updated Meta-Analysis of 13 Randomized Controlled Trials Involving 127 477 Participants. *J. Am. Heart Assoc.* **2019**, *8*, e013543. [CrossRef]

48. Roche, H.M. Unsaturated Fatty Acids. *Proc. Nutr. Soc.* **1999**, *58*, 397–401.

49. Mazidi, M.; Mikhailidis, D.P.; Sattar, N.; Toth, P.P.; Judd, S.; Blaha, M.J.; Hernandez, A.V.; Penson, P.E.; Banach, M. Association of types of dietary fats and all-cause and cause-specific mortality: A prospective cohort study and meta-analysis of prospective studies with 1,164,029 participants. *Clin. Nutr.* **2020**, *39*, 3677–3686. [CrossRef]

50. Jang, H.; Park, K. Omega-3 and omega-6 polyunsaturated fatty acids and metabolic syndrome: A systematic review and meta-analysis. *Clin. Nutr.* **2020**, *39*, 765–773. [CrossRef] [PubMed]

51. Wanders, A.J.; Blom, W.; Zock, P.; Geleijnse, J.M.; A Brouwer, I.; Alssema, M. Plant-derived polyunsaturated fatty acids and markers of glucose metabolism and insulin resistance: A meta-analysis of randomized controlled feeding trials. *BMJ Open Diabetes Res. Care* **2019**, *7*, e000585. [CrossRef]

52. Das, U.N. Essential fatty acids: Biochemistry, physiology and pathology. *Biotechnol. J.* **2006**, *1*, 420–439. [CrossRef] [PubMed]

53. Tocher, D.R.; Glencross, B.D. Lipids and Fatty Acids. In *Dietary Nutrients, Additives and Fish Health*; Wiley: Hoboken, NJ, USA, 2015; pp. 47–94. ISBN 9781119005568.

54. Schwab, U.; Lauritzen, L.; Tholstrup, T.; Halldorsson, T.; Riserus, U.; Uusitupa, M.; Becker, W. Effect of the amount and type of dietary fat on cardiometabolic risk factors and risk of developing type 2 diabetes, cardiovascular diseases, and cancer: A systematic review. *Food Nutr. Res.* **2014**, *58*, 1–26. [CrossRef] [PubMed]

55. Bozzetto, L.; Prinster, A.; Annuzzi, G.; Costagliola, L.; Mangione, A.; Vitelli, A.; Mazzarella, R.; Longobardo, M.; Mancini, M.; Vigorito, C.; et al. Liver Fat Is Reduced by an Isoenergetic MUFA Diet in a Controlled Randomized Study in Type 2 Diabetic Patients. *Diabetes Care* **2012**, *35*, 1429–1435. [CrossRef]

56. Qian, F.; Korat, A.A.; Malik, V.; Hu, F.B. Metabolic Effects of Monounsaturated Fatty Acid–Enriched Diets Compared With Carbohydrate or Polyunsaturated Fatty Acid–Enriched Diets in Patients With Type 2 Diabetes: A Systematic Review and Meta-analysis of Randomized Controlled Trials. *Diabetes Care* **2016**, *39*, 1448–1457. [CrossRef]

57. Dulf, F.V.; Andrei, S.; Bunea, A.; Socaciu, C. Fatty acid and phytosterol contents of some Romanian wild and cultivated berry pomaces. *Chem. Pap.* **2012**, *66*, 925–934. [CrossRef]

58. Piasecka, I.; Górska, A.; Ostrowska-Ligęza, E.; Kalisz, S. The Study of Thermal Properties of Blackberry, Chokeberry and Raspberry Seeds and Oils. *Appl. Sci.* **2021**, *11*, 7704. [CrossRef]

59. Bada, J.; León-Camacho, M.; Copovi, P.; Alonso, L. Characterization of Berry and Currant Seed Oils from Asturias, Spain. *Int. J. Food Prop.* **2013**, *17*, 77–85. [CrossRef]

60. Oomah, B.; Ladet, S.; Godfrey, D.V.; Liang, J.; Girard, B. Characteristics of raspberry (*Rubus idaeus* L.) seed oil. *Food Chem.* **2000**, *69*, 187–193. [CrossRef]

61. Knothe, G. Fuel properties of methyl esters of borage and black currant oils containing methyl γ-linolenate. *Eur. J. Lipid Sci. Technol.* **2013**, *115*, 901–908. [CrossRef]

62. Dobson, G.; Shrestha, M.; Hilz, H.; Karjalainen, R.; McDougall, G.; Stewart, D. Lipophilic components in black currant seed and pomace extracts. *Eur. J. Lipid Sci. Technol.* **2011**, *114*, 575–582. [CrossRef]

63. Johansson, A.; Laine, T.; Linna, M.-M.; Kallio, H. Variability in oil content and fatty acid composition in wild northern currants. *Eur. Food Res. Technol.* **2000**, *211*, 277–283. [CrossRef]

64. Piskernik, S.; Vidrih, R.; Demšar, L.; Koron, D.; Rogelj, M.; Žontar, T.P. Fatty acid profiles of seeds from different Ribes species. *LWT Food Sci. Technol.* **2018**, *98*, 424–427. [CrossRef]

65. Shahidi, F.; De Camargo, A.C. Tocopherols and Tocotrienols in Common and Emerging Dietary Sources: Occurrence, Applications, and Health Benefits. *Int. J. Mol. Sci.* **2016**, *17*, 1745. [CrossRef]

66. Kamal-Eldin, A. Effect of fatty acids and tocopherols on the oxidative stability of vegetable oils. *Eur. J. Lipid Sci. Technol.* **2006**, *108*, 1051–1061. [CrossRef]

67. Wagner, K.-H.; Elmadfa, I. Effects of tocopherols and their mixtures on the oxidative stability of olive oil and linseed oil under heating. *Eur. J. Lipid Sci. Technol.* **2000**, *102*, 624–629. [CrossRef]

68. Player, M.; Kim, H.; Lee, H.; Min, D. Stability of α-, γ-, or δ-Tocopherol during Soybean Oil Oxidation. *J. Food Sci.* **2006**, *71*, C456–C460. [CrossRef]

69. Ayyildiz, H.F.; Topkafa, M.; Kara, H.; Sherazi, S.T.H. Evaluation of Fatty Acid Composition, Tocols Profile, and Oxidative Stability of Some Fully Refined Edible Oils. *Int. J. Food Prop.* **2015**, *18*, 2064–2076. [CrossRef]

70. Górnaś, P.; Soliven, A.; Seglina, D. Seed oils recovered from industrial fruit by-products are a rich source of tocopherols and tocotrienols: Rapid separation of α/β/γ/δ homologues by RP-HPLC/FLD. *Eur. J. Lipid Sci. Technol.* **2014**, *117*, 773–777. [CrossRef]

71. Schneider, C. Chemistry and biology of vitamin E. *Mol. Nutr. Food Res.* **2004**, *49*, 7–30. [CrossRef]

72. Piironen, V.; Lindsay, D.G.; Miettinen, T.A.; Toivo, J.; Lampi, A.-M. Plant Sterols: Biosynthesis, Biological Function and Their Importance to Human Nutrition. *J. Sci. Food Agric.* **2000**, *80*, 939–966. [CrossRef]

73. Guillaume, C.; Ravetti, L.; Ray, D.L.; Johnson, J. Technological Factors Affecting Sterols in Australian Olive Oils. *J. Am. Oil Chem. Soc.* **2011**, *89*, 29–39. [CrossRef]

74. Verleyen, T.; Sosinska, U.; Ioannidou, S.; Verhe, R.; Dewettinck, K.; Huyghebaert, A.; De Greyt, W. Influence of the vegetable oil refining process on free and esterified sterols. *J. Am. Oil Chem. Soc.* **2002**, *79*, 947–953. [CrossRef]

75. Hu, Y.; Xu, J.; Huang, W.; Zhao, Y.; Li, M.; Wang, M.; Zheng, L.; Lu, B. Structure–activity relationships between sterols and their thermal stability in oil matrix. *Food Chem.* **2018**, *258*, 387–392. [CrossRef] [PubMed]

76. Deepam, L.S.A.; Sundaresan, A.; Arumughan, C. Stability of Rice Bran Oil in Terms of Oryzanol, Tocopherols, Tocotrienols and Sterols. *J. Am. Oil Chem. Soc.* **2010**, *88*, 1001–1009. [CrossRef]

77. Wang, T.; Hicks, K.B.; Moreau, R. Antioxidant activity of phytosterols, oryzanol, and other phytosterol conjugates. *J. Am. Oil Chem. Soc.* **2002**, *79*, 1201–1206. [CrossRef]

78. Chang, M.; Xu, Y.; Li, X.; Shi, F.; Liu, R.; Jin, Q.; Wang, X. Effects of stigmasterol on the thermal stability of soybean oil during heating. *Eur. Food Res. Technol.* **2020**, *246*, 1755–1763. [CrossRef]

79. Qianchun, D.; Jie, S.; Mingming, Z.; Jiqu, X.; Chuyun, W.; Qingde, H.; Qi, Z.; Pingmei, G.; Fenghong, H.; Lan, W.; et al. Thermal Stability of Rapeseed Oil Fortified with Unsaturated Fatty Acid Sterol Esters. *J. Am. Oil Chem. Soc.* **2014**, *91*, 1793–1803. [CrossRef]

80. Lin, Y.; Knol, D.; Menéndez-Carreño, M.; Baris, R.; Janssen, H.-G.; Trautwein, E.A. Oxidation of sitosterol and campesterol in foods upon cooking with liquid margarines without and with added plant sterol esters. *Food Chem.* **2018**, *241*, 387–396. [CrossRef] [PubMed]

81. AbuMweis, S.S.; Barake, R.; Jones, P.J. Plant sterols/stanols as cholesterol lowering agents: A meta-analysis of randomized controlled trials. *Food Nutr. Res.* **2008**, *52*, 1–17. [CrossRef]

82. de Jong, A.; Plat, J.; Mensink, R.P. Metabolic effects of plant sterols and stanols (Review). *J. Nutr. Biochem.* **2003**, *14*, 362–369. [CrossRef]

83. Cacace, J.; Mazza, G. Mass transfer process during extraction of phenolic compounds from milled berries. *J. Food Eng.* **2003**, *59*, 379–389. [CrossRef]

84. Brglez Mojzer, E.; Knez Hrnčič, M.; Škerget, M.; Knez, Ž.; Bren, U. Polyphenols: Extraction Methods, Antioxidative Action, Bioavailability and Anticarcinogenic Effects. *Molecules* **2016**, *21*, 901. [CrossRef]

85. Martakos, I.; Kostakis, M.; Dasenaki, M.; Pentogennis, M.; Thomaidis, N. Simultaneous Determination of Pigments, Tocopherols, and Squalene in Greek Olive Oils: A Study of the Influence of Cultivation and Oil-Production Parameters. *Foods* **2019**, *9*, 31. [CrossRef] [PubMed]

86. Tir, R.; Dutta, P.C.; Badjah-Hadj-Ahmed, A.Y. Effect of the extraction solvent polarity on the sesame seeds oil composition. *Eur. J. Lipid Sci. Technol.* **2012**, *114*, 1427–1438. [CrossRef]

87. Sicaire, A.-G.; Vian, M.; Fine, F.; Joffre, F.; Carré, P.; Tostain, S.; Chemat, F. Alternative Bio-Based Solvents for Extraction of Fat and Oils: Solubility Prediction, Global Yield, Extraction Kinetics, Chemical Composition and Cost of Manufacturing. *Int. J. Mol. Sci.* **2015**, *16*, 8430–8453. [CrossRef]

88. Wajs-Bonikowska, A.; Stobiecka, A.; Bonikowski, R.; Krajewska, A.; Sikora, M.; Kula, J. A comparative study on composition and antioxidant activities of supercritical carbon dioxide, hexane and ethanol extracts from blackberry (*Rubus fruticosus*) growing in Poland. *J. Sci. Food Agric.* **2017**, *97*, 3576–3583. [CrossRef]

89. Dimic, E.; Vujasinovic, V.; Radocaj, O.; Pastor, O. Characteristics of blackberry and raspberry seeds and oils. *Acta Period. Technol.* **2012**, *43*, 1–9. [CrossRef]

90. Šućurović, A.; Vukelić, N.; Ignjatović, L.; Brčeski, I.; Jovanović, D. Physical-chemical characteristics and oxidative stability of oil obtained from lyophilized raspberry seed. *Eur. J. Lipid Sci. Technol.* **2009**, *111*, 1133–1141. [CrossRef]

91. Radočaj, O.; Vujasinović, V.; Dimić, E.; Basić, Z. Blackberry (*Rubus fruticosus* L.) and raspberry (*Rubus idaeus* L.) seed oils extracted from dried press pomace after longterm frozen storage of berries can be used as functional food ingredients. *Eur. J. Lipid Sci. Technol.* **2014**, *116*, 1015–1024. [CrossRef]

92. Pieszka, M.; Gogol, P.; Pietras, M.; Pieszka, M. Valuable Components of Dried Pomaces of Chokeberry, Black Currant, Strawberry, Apple and Carrot as a Source of Natural Antioxidants and Nutraceuticals in the Animal Diet. *Ann. Anim. Sci.* **2015**, *15*, 475–491. [CrossRef]

93. Puganen, A.; Kallio, H.P.; Schaich, K.M.; Suomela, J.-P.; Yang, B. Red/Green Currant and Sea Buckthorn Berry Press Residues as Potential Sources of Antioxidants for Food Use. *J. Agric. Food Chem.* **2018**, *66*, 3426–3434. [CrossRef] [PubMed]

94. Ross, K.A.; Ehret, D.; Godfrey, D.; Fukumoto, L.; Diarra, M. Characterization of Pilot Scale Processed Canadian Organic Cranberry (*Vaccinium macrocarpon*) and Blueberry (*Vaccinium angustifolium*) Juice Pressing Residues and Phenolic-Enriched Extractives. *Int. J. Fruit Sci.* **2016**, *17*, 202–232. [CrossRef]

95. Vulić, J.J.; Tumbas, V.T.; Savatović, S.M.; Djilas, S.; Ćetković, G.S.; Čanadanović-Brunet, J.M. Polyphenolic content and antioxidant activity of the four berry fruits pomace extracts. *Acta Period. Technol.* **2011**, *42*, 271–279. [CrossRef]

96. Čanadanović-Brunet, J.; Vulić, J.; Ćebović, T.; Ćetković, G.; Čanadanović, V.; Djilas, S.; Šaponjac, V.T. Phenolic Profile, Antiradical and Antitumour Evaluation of Raspberries Pomace Extract from Serbia. *Iran. J. Pharm. Res.* **2017**, *16*, 142–152.

97. Laroze, L.E.; Díaz-Reinoso, B.; Moure, A.; Zúñiga, M.E.; Domínguez, H. Extraction of antioxidants from several berries pressing wastes using conventional and supercritical solvents. *Eur. Food Res. Technol.* **2010**, *231*, 669–677. [CrossRef]

98. Kosmala, M.; Zduńczyk, Z.; Kołodziejczyk, K.; Klimczak, E.; Juśkiewicz, J.; Zdunczyk, P. Chemical composition of polyphenols extracted from strawberry pomace and their effect on physiological properties of diets supplemented with different types of dietary fibre in rats. *Eur. J. Nutr.* **2013**, *53*, 521–532. [CrossRef] [PubMed]

99. Machado, A.P.D.F.; Pereira, A.L.D.; Barbero, G.F.; Martínez, J. Recovery of anthocyanins from residues of *Rubus fruticosus*, Vaccinium myrtillus and Eugenia brasiliensis by ultrasound assisted extraction, pressurized liquid extraction and their combination. *Food Chem.* **2017**, *231*, 1–10. [CrossRef]

100. Medina-Torres, N.; Ayora-Talavera, T.; Espinosa-Andrews, H.; Sánchez-Contreras, A.; Pacheco, N. Ultrasound Assisted Extraction for the Recovery of Phenolic Compounds from Vegetable Sources. *Agronomy* **2017**, *7*, 47. [CrossRef]

101. D'Alessandro, L.G.; Kriaa, K.; Nikov, I.; Dimitrov, K. Ultrasound assisted extraction of polyphenols from black chokeberry. *Sep. Purif. Technol.* **2012**, *93*, 42–47. [CrossRef]

102. Herrera, M.C.; de Castro, M.D.L. Ultrasound-assisted extraction for the analysis of phenolic compounds in strawberries. *Anal. Bioanal. Chem.* **2004**, *379*, 1106–1112. [CrossRef] [PubMed]

103. Wang, W.; Jung, J.; Tomasino, E.; Zhao, Y. Optimization of solvent and ultrasound-assisted extraction for different anthocyanin rich fruit and their effects on anthocyanin compositions. *LWT Food Sci. Technol.* **2016**, *72*, 229–238. [CrossRef]

104. Gayas, B.; Kaur, G.; Gul, K. Ultrasound-Assisted Extraction of Apricot Kernel Oil: Effects on Functional and Rheological Properties. *J. Food Process Eng.* **2017**, *40*, e1243. [CrossRef]

105. Samaram, S.; Mirhosseini, H.; Tan, C.P.; Ghazali, H.; Bordbar, S.; Serjouie, A. Optimisation of ultrasound-assisted extraction of oil from papaya seed by response surface methodology: Oil recovery, radical scavenging antioxidant activity, and oxidation stability. *Food Chem.* **2015**, *172*, 7–17. [CrossRef]

106. Samaram, S.; Mirhosseini, H.; Tan, C.P.; Ghazali, H.M. Ultrasound-Assisted Extraction (UAE) and Solvent Extraction of Papaya Seed Oil: Yield, Fatty Acid Composition and Triacylglycerol Profile. *Molecules* **2013**, *18*, 12474–12487. [CrossRef]

107. Krivokapić, S.; Vlaović, M.; Vratnica, B.D.; Perović, A.; Perović, S. Biowaste as a Potential Source of Bioactive Compounds—A Case Study of Raspberry Fruit Pomace. *Foods* **2021**, *10*, 706. [CrossRef]

108. Bamba, B.S.B.; Shi, J.; Tranchant, C.C.; Xue, S.J.; Forney, C.F.; Lim, L.-T. Influence of Extraction Conditions on Ultrasound-Assisted Recovery of Bioactive Phenolics from Blueberry Pomace and Their Antioxidant Activity. *Molecules* **2018**, *23*, 1685. [CrossRef]

109. Zafra-Rojas, Q.Y.; Cruz-Cansino, N.S.; Lira, A.Q.; Gómez-Aldapa, C.A.; Alanís-García, E.; Cervantes-Elizarrarás, A.; Güemes-Vera, N.; Ramírez-Moreno, E. Application of Ultrasound in a Closed System: Optimum Condition for Antioxidants Extraction of Blackberry (*Rubus fructicosus*) Residues. *Molecules* **2016**, *21*, 950. [CrossRef] [PubMed]

110. Xue, H.; Tan, J.; Li, Q.; Tang, J.; Cai, X. Ultrasound-Assisted Enzymatic Extraction of Anthocyanins from Raspberry Wine Residues: Process Optimization, Isolation, Purification, and Bioactivity Determination. *Food Anal. Methods* **2021**, *14*, 1369–1386. [CrossRef]

111. Ramić, M.; Vidović, S.; Zeković, Z.; Vladić, J.; Cvejin, A.; Pavlić, B. Modeling and optimization of ultrasound-assisted extraction of polyphenolic compounds from *Aronia melanocarpa* by-products from filter-tea factory. *Ultrason. Sonochem.* **2015**, *23*, 360–368. [CrossRef]

112. He, B.; Zhang, L.-L.; Yue, X.-Y.; Liang, J.; Jiang, J.; Gao, X.-L.; Yue, P.-X. Optimization of Ultrasound-Assisted Extraction of phenolic compounds and anthocyanins from blueberry (*Vaccinium ashei*) wine pomace. *Food Chem.* **2016**, *204*, 70–76. [CrossRef] [PubMed]

113. Lončarić, A.; Celeiro, M.; Jozinović, A.; Jelinić, J.; Kovač, T.; Jukić, D.; Babić, J.; Moslavac, T.; Zavadlav, S.; Lores, M. Green Extraction Methods for Extraction of Polyphenolic Compounds from Blueberry Pomace. *Foods* **2020**, *9*, 1521. [CrossRef]

114. Gođevac, D.; Tešević, V.; Vajs, V.; Milosavljević, S.; Stanković, M. Blackberry Seed Extracts and Isolated Polyphenolic Compounds Showing Protective Effect on Human Lymphocytes DNA. *J. Food Sci.* **2011**, *76*, C1039–C1043. [CrossRef]
115. Teng, H.; Chen, L.; Huang, Q.; Wang, J.; Lin, Q.; Liu, M.; Lee, W.Y.; Song, H. Ultrasonic-Assisted Extraction of Raspberry Seed Oil and Evaluation of Its Physicochemical Properties, Fatty Acid Compositions and Antioxidant Activities. *PLoS ONE* **2016**, *11*, e0153457. [CrossRef] [PubMed]
116. Sady, S.; Matuszak, L.; Błaszczyk, A. Optimisation of ultrasonic-assisted extraction of bioactive compounds from chokeberry pomace using response surface methodology. *Acta Sci. Pol. Technol. Aliment.* **2015**, *18*, 249–256. [CrossRef]
117. Cvetanović, A.; Švarc-Gajić, J.; Zeković, Z.; Mašković, P.; Đurović, S.; Zengin, G.; Delerue-Matos, C.; Lozano-Sánchez, J.; Jakšić, A. Chemical and biological insights on aronia stems extracts obtained by different extraction techniques: From wastes to functional products. *J. Supercrit. Fluids* **2017**, *128*, 173–181. [CrossRef]
118. Oszmiański, J.; Lachowicz, S. Effect of the Production of Dried Fruits and Juice from Chokeberry (*Aronia melanocarpa* L.) on the Content and Antioxidative Activity of Bioactive Compounds. *Molecules* **2016**, *21*, 1098. [CrossRef] [PubMed]
119. Kumari, B.; Tiwari, B.K.; Hossain, M.B.; Brunton, N.P.; Rai, D.K. Recent Advances on Application of Ultrasound and Pulsed Electric Field Technologies in the Extraction of Bioactives from Agro-Industrial By-products. *Food Bioprocess Technol.* **2018**, *11*, 223–241. [CrossRef]
120. Pataro, G.; Bobinaitė, R.; Bobinas, Česlovas; Satkauskas, S.; Raudonis, R.; Visockis, M.; Ferrari, G.; Viskelis, P. Improving the Extraction of Juice and Anthocyanins from Blueberry Fruits and Their By-products by Application of Pulsed Electric Fields. *Food Bioprocess Technol.* **2017**, *10*, 1595–1605. [CrossRef]
121. Bobinaitė, R.; Pataro, G.; Lamanauskas, N.; Šatkauskas, S.; Viskelis, P.; Ferrari, G. Application of pulsed electric field in the production of juice and extraction of bioactive compounds from blueberry fruits and their by-products. *J. Food Sci. Technol.* **2015**, *52*, 5898–5905. [CrossRef] [PubMed]
122. Zhou, Y.; Zhao, X.; Huang, H. Effects of Pulsed Electric Fields on Anthocyanin Extraction Yield of Blueberry Processing By-Products. *J. Food Process. Preserv.* **2015**, *39*, 1898–1904. [CrossRef]
123. Gagneten, M.; Leiva, G.; Salvatori, D.; Schebor, C.; Olaiz, N. Optimization of Pulsed Electric Field Treatment for the Extraction of Bioactive Compounds from Blackcurrant. *Food Bioprocess Technol.* **2019**, *12*, 1102–1109. [CrossRef]
124. Veggi, P.C.; Martinez, J.; Meireles, M.A.A. Fundamentals of Microwave Extraction. In *Microwave-Assisted Extraction for Bioactive 15 Compounds: Theory and Practice*; Food Engineering Series; Springer Science and Business Media LLC: Boston, MA, USA, 2012; pp. 15–52.
125. Pap, N.; Beszédes, S.; Pongrácz, E.; Myllykoski, L.; Gábor, M.; Gyimes, E.; Hodúr, C.; Keiski, R.L. Microwave-Assisted Extraction of Anthocyanins from Black Currant Marc. *Food Bioprocess Technol.* **2013**, *6*, 2666–2674. [CrossRef]
126. Davis, E.J.; Andreani, E.S.; Karboune, S. Production of Extracts Composed of Pectic Oligo/Polysaccharides and Polyphenolic Compounds from Cranberry Pomace by Microwave-Assisted Extraction Process. *Food Bioprocess Technol.* **2021**, *14*, 634–649. [CrossRef]
127. Raghavan, S.; Richards, M. Comparison of solvent and microwave extracts of cranberry press cake on the inhibition of lipid oxidation in mechanically separated turkey. *Food Chem.* **2007**, *102*, 818–826. [CrossRef]
128. Klavins, L.; Kviesis, J.; Klavins, M. Comparison of Methods of Extraction of Phenolic Compounds from American Cranberry (*Vaccinium macrocarpon* L.) Press Residues. *Agron. Res.* **2017**, *15*, 1316–1329.
129. Wrona, O.; Rafińska, K.; Możeński, C.; Buszewski, B. Supercritical Fluid Extraction of Bioactive Compounds from Plant Materials. *J. AOAC Int.* **2017**, *100*, 1624–1635. [CrossRef] [PubMed]
130. Campalani, C.; Amadio, E.; Zanini, S.; Dall'Acqua, S.; Panozzo, M.; Ferrari, S.; De Nadai, G.; Francescato, S.; Selva, M.; Perosa, A. Supercritical CO2 as a green solvent for the circular economy: Extraction of fatty acids from fruit pomace. *J. CO2 Util.* **2020**, *41*, 101259. [CrossRef]
131. Marić, B.; Abramović, B.; Ilić, N.; Krulj, J.; Kojić, J.; Perović, J.; Bodroža-Solarov, M.; Teslić, N. Valorization of red raspberry (*Rubus idaeus* L.) seeds as a source of health beneficial compounds: Extraction by different methods. *J. Food Process. Preserv.* **2020**, *44*, e14744. [CrossRef]
132. Milala, J.; Grzelak-Błaszczyk, K.; Sójka, M.; Kosmala, M.; Dobrzyńska-Inger, A.; Rój, E. Changes of bioactive components in berry seed oils during supercritical CO2 extraction. *J. Food Process. Preserv.* **2017**, *42*, e13368. [CrossRef]
133. Correa, M.D.S.; Fetzer, D.L.; Hamerski, F.; Corazza, M.L.; Scheer, A.D.P.; Ribani, R.H. Pressurized extraction of high-quality blackberry (*Rubus* spp. Xavante cultivar) seed oils. *J. Supercrit. Fluids* **2021**, *169*, 105101. [CrossRef]
134. Pavlić, B.; Pezo, L.; Marić, B.; Tukuljac, L.P.; Zeković, Z.; Solarov, M.B.; Teslić, N. Supercritical fluid extraction of raspberry seed oil: Experiments and modelling. *J. Supercrit. Fluids* **2020**, *157*, 104687. [CrossRef]
135. Basegmez, H.I.O.; Povilaitis, D.; Kitrytė, V.; Kraujalienė, V.; Šulniūtė, V.; Alasalvar, C.; Venskutonis, P.R. Biorefining of blackcurrant pomace into high value functional ingredients using supercritical CO$_2$, pressurized liquid and enzyme assisted extractions. *J. Supercrit. Fluids* **2017**, *124*, 10–19. [CrossRef]
136. Sapkale, G.N.; Patil, S.M.; Surwase, U.S.; Bhatbhage, P.K. Supercritical Fluid Extraction. *Int. J. Chem. Sci.* **2010**, *8*, 729–743.
137. Mustafa, A.; Turner, C. Pressurized liquid extraction as a green approach in food and herbal plants extraction: A review. *Anal. Chim. Acta* **2011**, *703*, 8–18. [CrossRef] [PubMed]

138. Brazdauskas, T.; Montero, L.; Venskutonis, P.; Ibañez, E.; Herrero, M. Downstream valorization and comprehensive two-dimensional liquid chromatography-based chemical characterization of bioactives from black chokeberries (*Aronia melanocarpa*) pomace. *J. Chromatogr. A* **2016**, *1468*, 126–135. [CrossRef] [PubMed]
139. Grunovaitė, L.; Pukalskienė, M.; Pukalskas, A.; Venskutonis, P.R. Fractionation of black chokeberry pomace into functional ingredients using high pressure extraction methods and evaluation of their antioxidant capacity and chemical composition. *J. Funct. Foods* **2016**, *24*, 85–96. [CrossRef]
140. Kryževičūtė, N.; Jaime, I.; Diez, A.M.; Rovira, J.; Venskutonis, P.R. Effect of raspberry pomace extracts isolated by high pressure extraction on the quality and shelf-life of beef burgers. *Int. J. Food Sci. Technol.* **2017**, *52*, 1852–1861. [CrossRef]
141. Puri, M.; Sharma, D.; Barrow, C.J. Enzyme-assisted extraction of bioactives from plants. *Trends Biotechnol.* **2012**, *30*, 37–44. [CrossRef]
142. Syrpas, M.; Valanciene, E.; Augustiniene, E.; Malys, N. Valorization of Bilberry (*Vaccinium myrtillus* L.) Pomace by Enzyme-Assisted Extraction: Process Optimization and Comparison with Conventional Solid-Liquid Extraction. *Antioxidants* **2021**, *10*, 773. [CrossRef]
143. Kitrytė, V.; Kraujalienė, V.; Šulniūtė, V.; Pukalskas, A.; Venskutonis, P.R. Chokeberry pomace valorization into food ingredients by enzyme-assisted extraction: Process optimization and product characterization. *Food Bioprod. Process.* **2017**, *105*, 36–50. [CrossRef]
144. Saad, N.; Louvet, F.; Tarrade, S.; Meudec, E.; Grenier, K.; Landolt, C.; Ouk, T.; Bressollier, P. Enzyme-Assisted Extraction of Bioactive Compounds from Raspberry (*Rubus idaeus* L.) Pomace. *J. Food Sci.* **2019**, *84*, 1371–1381. [CrossRef]

Article

Application of Chromatographic and Thermal Methods to Study Fatty Acids Composition and Positional Distribution, Oxidation Kinetic Parameters and Melting Profile as Important Factors Characterizing Amaranth and Quinoa Oils

Magdalena Wirkowska-Wojdyła *, Ewa Ostrowska-Ligęza, Agata Górska and Joanna Bryś

Department of Chemistry, Institute of Food Sciences, Warsaw University of Life Sciences, 02-787 Warsaw, Poland; ewa_ostrowska_ligeza@sggw.edu.pl (E.O.-L.); agata_gorska@sggw.edu.pl (A.G.); joanna_brys@sggw.edu.pl (J.B.)
* Correspondence: magdalena_wirkowska@sggw.edu.pl; Tel.: +48-22-593-7606

Citation: Wirkowska-Wojdyła, M.; Ostrowska-Ligęza, E.; Górska, A.; Bryś, J. Application of Chromatographic and Thermal Methods to Study Fatty Acids Composition and Positional Distribution, Oxidation Kinetic Parameters and Melting Profile as Important Factors Characterizing Amaranth and Quinoa Oils. *Appl. Sci.* **2022**, *12*, 2166. https://doi.org/10.3390/app12042166

Academic Editors: Alessandra Durazzo and Anabela Raymundo

Received: 29 November 2021
Accepted: 17 February 2022
Published: 18 February 2022

Publisher's Note: MDPI stays neutral with regard to jurisdictional claims in published maps and institutional affiliations.

Abstract: Amaranth and quinoa are classed as pseudocereals that do not belong to the grass family, meaning they are not technically a grain. Both of them are seeds with tremendous nutritional value; compared to other cereals, they contain much more fat. The aim of the study was to present the parameters characterizing thermal properties of amaranth and quinoa oils, such as: oxidation induction time, oxidation kinetic parameters, and melting profile. In isolated oils, the peroxide value, oxidative stability by the Rancimat test (in 120 °C) and the pressure differential scanning calorimetry (PDSC) method (at 100, 110, 120, 130, 140 °C), fatty acids composition, and their distribution between the triacylglycerol positions were determined. The kinetic parameters of the oxidation process (activation energy, pre-exponential factor, and reaction rate constants) were calculated using the Ozawa–Flynn–Wall method and the Arrhenius equation. To measure the melting profile, the differential scanning calorimetry (DSC) method was used. Both types of seeds are a good source of unsaturated fatty acids. Induction time of oxidation suggests that amaranth oil may have better resistance to oxidation than quinoa oil. The melting characteristics of the oils show the presence of low-melting triacylglycerol fractions, mainly containing unsaturated fatty acids, which means that a small amount of energy is required to melt the fats.

Keywords: amaranth oil; quinoa oil; oxidative stability; DSC; Rancimat

1. Introduction

Amaranth (*Amaranthum*) and quinoa (*Chenopodium quinoa* Willd) are pseudocereals because their seeds resemble real cereals in terms of their composition and function. Amaranth and quinoa cultivated from tropical to subtropical regions were important food crops to Aztec, Mayan and Incan civilizations [1]. Pseudocereals are gluten-free products, which represent a significant advance towards ensuring an adequate intake of nutrients in subjects with celiac disease. Interest in quinoa and amaranth stems from their nutritional value, higher than cereal grains like corn, oats, wheat, and rice. The beneficial nutritional value of these pseudocereals is due to the high protein content, as well as dietary fibre and bioactive compounds such as polyphenolic compounds [2–4]. The quinoa seeds contain exogenous amino acids such as: lysine, arginine, histidine, and methionine [5]. These amino acids are most often found in small amounts in cereal grains and legumes [6]. Quinoa seed protein is characterized by a more balanced amino acids composition than wheat protein. Amaranth seed protein contains all the amino acids essential for the human body, surpassing soybean protein. The biological value of amaranth protein is higher than milk proteins, so amaranth can be used to produce milk replacement products for people intolerant to lactose [7].

Traditional cereals are not a rich source of fat in comparison to oilseed crop materials. On the other hand, pseudocereals like quinoa or amaranth, in comparison to other cereals, contain much more fat, supplementing the diet with this ingredient in the daily food ration. According to Jancurova et al. [8], the fat content in quinoa seeds can range from 2 to 10%. Navruz-Varli and Sanlier [9] classify quinoa as an alternative oilseed. Although the lipid content of amaranthus seed is typically 6–9%, some species such as *A. spinosus* and *A. tenuifolius* have been reported to contain as much as 19.3% [5].

The high amount of fat in pseudocereals means that it can be susceptible to the oxidation process. Oxidation is one of the most important processes that takes place in oils during storage or heat treatment. Moreover, the oxidative stability of an oil is one of the most significant parameters from the point of view of oil safety [10].

Among several available methods to measure the lipid oxidation, differential scanning calorimetry (DSC) or pressure differential scanning calorimetry (PDSC) can be considered as the most effective. The oxidative stability of lipids is determined as oxidative induction time; in practice, the time required to begin the oxidative decomposition process of the oil sample [11,12]. Differential scanning calorimetry is an applicable tool to describe the thermal transitions of examined samples in micro-scale. It provides information about thermodynamic and kinetic characteristics dependent on the temperature of the material. In food products analysis, DSC helps in determination of the transitions resulting from the specific composition of foods because different groups of nutrients are specified by various transitions [13]. Apart from that, DSC has been suggested as a possible method for verifying quality, determining varieties, and detecting adulteration of oil. DSC is preferable to other methods because it is fast and the use of environmental damaging solvents is not required [14].

According to our knowledge, in the literature, there are no studies on the kinetics of oxidation process of oil from pseudocereals using the PDSC method. Therefore, the aim of this study was to determine the oxidative stability and kinetic parameters of oxidation of oil isolated from quinoa and amaranth seeds. The melting characteristics of these oils are also presented, as well as the analysis of fatty acid composition and their stereochemical distribution in triacyglycerols.

2. Materials and Methods

2.1. Materials

Two packs of quinoa (*Chenopodium quinoa* Willd) and amaranthus (*Amaranthus cruentus*) seeds were purchased in two different retail outlets in Poland in October 2019 in order to avoid the same lots of seeds. The seeds were ground in a grinder model and passed through a 60-mesh screen for the determination of the oil yield by the Soxhlet procedure according to the ISO method [15]. Oil yield was defined as the grams of oil in 100 g of dry basis. In order to perform the remaining determinations, the fat was isolated from the seeds by cold extraction (with hexane as a solvent) according to the procedure described by Bryś et al. [16]. Extraction was performed in duplicate for each part of seeds. The graphical scheme of the study approach is presented in Figure 1.

2.2. GC Analysis

Fatty acids composition in fats isolated from quinoa and amaranth was determined as fatty acid methyl esters by gas chromatography according to the ISO method [17]. The YL6100 (Young Lin Bldg., Anyang, Hogyedong, Korea) gas chromatograph equipped with a flame ionization detector and a BPX-70 capillary column (SGE Analytical Science, Milton Keynes, UK) was used. The procedure for analyzing fatty acid methyl esters has been described in previous studies [18].

The analysis of distribution of fatty acids between the position of triacylglycerols (TAG) was also done as the percentage content of a fatty acid in the sn-2 position, according to the procedure described by Bryś et al. [19].

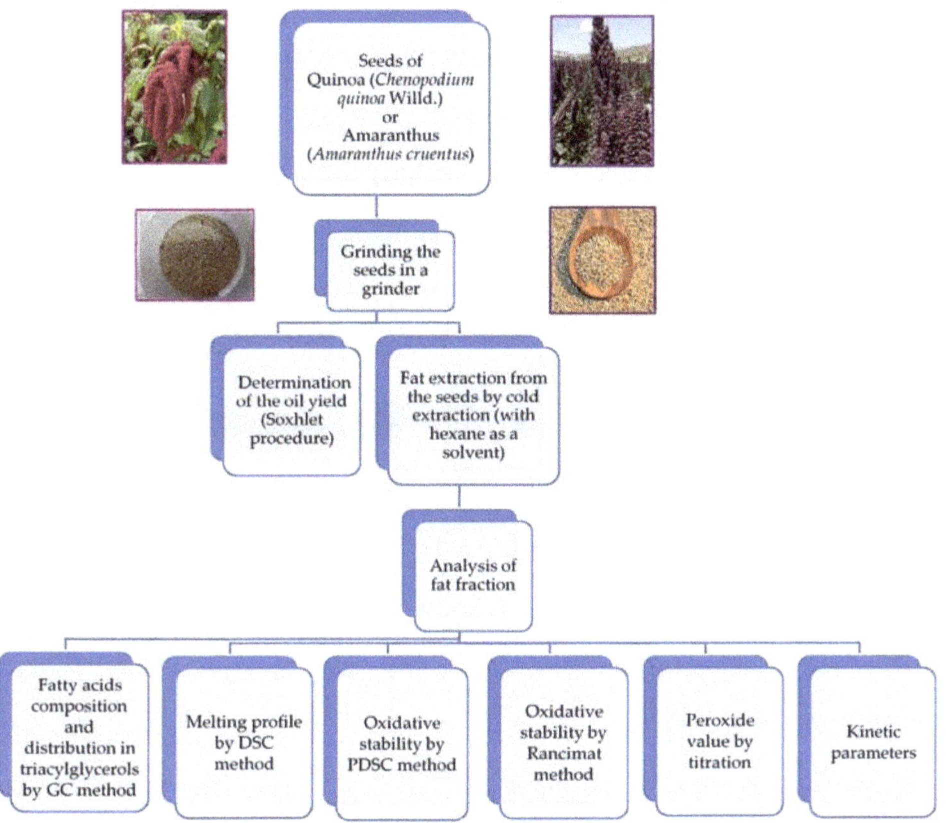

Figure 1. Graphical scheme of study approach (GC method—gas chromatography method).

2.3. Melting Profile

The instrument Q200 DSC (TA Instruments, Newcastle, DE, USA) was used to record the melting profile. Procedure was described by Aguedo et al. [20] and Wirkowska-Wojdyła et al. [21].

2.4. Peroxide Value

Peroxide values (PV) of oils were determined by the iodometric technique in correspondence with ISO standards [22].

2.5. Oxidative Stability by Rancimat Method

The oxidative stability of oils was determined using a Rancimat 743 Metrohm apparatus (Herisau, Switzerland), according to the ISO method [23], at constant temperature 120 °C. The exact procedure was described by Symoniuk et al. [10].

2.6. Oxidative Stability by PDSC Method

Pressure differential scanning calorimeter (DSC Q20 TA Instruments, Newcastle, WA, USA) was used to determine the oxidative stability of fats. The experiment was performed at constant temperatures: 100, 110, 120, 130, 140 °C under 1400 kPa pressure of oxygen. The procedure was described in previous studies [21].

2.7. Kinetic Parameters

The kinetic parameters of the oxidation process (activation energy, pre-exponential factor, and reaction rate constants) were calculated using the Ozawa–Flynn–Wall method and the Arrhenius equation.

Based on the results of the induction times obtained in the PDSC test, a graph of the logarithm of the induction time (τ) versus the reciprocal temperature (in absolute scale) was plotted. Regression lines with correlation coefficients >0.99 were determined according to the Equation (1):

$$log\ \tau = a\ T^{-1} + b \tag{1}$$

where a and b are adjustable coefficients. Reaction of fat oxidation proceeds in an excess of oxygen and can be treated as a first-order reaction. This fact can be used to determine the activation energy using the Ozawa–Flynn–Wall (Equation (2)):

$$Ea = 2.19 \times R \times a \tag{2}$$

where R is a gas constant and a is a coefficient from Equation (1). Based on the Arrhenius equation (Equation (3)):

$$k = Ze^{-Ea/RT} \tag{3}$$

activation energy—Ea, pre-exponential factor—Z, and reaction rate coefficient—k, for all temperatures were calculated.

2.8. Statistical Analysis

The Statgraphics Plus, version 5.1 (Statistical Graphics Corporation, Warrenton, VA, USA) program was used for statistical analysis of the results. Tuckey's multiple range test at a p-value of 0.05 was used to analyze significant differences.

3. Results and Discussion

3.1. Total Fat Content and Fatty Acid Composition

The quinoa seeds contained 5.43% of fat (Table 1). According to Villacrés et al. [24], yield in the oil extraction from quinoa seeds also reached about 5%. Rodriguez Gomez et al. [25], in the study of six quinoa varieties, obtained the fat content from 3.90 to 5.21 g/100 g of fresh weight. The flour from quinoa analyzed by Ascheri et al. [26], which was added to cereal products, contained 5.6% of lipids.

Table 1. Total fat content, peroxide value and Rancimat induction time (h) of fat extracted from quinoa and amaranth seeds.

Parameter	Quinoa	Amaranth
Fat content (g/100 g of seeds)	5.43 ± 0.51 [a]	7.30 ± 0.62 [b]
PV (meq O$_2$/kg of fat)	8.86 ± 1.38 [a]	6.57 ± 0.68 [b]
Induction time (h)	5.10 ± 0.47 [a]	6.90 ± 0.38 [b]

Data denoted by the same letter are not statistically different (α = 0.05).

In the study, oil yield recovery from amaranth seeds reached 7.30% (Table 1). According to Caselato-Sous and Amaya-Farfán [27], the fat content in amaranth was 7%. Based on the studies conducted by Sanz-Penell et al. [28], the more than six-times higher fat content in amaranth than in wheat, i.e., about 10 g/100 g of fresh weight, affected the functionality of the flour as a stabilizing agent for the gas release during baking, which probably makes the dough elastic.

The total percentage of individual fatty acids and fatty acid groups (SFA—saturated fatty acids, MUFA—monounsaturated fatty acids, PUFA—polyunsaturated fatty acids) is shown in Table 2. The quinoa lipid fraction contained 10.98% of SFA, 30.82% of MUFA and 58.19% of PUFA. The content of fatty acid groups in amaranth seeds was: 24.94% SFA, 25.46% MUFA, 49.61% PUFA. The experiment of Ryan et al. [29] on quinoa oil showed the content of fatty acids at the level of: SFA 11.2%, MUFA 32.8% and PUFA 56.1%. The total content of SFA, MUFA and PUFA in the experiment conducted by Palombini et al. [30] for quinoa was 13.55%, 28.55%, 57.90%, respectively. For amaranth oil, the same researchers determined the content of SFA, MUFA, and PUFA at the levels of 25.28%, 33.82% and 40.90%. In the studies conducted by León-Camacho and García-González [31] on *Amaranthus cruentus*, the amount of SFA was 26.4%, MUFA was 34.7%, and PUFA was 38.9%. The saturated fatty acid present in the predominant amount in quinoa oil was palmitic acid, at 9.59%. Stearic acid was detected in a smaller amount, at 0.59% (Table 2). Similar amounts were obtained by Ando et al. [32]: about 10% of palmitic acid, and 0.75% of stearic acid. In amaranth fat, palmitic acid content was 18.92%, and stearic acid was 4.50% (Table 2). The experiment of Jahaniaval et al. [5] on *A. cruentus* showed the presence of palmitic acid at the level of 22.2%, and of stearic acid at the level of 3.57%. The tested quinoa oil showed a high proportion (52.59%) of C18: 2n−6 (linoleic) acid. The content of C18: 3n-3 (α-linolenic) acid reached the value of 5.42% (Table 2). Tang et al. [33] experimented with white quinoa, and showed the presence of C18: 2n−6 acid at the level of 47.39%, and the content of C18: 3n-3 acid at the level of 8.44%. The total content of unsaturated fatty acids in analyzed pseudocereals can help in maintaining health, as linoleic acid reduces levels of cholesterol and LDL in serum, while oleic acid presents a neutral behavior with respect to LDL, but moderately increases the level of high-density lipoproteins (HDL) [24]. Generally, this distinctive fatty acid profile is suitable for human health as it reduces risk factors related to cardiovascular diseases [25].

Table 2. Fatty acids composition (%) of oil extracted from quinoa and amaranth seeds.

Type of Fatty Acids	Quinoa	Amaranth
C14:0	0.14 ± 0.08 [a]	0.25 ± 0.08 [a]
C15:0	0.05 ± 0.01 [a]	0.09 ± 0.02 [a]
C16:0	9.59 ± 1.21 [b]	18.92 ± 1.15 [b]
C16:1	0.07 ± 0.01 [a]	0.11 ± 0.04 [a]
C17:0	0.04 ± 0.01 [a]	0.08 ± 0.03 [a]
C18:0	0.59 ± 0.10 [b]	4.50 ± 0.59 [b]
C18:1 cis	28.84 ± 2.15 [a]	25.03 ± 1.87 [a]
C18:2 n-6	52.59 ± 2.67 [a]	48.55 ± 2.48 [a]
C18:3 n-3	5.42 ± 1.26 [a]	1.06 ± 0.36 [b]
C20:0	0.57 ± 0.16 [a]	1.10 ± 0.24 [b]
C20:1	1.91 ± 0.34 [a]	0.32 ± 0.11 [b]
C20:2	0.18 ± 0.08	-
SFA	10.98 ± 1.57 [a]	24.94 ± 2.11 [b]
MUFA	30.82 ± 2.50 [a]	25.46 ± 2.02 [a]
PUFA	58.19 ± 4.01 [a]	49.61 ± 2.84 [b]

Data denoted by the same letter are not statistically different (α = 0.05). SFA—saturated fatty acids, MUFA—monounsaturated fatty acids, PUFA—polyunsaturated fatty acids.

As for the distribution of fatty acids between the positions of triacylglycerols in quinoa and amaranth lipid fraction, it was typical for vegetable oils. In both quinoa and amaranth oils, unsaturated fatty acids tended to be located in the internal position sn-2 of triacylglycerols (Figure 2). The proportion of saturated fatty acids in the sn-2 position was below 33.3%, which means that saturated fatty acids were mainly located in the external sn-1,3 positions.

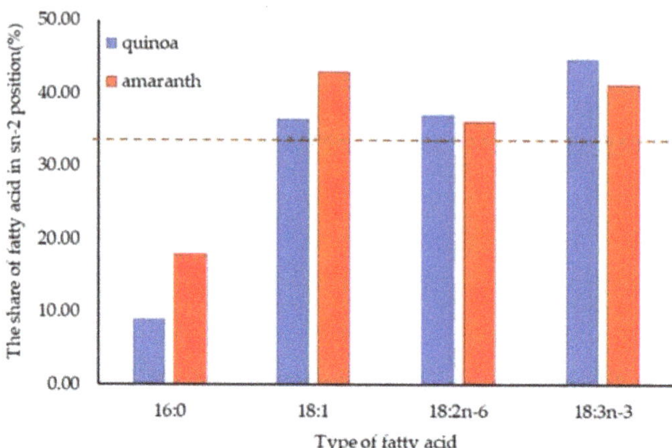

Figure 2. The percentage of palmitic, oleic, linoleic and α-linolenic acids in the sn-2 position of triacylglycerols (TAG) of amaranth and quinoa oil.

In the literature, influence of the TAG structure on the oxidative stability of oils is not always consistent. Some researchers reported that TAG with unsaturated fatty acids located at the sn-2 positions were more stable than TAG with unsaturated fatty acids in external positions [34,35]. Saturated fatty acids coexisting in TAG with unsaturated fatty acids, did not affect the oxidative of fat, whereas short chain saturated fatty acids can enhance oxidative stability of unsaturated TAG. On the other hand, Martin et al. [36] discussed that location of the fatty acids in the TAG position did not seem to be conclusive of the oxidative stability of fat.

3.2. Melting Profile

The melting point is one of the main indicators characterizing the consistency of fats and oils, and consequently, also their possible uses. DSC melting curves of amaranth and quinoa oils are presented in Figure 3. In the quinoa oil, three endothermic peaks were observed at temperatures: −44.81 °C, −20.34 °C, and 9.08 °C. For amaranth oil, three endothermic peaks can also be observed, but in a different temperature range. The maximum of the first peak was recorded at temperature of −26.10 °C, the second at −7.05 °C, and the third at 3.58 °C. The occurrence of maximum peaks at low temperatures was attributed to the high content of low-melting triacylglycerols with a high proportion of mono- and polyunsaturated fatty acids [21]. The performed determination of the fatty acid composition of the analyzed amaranth and quinoa oils confirmed the predominant share of unsaturated fatty acids. Generally, vegetable oil with a high degree of unsaturation can remain liquid over a wide temperature range [37]. Our results are in agreement with those reported by Wirkowska et al. [21], Wirkowska et al. [38], and Rezig et al. [39], who observed that vegetable oils were characterized by the presence of peaks in the low temperature range on the melting curve, which corresponded to the presence of low-melting TAG fractions.

3.3. Oxidative Stability

In both oils, peroxide value did not exceed the value specified in Codex Alimentarius for refined oils (<10 meq O_2/kg of fat), although in the quinoa oil, the level of peroxide value was higher. This may mean that quinoa oil was less resistant to oxidation than amaranth oil. The amount of peroxides, as the only indicator, did not clearly indicate the fat's resistance to oxidation. It is advisable to perform other tests characterizing the degree of oxidation. Oil extracted from amaranth seeds was characterized by a longer induction time measured wth the Rancimat test than fat isolated from quinoa seeds (Table 1). Induction time obtained for amaranth oil (6.9 h) was consistent with the results obtained by

Szterk et al. [40] (6.14 h). In tested amaranth oils, a longer induction time was noticed than for linseed oil—0.39–0.76 h [10], rapeseed oil—4.76–5.84, and olive oil—4.79–5.26 h [41]. By contrast, quinoa oil was characterized by shorter induction time than hazelnut oil—5.19–8.94 h [41].

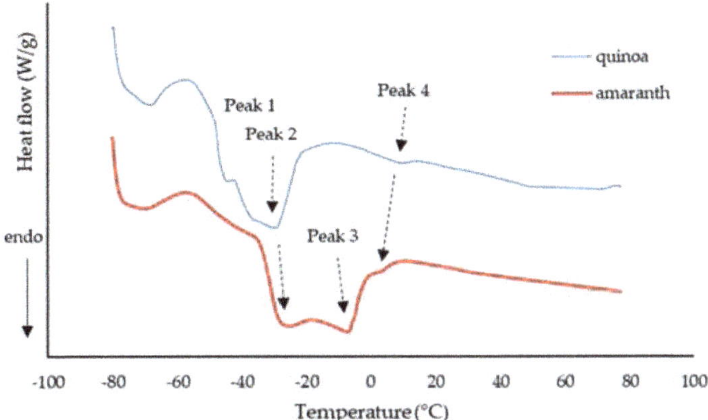

Figure 3. Melting DSC curves, determined in the temperature range −80–80 °C of amaranth (red line) and quinoa (blue line) oils.

The results for oxidative stability of amaranth and quinoa oils measured in the PDSC test at five different temperatures are summarized in Table 3. As expected, the induction time of the tested oils decreased with increase in temperature. The induction time for amaranth oil ranged from 435.26 min (at 100 °C) to 32.20 min (at 140 °C), and for quinoa oil ranged from 300.15 min to 20.51 min. The oxidative stability of oils assessed by the PDSC method had only been the subject of a few studies. Irwandi et al. [42] obtained an induction time of 71.4 min in a study of the cultivar *Amarantus gangeticus*. The PDSC result in an experiment carried out by Ando et al. [32] for quinoa oil was 71 min at 120 °C. In our study, the PDSC induction time of investigated amaranth and quinoa oils was much longer than in results obtained by Symoniuk et al. [10] in a similar experiment for linseed oil: 104.20–111.19 min at 100 °C, 46.19–51.93 min at 110 °C, 21.20–24.72 min at 120 °C, 10.48–11.30 min at 130 °C, 4.33–4.97 min at 140 °C. On the other hand, studies conducted by Ciemniewska et al. [41], also in similar experiment, indicated that rapeseed oils, olive oils and hazelnut oils were characterized by a longer induction time in the temperature range of 100–130 °C.

Table 3. PDSC induction time (min.) in the temperature range 100–140 °C of oils extracted from quinoa and amaranth seeds.

Temperature (°C)	Induction Time (min)	
	Quinoa	Amaranth
100	300.15 ± 4.98 [a]	435.26 ± 6.50 [b]
110	169.68 ± 3.21 [a]	200.50 ± 4.60 [b]
120	82.82 ± 2.10 [a]	112.00 ± 3.20 [b]
130	47.36 ± 2.06 [a]	62.40 ± 2.90 [b]
140	20.51 ± 1.13 [a]	32.20 ± 2.20 [b]

Data denoted by the same letter are not statistically different ($\alpha = 0.05$).

PDSC induction times in whole temperatures range (100–140 °C) were significantly longer for amaranth oil than for quinoa oil. The free fatty acids present in the oil made the fat more susceptible to oxidation. The investigated quinoa oil was characterized by a higher level of free fatty acids (measured by acid value), which resulted in lower oxidative stability compared to amaranth oil with lower level of free fatty acids (data not presented).

If we compare the oxidative stability measured at the same temperature (120 °C) by the Rancimat and PDSC tests, it turned out that induction time values measured by PDSC were four times shorter compared to those measured by the Rancimat method. The differences could be related to the smaller sample size used in PDSC measurement (3–4 mg) in comparison to the Rancimat quantity of the sample (2.5 g) [41]. According to Tan et al. [43], higher surface-volume ratio of the PDSC oil sample also played an important role leading to the shortened analyzed time. It may also be due to the fact that the oil in the PDSC test was oxidized with pure oxygen at a pressure of 1400 kPa, while in the Rancimat test, air containing approx. 21% oxygen at a flow of 20 L/h [41] was used. Ciemniewska et al. [41] managed to obtain a statistically significant linear correlation (R > 0.99) between PDSC and the Rancimat values for hazelnut oil. The authors recommended PDSC as an appropriate objective method for assessing the oxidative stability of oil.

3.4. Kinetics Analysis of Oxidation

Oxidation time measured in isothermal conditions (100–140 °C) enabled preparation of graphical dependence between the logarithm of the induction time (τ) and the reciprocal temperature (in absolute scale) (Figure 4).

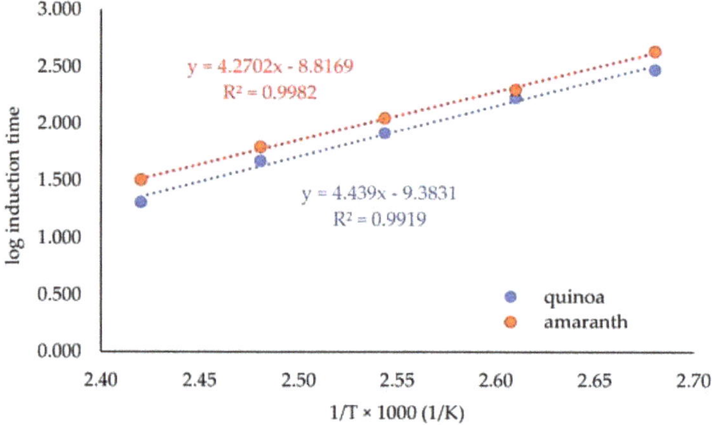

Figure 4. Log PDSC induction time (τ) versus reciprocal temperature for oxidation of amaranth (red line) and quinoa (blue line) oil.

Due to the fact that high R^2 coefficients were obtained (>0.99), the data from the equation describing the linear relationship were the starting point for calculating the kinetic parameters (Table 4). In Table 5, the oxidation parameters of amaranth and quinoa oil (obtained in this study) with the available references data on the oxidation of other vegetable oils were presented. Obtained values of activation energy were lower than values obtained by Symoniuk et al. [10] for linseed oil (93.14–94.53 kJ), Ratusz et al. [44] for *Camelina sativa* oil (87.63–93.61 kJ), and Ciemniewska et al. [41] for hazelnut oil (89.06 kJ), olive oil (92.81 kJ), and rapeseed oil (92.68 kJ). Due to the high content of polyunsaturated fatty acids in the analyzed amaranth and quinoa oils, a small amount of energy was needed to initiate the oxidation reaction. According to Adhvaryu et al. [45], high content of MUFA and SFA would increase the activation energy, whereas the high content of PUFA would decrease the activation energy value for lipid oxidation.

Table 4. Kinetics parameters (*a* and *b*—adjustable coefficients, *Ea*—activation energy, *Z*—pre-exponential factor, *k*—reaction rate coefficient) of oxidation of oils extracted from quinoa and amaranth seeds.

Parameters	Quinoa	Amaranth
a	4.27	4.44
b	8.82	9.38
R^2	0.9919	0.9982
Ea (kJ/mol)	77.75 ± 3.24 [a]	80.82 ± 2.87 [a]
Z (1/min)	1.44×10^7	5.13×10^7
k at 100 °C	1.89×10^{-4}	2.48×10^{-4}
k at 110 °C	3.63×10^{-4}	4.91×10^{-4}
k at 120 °C	6.76×10^{-4}	9.36×10^{-4}
k at 130 °C	1.22×10^{-3}	1.72×10^{-3}
k at 140 °C	2.14×10^{-3}	3.10×10^{-3}

Data denoted by the same letter are not statistically different ($\alpha = 0.05$).

Table 5. Comparison of oxidation parameters (activation energy—*Ea*, pre-exponential factor—*Z*, and reaction rate coefficient—*k*) of pseudo-cereal oils with camelina, linseed, hazelnut, rapeseed and olive oil.

Typ of Oil	*Ea* (kJ/mol)	*Z* (1/min)	*k* ange	References
quinoa seeds oil	77.75 ± 3.24	1.44×10^7	1.89×10^{-4}–2.14×10^{-3} (at 100–140 °C)	this study
amaranth seeds oil	80.82 ± 2.87	5.13×10^7	2.48×10^{-4}–3.10×10^{-3} (at 100–140 °C)	this study
camelina oil	88.37–93.61	1.40×10^{11}–4.12×10^{11}	1.17×10^{-2}–3.06×10^{-1} (at 90–130 °C)	[42]
linseed oil	93.14–94.53	3.24×10^{11}–5.38×10^{11}	1.25×10^{-2}–7.43×10^{-1} (at 90–140 °C)	[10]
hazelnut oil	89.06	2.08×10^{10}	7.10×10^{-3}–1.14×10^{-1} (at 100–140 °C)	[39]
rapeseed oil	92.68	1.02×10^{11}	1.08×10^{-2}–1.95×10^{-1} (at 100–140 °C)	[39]
olive oil	92.81	6.5×10^{10}	6.60×10^{-3}–1.19×10^{-1} (at 100–140 °C)	[39]

Taking into account the rate constant of the oxidation reaction, it can be observed that the reaction rate increased with increasing temperature, both for amaranth oil and quinoa oil. The k values obtained by other research [10,41,44] demonstrated the same pattern.

This study had limitations. In the present study we did not analyze the oxidative stability by the Rancimat method in wide range of temperatures (only at 120 °C), as well as in total phenolic compounds. Interest in quinoa and amaranth has markedly increased, due to their significant amount of bioactive compounds, such as phenolic compounds, flavonoids and their glycosides, betanins, and carotenoids [46]. Total phenolic compounds were positively correlated with antioxidant activities of quinoa and amaranth seeds [46,47]. Phytochemical nutrients in quinoa and amaranth may help restore the balance between oxidative stress and antioxidant defense. Diet supplemented with quinoa and amaranth seeds reduced oxidative stress in the plasma, heart, kidney, liver, spleen, lung, testis and pancreas of fructose administered rats [46]. Cisneros-Yupanqui et al. [48] reported that activity of superoxide dismutase (an enzyme that catalyzes the dissociation of the free radical in water and hydrogen peroxide) was increased when quinoa seeds were supplied to hypertension-induced rats. In addition, the free phenolic and PUFA fractions of cooked quinoa showed strong antioxidant ability based on Caco-2 cell-based antioxidant activity assay. In addition, the phenolics and unsaturated fatty acids exhibited protective effects on H_2O_2-induced Caco-2 cell oxidative injury [46].

4. Conclusions

Our results showed that quinoa oil contained more polyunsaturated fatty acids and less saturated fatty acids than amaranth oil. Both oils were a very good source of unsaturated fatty acids. The distribution of fatty acids between the triacylglycerol positions in amaranth and quinoa oil was typical for vegetable oils. Saturated fatty acids were found mainly in the external positions, while unsaturated fatty acids were located in the internal position. The PDSC test was a very useful method for determining the oxidative stability of oils due to the short time of the analysis, the small amount of sample required, and the fact that it did not require chemical reagents for the test. PDSC has the potential to be used as an alternative method for evaluation of the stability of edible oils. Induction time of oxidation of quinoa and amaranth oils suggested that amaranth oil may present better resistance to oxidation, while activation energy values indicate that a small amount of energy should be supplied to initiate oxidation of both amaranth and quinoa oils. The obtained results of the kinetics of the oxidation reaction allowed prediction of the oxidation process under various conditions and may be useful for the assessment of the oxidation rate of oils from other pseudocereals.

The presented paper contains the preliminary study, and further investigations are needed in order to confirm the suitability of the DSC method to determine the thermal properties of pseudocereal oils.

Author Contributions: Conceptualization, M.W.-W.; methodology, M.W.-W. and E.O.-L.; formal analysis, M.W.-W., E.O.-L. and J.B.; data curation, E.O.-L. and J.B.; writing—original draft preparation, M.W.-W.; writing—review and editing, A.G. All authors have read and agreed to the published version of the manuscript.

Funding: The study was financially supported by sources of the Ministry of Education and Science within funds of the Institute of Food Sciences of Warsaw University of Life Sciences (WULS), for scientific research.

Institutional Review Board Statement: Not applicable.

Informed Consent Statement: Not applicable.

Data Availability Statement: The data generated or analyzed during this study are available from the corresponding author on reasonable request.

Acknowledgments: We would like to thank Lena Dąbrowska for help in carrying out the analyses and technical support.

Conflicts of Interest: The authors declare no conflict of interest.

References

1. Alvarez-Jubete, L.; Arendt, E.K.; Gallagher, E. Nutritive value of pseudocereals and their increasing use as functional gluten-free ingredients. *Trends Food Sci. Technol.* **2010**, *21*, 106–113. [CrossRef]
2. Nowak, V.; Du, J.; Charrondière, U.R. Assessment of the nutritional composition of quinoa (*Chenopodium quinoa* Willd). *Food Chem.* **2016**, *193*, 47–54. [CrossRef] [PubMed]
3. Pellegrini, M.; González, R.L.; Ricci, A.; Fontecha, J.; Fernández-López, J.; Pérez-Alvarez, J.A.; Viuda-Martos, M. Chemical, fatty acid, polyphenolic profile, techno-functional and antioxidant properties of flours obtained from quinoa (*Chenopodium quinoa* Willd) seeds. *Ind. Crop. Prod.* **2018**, *111*, 38–46. [CrossRef]
4. Nasir, S.; Allai, F.M.; Gani, M.; Ganaie, S.; Gul, K.; Jabeen, A.; Majeed, D. Physical, Textural, Rheological, and Sensory Characteristics of Amaranth-Based Wheat Flour Bread. *Int. J. Food Sci.* **2020**, *2020*, 1–9. [CrossRef] [PubMed]
5. Jahaniaval, F.; Kakuda, Y.; Marcone, M.F. Fatty acid and triacylglycerol compositions of seed oils of five Amaranthus accessions and their comparison to other oils. *J. Am. Oil Chem. Soc.* **2000**, *77*, 847–852. [CrossRef]
6. Ruas, P.M.; Bonifacio, A.; Ruas, C.F.; Fairbanks, D.J.; Andersen, W.R. Genetic relationship among 19 accessions of six species of *Chenopodium*, L., by Random Amplified Polymorphic DNA fragments (RAPD). *Euphytica* **1999**, *105*, 25–32. [CrossRef]
7. Escudero, N.L.; De Arellano, M.L.; Luco, J.; Giménez, M.S.; Mucciarelli, S.I. Comparison of the Chemical Composition and Nutritional Value of Amaranthus cruentus Flour and Its Protein Concentrate. *Plant Foods Hum. Nutr.* **2004**, *59*, 15–21. [CrossRef]
8. Jancurowa, M.; Minarovicova, L.; Dandar, A. Quinoa—A review. *Czech J. Food Sci.* **2009**, *27*, 71–79. [CrossRef]
9. Navruz-Varli, S.; Sanlier, N. Nutritional and health benefits of quinoa (*Chenopodium quinoa* Willd.). *J. Cereal Sci.* **2016**, *69*, 371–376. [CrossRef]

10. Symoniuk, E.; Ratusz, K.; Krygier, K. Comparison of the oxidative stability of linseed (*Linum usitatissimum* L.) oil by pressure differential scanning calorimetry and Rancimat measurements. *J. Food Sci. Technol.* **2016**, *53*, 3986–3995. [CrossRef]

11. Vicente, J.; Cappato, L.P.; Calado, V.M.D.A.; de Carvalho, M.G.; Garcia-Rojas, E.E. Thermal and oxidative stability of Sacha Inchi oil and capsules formed with biopolymers analyzed by DSC and 1H NMR. *J. Therm. Anal.* **2017**, *131*, 2093–2104. [CrossRef]

12. Raczyk, M.; Bryś, J.; Brzezińska, R.; Ostrowska-Ligęza, E.; Wirkowska-Wojdyła, M.; Górska, A. Quality assessment of cold-pressed strawberry, raspberry and blackberry seed oils intended for cosmetic purposes. *Acta Sci. Pol. Technol. Aliment.* **2021**, *20*, 127–133. [PubMed]

13. Piasecka, I.; Górska, A.; Ostrowska-Ligęza, E.; Kalisz, S. The Study of Thermal Properties of Blackberry, Chokeberry and Raspberry Seeds and Oils. *Appl. Sci.* **2021**, *11*, 7704. [CrossRef]

14. van Wetten, I.; van Herwaarden, A.; Splinter, R.; van Ruth, S. Oil Analysis by Fast DSC. *Procedia Eng.* **2014**, *87*, 280–283. [CrossRef]

15. *ISO 659*; Oilseeds—Determination of Oil Content (Reference Method). International Organization for Standardization: Geneva, Switzerland, 2009.

16. Bryś, J.; Obranović, M.; Repajić, M.; Kraljić, K.; Škevin, D.; Bryś, A.; Górska, A.; Ostrowska-Ligęza, E.; Wirkowska-Wojdyła, M. Comparison of Different Methods of Extraction for Pomegranate Seeds. *Proceedings* **2020**, *70*, 91. [CrossRef]

17. *ISO 5509*; Animal and Vegetable Fats and Oils—Preparation of Methyl Esters of Fatty Acids. International Organization for Standardization: Geneva, Switzerland, 2001.

18. Wirkowska-Wojdyła, M.; Bryś, J.; Ostrowska-Ligęza, E.; Górska, A.; Chmiel, M.; Słowiński, M.; Piekarska, J. Quality and oxidative stability of model meat batters as affected by interesterified fat. *Int. J. Food Prop.* **2019**, *22*, 607–617. [CrossRef]

19. Bryś, J.; Flores, L.F.V.; Górska, A.; Wirkowska-Wojdyła, M.; Ostrowska-Ligęza, E.; Bryś, A. Use of GC and PDSC methods to characterize human milk fat substitutes obtained from lard and milk thistle oil mixtures. *J. Therm. Anal.* **2017**, *130*, 319–327. [CrossRef]

20. Aguedo, M.; Giet, J.; Hanon, E.; Lognay, G.; Wathelet, B.; Destain, J.; Brasseur, R.; Vandenbol, M.; Danthine, S.; Blecker, C.; et al. Calorimetric study of milk fat/rapeseed oil blends and their interesterification products. *Eur. J. Lipid Sci. Technol.* **2009**, *111*, 376–385. [CrossRef]

21. Wirkowska-Wojdyła, M.; Bryś, J.; Górska, A.; Ostrowska-Ligęza, E. Effect of enzymatic interesterification on physiochemical and thermal properties of fat used in cookies. *LWT* **2016**, *74*, 99–105. [CrossRef]

22. *ISO 3960*; Animal and Vegetable Fats and Oils—Determination of Peroxide Value—Iodometric (Visual) Endpoint Determination. International Organization for Standardization: Geneva, Switzerland, 2007.

23. *ISO 6886*; Animal and Vegetable Fats and Oils—Determination of Oxidation Stability (Accelerated Oxidation Test). International Organization for Standardization: Geneva, Switzerland, 2009.

24. Villacrés, E.; Pástor, G.; Quelal, M.B.; Zambrano, I.; Morales, S.H. Effect of processing on the content of fatty acids, tocopherols and sterols in the oils of quinoa (*Chenopodium quinoa* Willd), lupine (*Lupinus mutabilis* Sweet), amaranth (*Amaranthus caudatus* L.) and sangorache (*Amaranthus quitensis* L.). *Glob. Adv. Res. J. Food Sci. Technol.* **2013**, *2*, 44–53.

25. Gómez, M.J.R.; Prieto, J.M.; Sobrado, V.C.; Magro, P.C. Nutritional characterization of six quinoa (*Chenopodium quinoa* Willd) varieties cultivated in Southern Europe. *J. Food Compos. Anal.* **2021**, *99*, 103876. [CrossRef]

26. Ascheri, J.L.R.; Nascimento, R.E.; Spehar, C.R. Composição química comparativa de farinha instantânea de quinoa, arroz e milho. *Comun. Tec.* **2002**, *52*, 1–4.

27. Caselato-Sousa, V.M.; Amaya-Farfán, J. State of Knowledge on Amaranth Grain: A Comprehensive Review. *J. Food Sci.* **2012**, *77*, R93–R104. [CrossRef]

28. Sanz-Penella, J.; Wronkowska, M.; Soral-Smietana, M.; Haros, M. Effect of whole amaranth flour on bread properties and nutritive value. *LWT* **2013**, *50*, 679–685. [CrossRef]

29. Ryan, E.; Galvin, T.P.; Connor, A.R.; Maguire, A.R. Phytosterol, squalene, tocopherol content and fatty acid profile of selected seeds, grains, and legumes. *Plant Foods Hum. Nutr.* **2007**, *62*, 85–91. [CrossRef]

30. Palombini, S.V.; Claus, T.; Maruyama, S.A.; Gohara, A.K.; Souza, A.H.P.; De Souza, N.E.; Visentainer, J.V.; Gomes, S.T.M.; Matsushita, M. Evaluation of nutritional compounds in new amaranth and quinoa cultivars. *Food Sci. Technol.* **2013**, *33*, 339–344. [CrossRef]

31. León-Camacho, M.; González, D.G.; Aparicio, R. A detailed and comprehensive study of amaranth (*Amaranthus cruentus* L.) oil fatty profile. *Eur. Food Res. Technol.* **2001**, *213*, 349–355. [CrossRef]

32. Ando, H.; Chen, Y.-C.; Tang, H.; Shimizu, M.; Watanabe, K.; Mitsunaga, T. Food Components in Fractions of Quinoa Seed. *Food Sci. Technol. Res.* **2002**, *8*, 80–84. [CrossRef]

33. Tang, Y.; Xihong, L.; Peter, X.; Zhang, B.; Hernandez, M.; Zhang, H.; Massimo, F. Bound phenolics of quinoa seeds released by acid, alkaline, and enzymatic treatments and their antioxidant and α-glucosidase and pancreatic lipase inhibitory effects. *J. Agric. Food Chem.* **2015**, *64*, 1712–1719. [CrossRef]

34. Endo, Y.; Hoshizaki, S.; Fujimoto, K. Autoxidation of synthetic isomers of triacylglycerol containing eicosapentaenoic acid. *J. Am. Oil Chem. Soc.* **1997**, *74*, 543–548. [CrossRef]

35. Endo, Y.; Hoshizaki, S.; Fujimoto, K. Oxidation of synthetic triacylglycerols containing eicosapentaenoic and docosahexaenoic acids: Effect of oxidation system and triacylglycerol structure. *J. Am. Oil Chem. Soc.* **1997**, *74*, 1041–1045. [CrossRef]

36. Martin, D.; Reglero, G.; Señoráns, F.J. Oxidative stability of structured lipids. *Eur. Food Res. Technol.* **2010**, *231*, 635–653. [CrossRef]

37. Cheong, L.-Z.; Zhang, H.; Nersting, L.; Jensen, K.; Haagensen, J.A.J.; Xu, X. Physical and sensory characteristics of pork sausages from enzymatically modified blends of lard and rapeseed oil during storage. *Meat Sci.* **2010**, *85*, 691–699. [CrossRef] [PubMed]

38. Wirkowska-Wojdyła, M.; Chmiel, M.; Ostrowska-Ligęza, E.; Górska, A.; Bryś, J.; Słowiński, M.; Czerniszewska, A. The Influence of Interesterification on the Thermal and Technological Properties of Milkfat-Rapeseed Oil Mixture and Its Potential Use in Incorporation of Model Meat Batters. *Appl. Sci.* **2020**, *11*, 350. [CrossRef]

39. Rezig, L.; Chouaibi, M.; Msaada, K.; Hamdi, S. Chemical composition and profile characterisation of pumpkin (*Cucurbita maxima*) seed oil. *Ind. Crop. Prod.* **2012**, *37*, 82–87. [CrossRef]

40. Szterk, A.; Roszko, M.; Sosińska, E.; Derewiaka, D.; Lewicki, P.P. Chemical Composition and Oxidative Stability of Selected Plant Oils. *J. Am. Oil Chem. Soc.* **2010**, *87*, 637–645. [CrossRef]

41. Ciemniewska-Żytkiewicz, H.; Ratusz, K.; Bryś, J.; Reder, M.; Koczoń, P. Determination of the oxidative stability of hazelnut oils by PDSC and Rancimat methods. *J. Therm. Anal.* **2014**, *118*, 875–881. [CrossRef]

42. Irwandi, J.; Fitri, O.; Ridar, H.; Khatib, A. Oxidative behaviour of four Malaysian edible plant extracts in model and food oil systems. *J. Med. Plants Res.* **2011**, *6*, 1556–1561. [CrossRef]

43. Tan, C.P.; Man, Y.C.; Selamat, J.; Yusoff, M. Comparative studies of oxidative stability of edible oils by differential scanning calorimetry and oxidative stability index methods. *Food Chem.* **2002**, *76*, 385–389. [CrossRef]

44. Ratusz, K.; Popis, E.; Ciemniewska-Żytkiewicz, H.; Wroniak, M. Oxidative stability of camelina (*Camelina sativa* L.) oil using pressure differential scanning calorimetry and Rancimat method. *J. Therm. Anal.* **2016**, *126*, 343–351. [CrossRef]

45. Adhvaryu, A.; Erhan, S.; Liu, Z.; Perez, J. Oxidation kinetic studies of oils derived from unmodified and genetically modified vegetables using pressurized differential scanning calorimetry and nuclear magnetic resonance spectroscopy. *Thermochim. Acta* **2000**, *364*, 87–97. [CrossRef]

46. Tang, Y.; Tsao, R. Phytochemicals in quinoa and amaranth grains and their antioxidant, anti-inflammatory, and potential health beneficial effects: A review. *Mol. Nutr. Food Res.* **2017**, *61*, 1600767. [CrossRef] [PubMed]

47. Souza, S.P.; Roos, A.A.; Gindri, A.L.; Domingues, V.O.; Ascari, J.; Guerra, G.P. Neuroprotective effect of red quinoa seeds extract on scopolamine-induced declarative memory deficits in mice: The role of acetylcholinesterase and oxidative stress. *J. Funct. Foods* **2020**, *69*, 103958. [CrossRef]

48. Cisneros-Yupanquil, M.; Lante, A.; Mihaylova, D.; Krastanov, A.I.; Vílchez-Perales, C. Impact of consumption of cooked red and black Chenopodium quinoa Willd. over blood lipids, oxidative stress, and blood glucose levels in hypertension-induced rats. *Cereal Chem.* **2020**, *97*, 1254–1262. [CrossRef]

 *applied sciences*

Article

The Influence of a Chocolate Coating on the State Diagrams and Thermal Behaviour of Freeze-Dried Strawberries

Ewa Ostrowska-Ligęza [1,*], Karolina Szulc [2], Ewa Jakubczyk [2], Karolina Dolatowska-Żebrowska [1], Magdalena Wirkowska-Wojdyła [1], Joanna Bryś [1] and Agata Górska [1]

[1] Department of Chemistry, Institute of Food Sciences, Warsaw University of Life Sciences, 159c Nowoursynowska Street, 02-776 Warsaw, Poland; k.dolatowska@gmail.com (K.D.-Ż.); magdalena_wirkowska@sggw.edu.pl (M.W.-W.); joanna_brys@sggw.edu.pl (J.B.); agata_gorska@sggw.edu.pl (A.G.)

[2] Department of Food Engineering and Process Management, Institute of Food Sciences, Warsaw University of Life Sciences, 159c Nowoursynowska Street, 02-776 Warsaw, Poland; karolina_szulc@sggw.edu.pl (K.S.); ewa_jakubczyk@sggw.edu.pl (E.J.)

* Correspondence: ewa_ostrowska_ligeza@sggw.edu.pl; Tel.: +48-22-5937635

Citation: Ostrowska-Ligęza, E.; Szulc, K.; Jakubczyk, E.; Dolatowska-Żebrowska, K.; Wirkowska-Wojdyła, M.; Bryś, J.; Górska, A. The Influence of a Chocolate Coating on the State Diagrams and Thermal Behaviour of Freeze-Dried Strawberries. *Appl. Sci.* **2022**, *12*, 1342. https://doi.org/10.3390/app12031342

Academic Editor: Francisco Artés-Hernández

Received: 13 December 2021
Accepted: 25 January 2022
Published: 27 January 2022

Publisher's Note: MDPI stays neutral with regard to jurisdictional claims in published maps and institutional affiliations.

Abstract: Chocolate-coated fruit is becoming more and more popular as a tasty snack. The subjects of the research were freeze-dried strawberries and dark and milk chocolate-coated freeze-dried strawberries. The DSC curves, sorption isotherms, and glass transition temperature were determined. The state diagrams of the freeze-dried strawberries and dark and milk chocolate-coated freeze-dried strawberries were investigated. The modulated differential scanning calorimetry (MDSC) technique was used to determine the glass transition temperature. The DSC diagrams of the studied samples showed differences in shape and course. The sorption isotherms of the freeze-dried strawberries and dark and milk chocolate-coated strawberries belonged to type II according to BET classification. A coating of milk or dark chocolate resulted in a significant reduction in the hygroscopic behaviour of the freeze-dried strawberries and could be considered a promising way to improve the shelf life of the product and improve the nutritional value for commercial production purposes.

Keywords: glass transition temperature; freeze-dried strawberries; milk and dark chocolate; MDSC

1. Introduction

New products that act as snacks are in great demand in the snack market. The development of novel snacks with nutritious ingredients has an effective role in improving diet quality. Eating between the main meals or snacking are popular behaviours throughout the world and are especially appreciated by children [1]. Therefore, the nutritional quality of such snacks has to be greatly considered [2]. Consumption of fruit-based snacks increases the intake of nutrients and phytochemicals, which leads to positive health effects [3,4]. Fruit snacks may be prepared so they are comprised of a core portion coated with an outer shell. A wide range of edible ingredients could be used as cores and coatings. Nowadays, natural ingredients such as fruits have gained interest as a suitable material for core formulation [5,6].

Chocolates and confectionery items are food products favoured by many people, especially children. Confectionery products are usually chosen for consumption because of their flavour and textural properties and are often treated as sources of fats and carbohydrates. However, confectionery products can be also considered a source of vitamins, minerals, and even polyphenols and tocopherols, which are believed to have beneficial effects in the prevention of heart disease and possibly some cancers [5,7]. Chocolate, the most popular among cocoa products, is considered a luxury good [8], with the composition varying with the type of product. In addition to cocoa, the final composition of the product is influenced by many extra ingredients [7]. Dark chocolates with a high proportion of cocoa solids are generally treated as a rich source of magnesium (Mg) and copper (Cu), whereas milk and white chocolates are relatively good sources of calcium (Ca) [9].

Water activity has usually been related to food stability, assuming internal thermodynamic equilibrium in the product and its equilibrium with the environment while avoiding kinetic aspects. Nevertheless, equilibrium may not be achieved in complex food systems such as dehydrated or intermediate moisture products [10]. The removal of water during processing of many products often results in the formation of an amorphous state, which is a nonequilibrium state with time-dependent properties. The physical state of amorphous materials may change from a solid glassy state to a liquid-like rubbery one when the glass transition temperature (T_g) is reached. As the T_g is dependent on the water content, a change from a rubbery to a glassy state can also occur as a consequence of a water content decrease in the product during its processing or storage. The glass transition is supposed to cause dramatic changes in some physical properties of the product. It is known that stickiness and the collapse of dehydrated powdered products occur due to a drastic decrease in the viscosity above the T_g [11–13]. On the other hand, an increase in the molecular mobility above the T_g may allow the crystallization of amorphous compounds, especially in food products containing low molecular weight sugars such as strawberries. In products with crystallizing components, the determination of the critical water activity or critical water content at a given temperature to avoid crystallization will allow a prolonged product shelf life. State diagrams showing the relationships between a product's water content and its physical state as a function of the temperature together with sorption isotherms are useful tools in process optimisation and food formulation for defining processing equipment and operation variables and designing package and storage conditions [14].

Removal of water from food products in order to increase their shelf life can be achieved by different dehydration techniques. However, shrinkage phenomena which depend on the interstitial mobility [15] affect the quality of the final product in classical dehydration methods markedly [16]. Berries, for example, with delicate structures and high levels of water content are very difficult to dehydrate by classical methods [17,18], especially during air-drying, when a collapse causes considerable damage to their physical structures [16]. Vacuum freeze-drying of biological products is the best method of water removal, with end products of the highest quality compared with other dehydration techniques [19,20]. Freeze-dried strawberries were found to be of excellent colour and flavour [18,21] with a high rehydration capacity [18,22].

The chocolate panned products represent a type of food product exhibiting high storage stability. Preservation of high-quality food products during their storage requires monitoring such parameters as the temperature, air humidity, oxygen content in the storage room, as well as the proximate chemical compositions of the products [23–25].

This presented study aimed to determine the influence of the chocolate coating on the state diagrams of strawberry snacks. The water activity of the environment, water content, and glass transition temperature of freeze-dried strawberries and dark and milk chocolate-coated strawberries are critical parameters that enable predicting the strawberry snacks' storage conditions and can be treated as an attempt to define the shelf life of the products.

2. Materials and Methods

2.1. Material

Fresh Senga Sengana (*Fragaria ananassa*) strawberries were purchased from a local market in Warsaw (Poland). Then, the strawberries were cleaned and washed thoroughly in tap water. After this process, they were thoroughly dried on paper-lined trays. After drying, the strawberries were cut into 2–3-mm thick slices.

2.2. Freeze-Drying of Strawberries

The strawberry slices were prepared according to the methodology of Jakubczyk et al [26]

Samples were placed on an aluminum tray and frozen at −40 °C for 4 h using a shock freezer (Irinox, Corbanese, Italy). Then, the material was freeze-dried for 24 h with the application of a Christ Gamma 1–16 LSC freeze-dryer (Martin Christ Gefriertrock-nungsanlagen GmbH, Osterode am Harz, Germany) under a pressure of 63 Pa and at a shelf temperature of 20 °C.

2.3. Chocolate Coating of Freeze-Dried Strawberries

The freeze-dried strawberry slices were coated with dark or milk couverture choco-late. Then, the coated strawberries were taken out and put on sieves. The sieves were placed on paper trays in order to remove the excess chocolate. After drying, the dark and milk couverture chocolate-coated strawberries were placed in tightly closed polyethylene bags [25]. The dark couverture chocolate composition (according to the manufacturer's declaration) was as follows: cocoa mass: 70%; sugar: 32.5%; and fat: 44.9%. The milk couverture chocolate composition (according to the manufacturer's declaration) was as follows: cocoa mass: 35%; sugar: 53.5%; and fat: 37.3%.

2.4. Sorption Isotherms

The water vapour sorption isotherms were determined according to the method of Jakubczyk et al. [26] and Peleg [27]. The water vapour sorption isotherms were determined using the static gravimetric method. The different saturated salt solutions (LiCl, CH_3COOK, $MgCl_2$, K_2CO_3, $Mg(NO_3)_2$, $NaNO_2$, and NaCl) and calcium chloride ($CaCl_2$) were prepared to obtain a water activity in the range of 0.0–0.753. The triplicate samples of the same variant of freeze-dried strawberries without or with chocolate coatings were stored in desiccators at a stable water activity and a temperature of 25 °C for 3 months. A small amount of thymol was placed in the desiccator with the NaCl to avoid microbial growth in the samples.

The Peleg model (Equation (1)) was evaluated by determining the best fit to the experimental data [26,27].

The Peleg model was chosen based on the best fit to the experimental data [26,27]:

$$u = Aa_w^B + Ca_w^D \tag{1}$$

where u is the equilibrium moisture content, a_w is the water activity, and A, B, C, and D areconstants.

The regression analysis was performed according to the method of Jakubczyk et al. [26]. The determination coefficient (R^2), the adjusted coefficient of determination (R^2 adj), and the root mean square error (RMSE) were used to estimate the compliance of the model with the empirical data. The regression analysis was performed with the application of Table Curve 2D v 5.01 (Systat Software Inc., San Jose, CA, USA).

2.5. Differential Scanning Calorimetry

The calorimetric measurements were performed with a Q200 DSC (TA Instruments, New Castle, DE, USA) according to the methods of Jakubczyk et al. [26] and Ostrowska-Ligeza et al. [28,29].

DSC curves are presented for three levels of water activity: 0.001, 0.329, and 0.753.

The water activity of 0.001 was a control, being basic and the first level of activity. It is the lowest level water activity for food, and in the real world, no food has a water activity level of 0.001.

The water activity 0.329 is the level corresponding to the water activity of many food products stored under appropriate conditions.

The water activity 0.753 is the level corresponding to the critical water activity of many food products stored under inappropriate conditions, such as at a high temperature and air humidity. With such water activity, pathogenic microorganisms can develop, and food products and ingredients can spoil very quickly (e.g., hydrolysis of fats occurs).

2.6. Glass Transition Temperature

The glass transition temperatures of the freeze-dried strawberries and dark and milk chocolate-coated strawberries were determined by modulated differential scanning calorimetry (MDSC) as described by Jakubczyk et al. [26] and Ostrowska-Ligęza et al. [29]. The Gordon–Taylor model (Equation (2)) was used to describe the changes in T_g with the water content and water activity [30]. The goodness of fit was estimated based on R^2:

$$T_g = \frac{(1 - x_w)T_{gs} + kx_w T_{gw}}{(1 - x_w) + kx_w} \tag{2}$$

where T_{gs}, T_{gw}, and T_g are the glass transition temperatures of the solids, water, and samples, respectively, x_w is the mass fraction of water, and k is the Gordon–Taylor model parameter.

2.7. Statistical Analysis

The statistical analysis was conducted as described by Ostrowska-Ligęza et al. [29].

3. Results

3.1. DSC Studies of Freeze-Dried Strawberries and Dark and Milk Chocolate-Coated Strawberries at Different Levels of Water Activity

The subjects of the research were freeze-dried strawberries and dark and milk chocolate-coated strawberries. Fructose, glucose, and sucrose is the mixture of sugars present in strawberries [31,32]. The DSC diagrams of the samples stored at water activities of 0.001, 0.329, and 0.753 are presented in Figures 1–3, respectively.

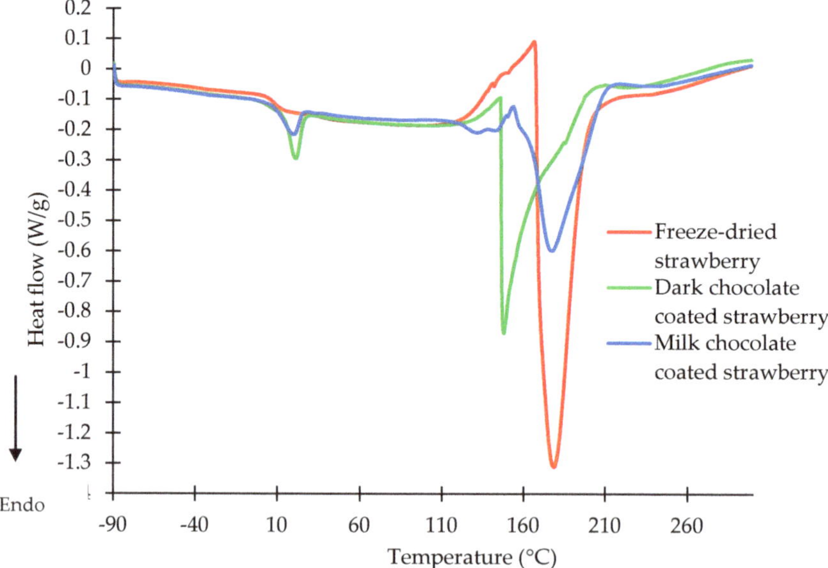

Figure 1. DSC curves of freeze-dried strawberries and strawberries coated with dark and milk chocolate stored at a water activity of 0.001.

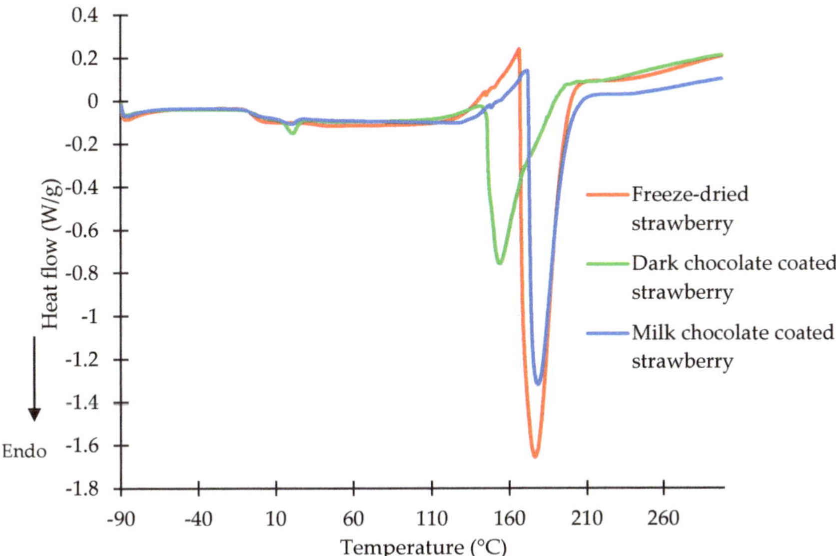

Figure 2. DSC curves of freeze-dried strawberries and strawberries coated with dark and milk chocolate at a water activity of 0.329.

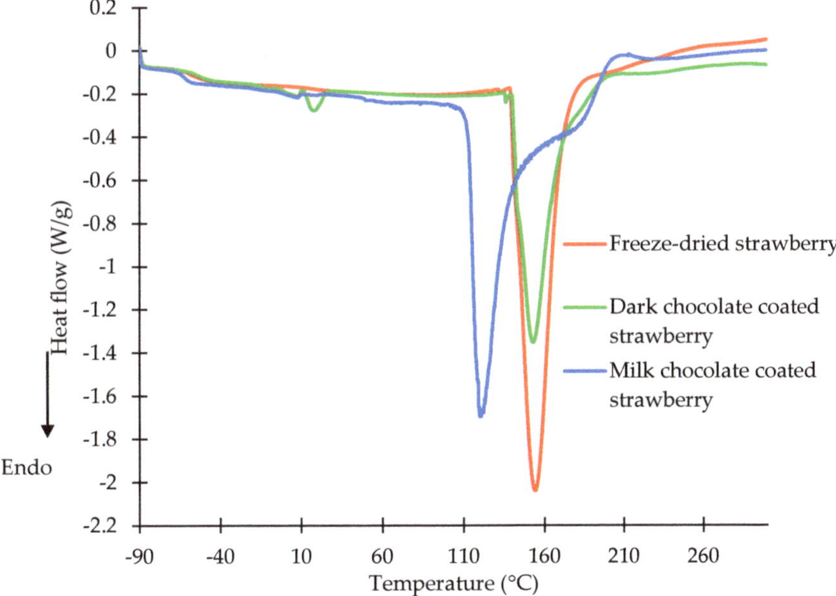

Figure 3. DSC curves of freeze-dried strawberries and strawberries coated with dark and milk chocolate at a water activity of 0.753.

Mild endothermic peaks at temperatures ranging from 0 to 30 °C were observed on the DSC diagrams of the dark and milk chocolate-coated strawberries. The maximum temperature of the first endothermic peak was observed at 21.21 and 20.15 °C for the dark and milk chocolate-coated strawberries, respectively (Figure 1). The peak corresponds to the melting of the chocolate that covered the strawberry. Cocoa butter was present in the dark chocolate, and a mixture of cocoa butter and milk fat was present in the milk chocolate. The melting point of the chocolates was higher, but it should be mentioned that the chocolate coatings were cooled to -90 °C, and therefore the fat memory was erased [33,34]. Ostrowska-Ligęza et al. [35] investigated three types of chocolates: dark, milk, and white. The authors found that the course of the DSC melting curves and melting temperatures of the chocolates depended on the quality of the fat, addition of sugar or emulsifiers, and particle size distribution in the chocolate.

The DSC diagrams of the freeze-dried strawberries and dark and milk chocolate-coated strawberries were characterised by distinct endothermic peaks at a water activity of 0.001. The peaks were characterised by a very sharp course at a temperature range from about 130 to 210 °C (Figure 1). For the DSC curve of the freeze-dried strawberries, one endothermic peak was observed, with the maximum temperature found at 178.41 °C. The maximum temperature of the second peak for the milk chocolate-coated strawberries was characterised by a similar value of 177.68 °C. For the dark chocolate-coated strawberries, the presence of a second endothermic peak on the DSC curve was observed at a maximum temperature of 147.25 °C. The values of the transition temperatures corresponded to the melting of the sugars contained in the snacks. The mixtures of sugars in the compositions of the freeze-dried and coated strawberries may be in crystalline or amorphous states. The melting temperature was dependent on the sugar state and composition of its mixtures. Wang et al. [36] studied the temperatures of the melting transitions of sucrose, glucose, fructose, and their mixtures at a water activity of about zero. The melting maximum temperature of sucrose was observed at 190.6 °C, while that of glucose was 161.1 °C and fructose's was 128.2 °C. The melting transitions of the mixtures of glucose–sucrose, fructose–sucrose, and glucose–fructose–sucrose were investigated by DSC. The melting temperature of sucrose was found to decrease in the presence of either fructose or glucose [36]. Lee et al. [37] determined the influence of the heating rate level on the melting transitions of sucrose, glucose, and fructose. The onset melting temperature was investigated. The values of the melting temperature were in agreement with the results published by Wang [36]. The dark chocolate was characterised by the lowest content of sugar. Coating the strawberry with dark chocolate probably lowered the melting temperature of the sugar. This melting point of the sugars may prove that the chocolate coating of the strawberry was carefully carried out.

The DSC diagrams of the freeze-dried strawberries and dark and milk chocolate-coated strawberries at a water activity of 0.329 are shown in Figure 2. The shapes, diagrams, and values of the melting temperatures were similar to those in the DSC curves with a water activity of 0.001. The intensity of the chocolate melting peaks changed, being lower for the dark and milk chocolate-coated strawberries at a water activity of 0.329. An endothermic peak with a maximum temperature of 21.25 °C for the dark chocolate-coated strawberries and 20.23 °C for the milk chocolate-coated strawberries was observed (Figure 2). Water activity in the range from 0.3 to 0.4 corresponds to the water activity during storage [28]. The maximum temperature of the carbohydrate transitions of the freeze-dried strawberries and dark and milk chocolate-coated strawberries at a water activity of 0.329 were observed at 175.87, 152.98, and 178.01 °C, respectively. Based on the obtained results, it can be stated that the thermal properties were not influenced by the level of water activity of the studied freeze-dried strawberry or dark or milk chocolate-coated strawberry samples.

The DSC curves of the freeze-dried strawberries and dark and milk chocolate-coated strawberries at a water activity of 0.753 are shown in Figure 3. The first endothermic peak corresponds to the melting transition of the dark chocolate coating, with the value of the maximum temperature at 17.95 °C. On the DSC diagrams of the freeze-dried strawberries and milk chocolate-coated strawberries, no peaks were observed in the range of temperatures from 0 to 30 °C. The milk chocolate coating was characterised by a lower content of fat, and this could cause the coating to partially dissolve. The peaks observed between 100 and 210 °C correspond to the melting transition of the carbohydrates for all samples. The maximum temperatures of the melting peaks for the freeze-dried strawberries and milk chocolate-coated strawberries were determined to be 153.76 and 118.8 °C, respectively. The increase in water content in the strawberries caused a decrease in the melting points of the sugars contained in the strawberries and the milk chocolate coating. An endothermic peak with a maximum temperature of 153.46 °C was observed for the dark chocolate-coated strawberries. The increased water activity of the carbohydrates lowered the melting points [38,39]. The freeze-dried strawberries and dark and milk chocolate-coated strawberries at a water activity of 0.753 could undergo permanent transformations. At a water activity of 0.753, oxidation and hydrolysis of fats can take place, as well as rapid rehydration of dried strawberries, the decomposition of sugars, and the development of pathogenic microorganisms, which lead to a reduction in the quality of the product. Kita et al. [25] investigated the influence of the packaging methods and storage time on chocolate-panned figs, cherries, hazelnuts, and almonds. They found that the right choice of package type allowed for minimizing transformations proceeding in chocolate-panned products during their long-term storage.

3.2. Sorption Isotherms of Freeze-Dried Strawberries and Dark and Milk Chocolate-Coated Strawberries

The moisture sorption isotherm of the freeze-dried strawberries and dark and milk chocolate-coated strawberries measured at 25 °C is shown in Figure 4. Adding dried fruits to chocolates has been widely used as a way of bringing nutritional and sensory benefits to these products [40]. The moisture sorption isotherm of a complex food is one of the most important measures affecting the acceptability, shelf life, and packaging and storage requirements [41]. The sorption isotherms as experimental data and the Peleg fitted model for freeze-dried strawberries with and without chocolate coatings at temperatures of 25 °C are presented in Figure 4. The sorption curves obtained for the samples were sigmoidal in shape, which is typical for type II isotherms [14,42]. The data showed typical behaviour of rich sugar foods: a slight increase in the equilibrium moisture content in the low water activity and a sharp increase above the water activity of 0.648 due to the prevailing effect of solute–solvent interactions associated with sugar dissolution [14,43,44]. Chocolate coating of freeze-dried strawberries influenced the course of the water vapour sorption isotherms. The equilibrium water content of the coated samples was lower than that of the freeze-dried fruit in the whole range of water activity. Lower values were obtained for strawberries coated with milk than dark chocolate, indicating that the chocolate coating protected the core against moisture absorption and the composition of the chocolate had a significant impact on the course of the sorption isotherms. Ghosh et al. [45] showed that the water vapour permeability of the chocolate coatings increased with the addition of cocoa powder and lecithin but decreased with the addition of sugar. Along with the diffusion of moisture, structural changes in the coating were observed by Ghost et al. [45] which altered the diffusivity of the moisture through the coating. These structural changes can occur due to the swelling of the cocoa particles or dissolving of the sucrose particles in the moisture and subsequent migration to the surface. The mechanism of diffusion through the two dispersed phases—sucrose and cocoa powder—is totally different. Moisture diffusion through the cocoa powder occurs through the particles, which is influenced by the porous nature of powder. The crystalline structure of sucrose makes it so that moisture cannot diffuse through sugar crystals, and therefore, moisture diffuses along the surface of the sugar particles [45]. The presence of a moisture layer on the surface can contribute to dissolving the sugar, and sugar can seep into the surface of the coating to cause sugar bloom.

The goodness of fit for Peleg model was estimated based on the determination coefficient (R^2), the adjusted coefficient of determination (R^2 *adj*), and the root mean square error (*RMSE*) (Table 1). The Peleg model was described by high values for R^2 (from 0.9962 to 0.9995) and low values for the *RMSE* (from 2.94 to 7.17%) for all samples (Table 2).

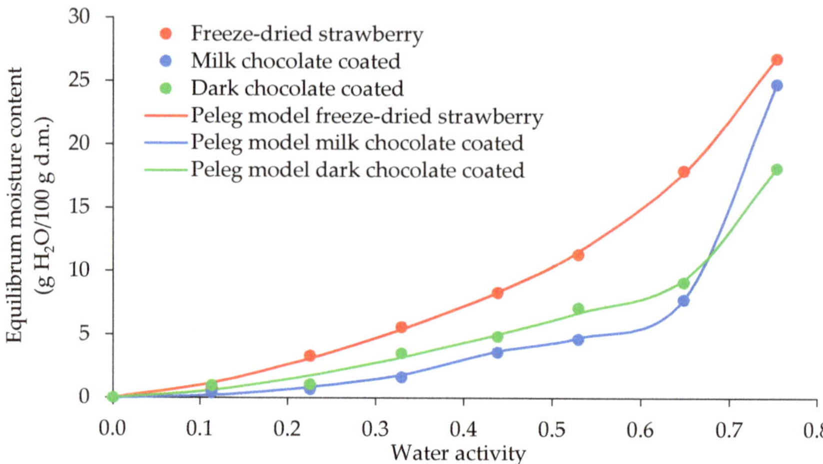

Figure 4. Water sorption isotherms of freeze-dried and dark and milk chocolate-coated strawberry.

Table 1. Fitting of Peleg model equation for obtained data.

Sample	A	B	C	D	R^2	R^2 *adj*	*RMSE* (%)
Freeze-dried strawberry	5.371 ± 0.899	0.599 ± 0.025	2.543 ± 0.032	0.141 ± 0.006	0.9995	0.9961	2.94
Milk chocolate-coated strawberry	547.074 ± 1.018	2.074 ± 0.105	1.682 ± 0.012	0.198 ± 0.027	0.9995	0.9953	3.38
Dark chocolate-coated strawberry	1017.400 ± 1.865	2.591 ± 0.241	1.808 ± 0.101	0.155 ± 0.008	0.9962	0.9896	7.17

Values represent means $\pm$ standard deviations. A, B, C, D = constants of Peleg model; R^2 = determination coefficient; R^2 *adj* = adjusted coefficient of determination; *RMSE* = root mean square error. Mean values $\pm$ standard deviation (n = 3).

Table 2. Gordon–Taylor model fitting for experimental data.

Sample	T_{gs} (°C)	k	R^2	R^2 *adj*	*RMSE* (%)
Freeze-dried strawberry	18.18 ± 0.23	2.73 ± 0.05	0.9811	0.9735	9.53
Milk chocolate-coated strawberry	9.07 ± 0.17	3.18 ± 0.13	0.8994	0.8592	24.72
Dark chocolate-coated strawberry	11.54 ± 0.42	3.71 ± 0.21	0.9232	0.8925	39.19

Values represent means $\pm$ standard deviations. T_{gs} = glass transition temperature of solids; k = Gordon-Taylor model parameter; R^2 = determination coefficient; R^2 *adj* = adjusted coefficient of determination; *RMSE* = root mean square error. Mean values $\pm$ standard deviation (n = 3).

Ciurzyńska and Lenart [42] confirmed that the highest probability of fitting experimental data for freeze-dried strawberries is given by Peleg's model (R^2 0.990–0.997). The Peleg model was also the best fitting observed for freeze-dried *Syzygium cumini* fruit (jambolan) [46] and freeze-dried apple puree gels [26] sorption values.

3.3. Influence of Water Activity on the Glass Transition Temperature of Freeze-Dried Strawberries and Dark and Milk Chocolate-Coated Strawberries

The relationships between the water activity and glass transition temperature as a function of the equilibrium moisture content are shown in Figures 5–7 for the freeze-dried strawberries and dark and milk chocolate-coated strawberries, respectively. The experimental data of the glass transition temperature obtained for the freeze-dried and dark and milk chocolate-coated strawberries are presented in Table 3. The fitting of the Gordon–Taylor model for data obtained for the freeze-dried and dark and milk chocolate-coated strawberries is shown in Table 2.

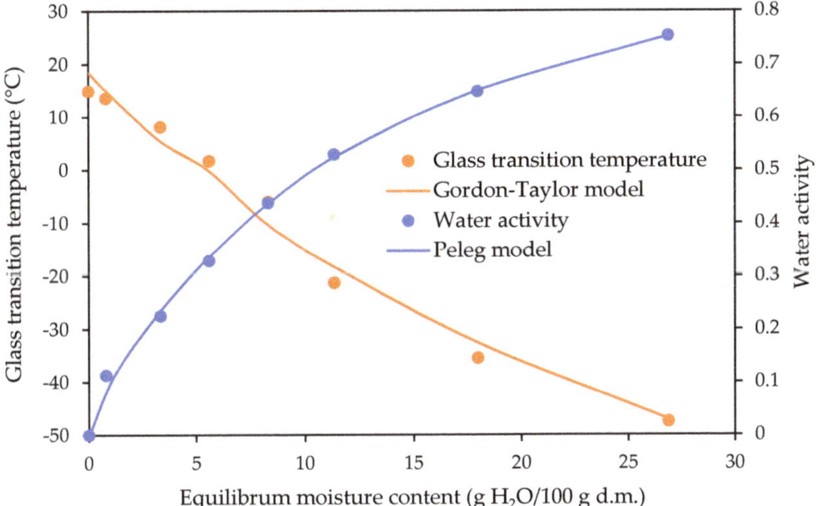

Figure 5. State diagram of freeze-dried strawberries.

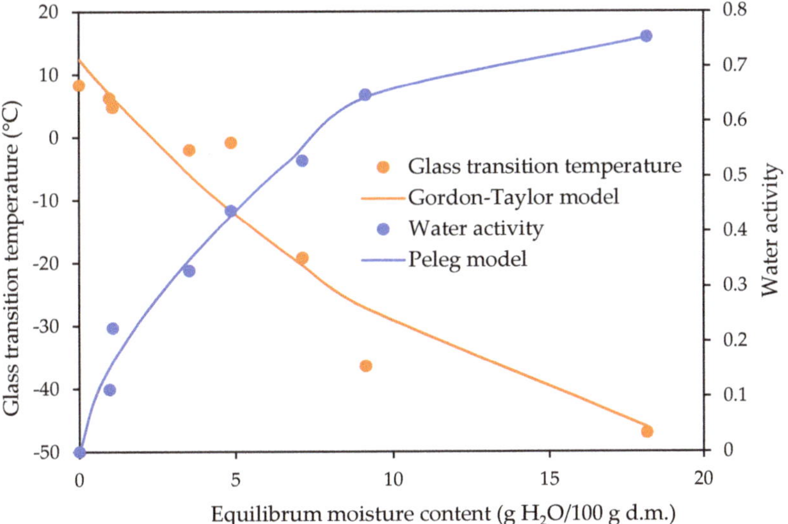

Figure 6. State diagram of dark chocolate-coated strawberries.

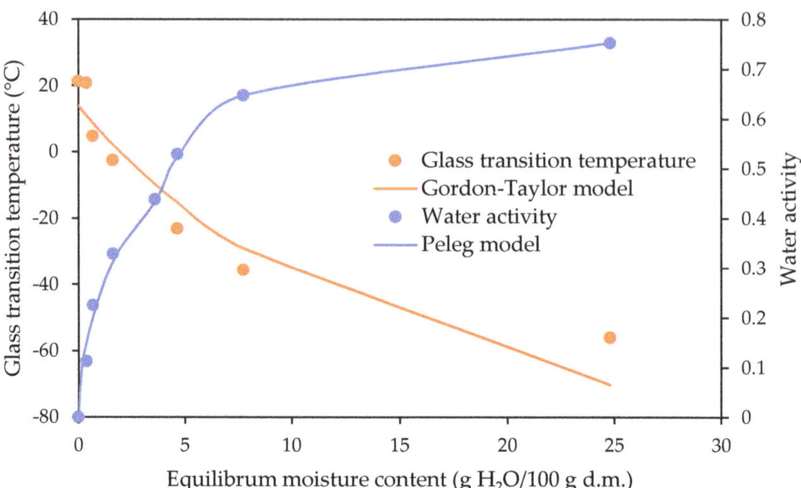

Figure 7. State diagram of milk chocolate-coated strawberries.

Table 3. Glass transition temperature of strawberries (freeze-dried and coated with dark chocolate and milk chocolate).

Water Activity	Freeze-Dried			Dark Chocolate-Coated			Milk Chocolate-Coated		
	$T_{gi}/°C$	$T_{gm}/°C$	$T_{ge}/°C$	$T_{gi}/°C$	$T_{gm}/°C$	$T_{ge}/°C$	$T_{gi}/°C$	$T_{gm}/°C$	$T_{ge}/°C$
0.000	9.15 ± 3.35	14.62 ± 3.22 [a]	20.02 ± 3.14	3.50 ± 2.12	7.99 ± 1.17 [b]	12.54 ± 0.31	18.58 ± 1.78	21.08 ± 0.08 [c]	23.59 ± 1.63
0.113	8.70 ± 1.13	14.02 ± 0.23 [a]	19.34 ± 0.66	0.57 ± 0.92	6.12 ± 0.12 [b]	11.65 ± 1.16	16.69 ± 3.97	18.92 ± 3.71 [c]	21.41 ± 3.05
0.225	2.45 ± 0.04	8.13 ± 0.18 [a]	13.79 ± 0.35	−0.34 ± 1.49	4.56 ± 1.30 [b]	9.46 ± 1.12	4.65 ± 0.86	11.27 ± 0.24 [c]	17.76 ± 1.63
0.329	−3.98 ± 0.86	1.89 ± 0.11 [a]	7.74 ± 0.63	−7.94 ± 0.45	−2.04 ± 1.74 [b]	3.53 ± 3.03	−6.44 ± 0.45	−3.62 ± 2.59 [c]	−0.74 ± 5.54
0.438	−10.59 ± 3.53	−4.79 ± 3.71 [a]	0.94 ± 3.85	−11.37 ± 1.99	−3.73 ± 0.54 [b]	4.00 ± 2.72	−19.68 ± 0.57	−15.40 ± 0.39 [c]	−11.14 ± 0.21
0.529	−26.98 ± 0.31	−21.36 ± 0.57 [a]	−15.71 ± 0.85	−24.52 ± 2.49	−16.44 ± 1.69 [b]	−8.90 ± 5.09	−28.37 ± 0.52	−23.01 ± 0.79 [c]	−17.65 ± 1.06
0.648	−40.75 ± 2.16	−35.62 ± 1.73 [a]	−30.48 ± 1.29	−42.41 ± 0.35	−37.07 ± 0.57 [b]	−31.72 ± 0.78	−40.69 ± 1.32	−35.73 ± 0.70 [a]	−30.77 ± 0.07
0.753	−52.24 ± 3.67	−47.48 ± 3.34 [a]	−42.75 ± 3.08	−53.42 ± 1.85	−47.79 ± 0.88 [a]	−42.14 ± 0.08	−60.00 ± 2.53	−55.44 ± 2.72 [b]	−50.89 ± 2.90

Values represent means ± standard deviations. Different letters in the rows indicate that the samples are considered significantly different at the 5% level ($p < 0.05$).

The sugar mixture (fructose, sucrose, and glucose) in the freeze-dried strawberries is in an amorphous state. These materials are not stable due to a lack of thermodynamic equilibrium [26,47]. With the values of the water contents increasing, the values of the glass transition decreased, as presented in the curves of the variations in glass transition temperature and water activity with the equilibrium moisture content of the freeze-dried strawberries and dark and milk chocolate-coated strawberries (Figures 5–7). The temperature of determination of the sorption isotherm was 25 °C, and according to the glass transition temperature conception, the values of the critical water activity and equilibrium moisture content should be determined at exactly this temperature. This is not possible because in the study, the highest glass transition temperature was determined for the milk chocolate-coated strawberries and reached 21.08 °C, with an equilibrium moisture content value of 0.00 at a water activity of 0.001 (Table 3).

The values of the glass transition temperature for the freeze-dried strawberries and dark chocolate-coated strawberries were lower, being 14.62 and 7.99 °C, respectively (Table 3). Jakubczyk et al. [26] researched the effect of the incorporation of apple puree and maltodextrin to agar sol on the sorption properties and structure of the dried gel. The authors investigated the relations between the glass transition temperature, water activity, and water content for apple snacks. They found that a higher value for the critical water content and critical water activity at the glass transition temperature of 25 °C, determined from the relations of water content–water activity and water content–glass transition temperature, can result in better stability for the physical properties of stored products. Dried apple puree gels with maltodextrin were characterised by a significantly higher critical water activity than dried gel without a carrier addition. Sà and Sereno [39] investigated phase transitions and unfreezable water in fresh and freeze-dried samples of onions, grapes, and strawberries after equilibration at different relative humidities. Freeze-dried strawberries and onions were obtained in the form of powder. From the DSC trace for each product, the glass transition temperature and melting temperatures were determined. The onset and end temperature of the glass transition were defined. The glass transition temperature ranges obtained by Sà and Sereno [39] differed from those obtained in this article. The values of the onset and end glass transition temperature at a water activity of 0.33 obtained by Sà and Sereno [39] reached −22.6 and −14.8 °C, respectively, and the results obtained in this study at the same water activity were −3.9 and 7.7 °C, respectively (Table 2). This could be due to different sample preparation methods and a different methodology of using DSC. The Gordon–Taylor equation was able to predict the glass transition temperature for the water–food systems studied from the corresponding pure component values [39]. The freeze-dried entire tissue and homogenised tissue of the strawberries were used to obtain the moisture sorption isotherms and glass transition temperature by Moraga et al. [14]. Strawberry pretreatments cause changes in the tissue structure that affect the water binding capacity of the different product phases at equilibrium with a determined aw value. They found that freeze-drying was the method by which products with a very low water content were obtained. This ensures the stability of the product during storage.

Nightingale et al. [48] investigated the influence of the fluctuation temperature and relative humidity on the storage of dark chocolate. The impact of the storage conditions on the quality of the dark chocolate by sensory and, among others, DSC measurements was determined. The dark chocolate was kept under various conditions and analysed at 0, 4, and 8 weeks of storage. They found that varying the storage conditions (e.g., high or low temperature and varying humidity) caused changes in the texture of the chocolate. The cocoa butter changed from the V to VI polymorphic form, which caused blooms on the surface of the dark chocolate. The temperature variation had the worst effect on the quality of the dark chocolate. The authors recommended storage of dark chocolate at a constant temperature and relative humidity. According to Kita et al. [25], the selection of appropriate storage and packaging conditions for chocolate-covered fruit and nuts will ensure their appropriate quality.

In this study, phase diagrams for the dark and milk chocolate-coated freeze-dried strawberries were determined for the first time.

4. Conclusions

Differences in shape and course were observed in the DSC diagrams of the freeze-dried strawberries and dark and milk chocolate-coated strawberries. The sorption isotherms of the freeze-dried strawberries and dark and milk chocolate-coated strawberries were classified as type II according to BET. Based on the obtained results, it can be stated that the shape and course of the sorption isotherms of the freeze-dried strawberries and dark and milk chocolate-coated strawberries were influenced by the method of snack preparation. The glass transition temperature decreased with the increase in the moisture content, which could be due to the strong plasticizing effect of water on this parameter. Coating with milk or dark chocolate resulted in a significant reduction in the hygroscopic behaviour of the freeze-dried strawberries and could be considered a promising way to improve the shelf life of a product.

Author Contributions: Conceptualization, E.O.-L.; methodology, E.O.-L., K.S. and E.J.; investigation, K.D.-Ż., K.S., J.B., M.W.-W. and A.G.; formal analysis, E.O.-L., A.G., M.W.-W. and E.J.; writing—original draft preparation, E.O.-L. and K.S.; writing—review and editing, E.O.-L., A.G. and M.W.-W. All authors have read and agreed to the published version of the manuscript.

Funding: The study was financially supported by sources of the Ministry of Education and Science within funds of the Institute of Food Sciences of Warsaw University of Life Sciences (WULS) for scientific research.

Institutional Review Board Statement: Not applicable.

Informed Consent Statement: Not applicable.

Data Availability Statement: The data generated or analysed during this study are available from the corresponding author on reasonable request.

Conflicts of Interest: The authors declare no conflict of interest.

References

1. Adams, E.L.; Savage, J.S. From the children's perspective: What are candy, snacks, and meals? *Appetite* **2017**, *116*, 215–222. [CrossRef] [PubMed]
2. Green, H.; Siwajek, P.; Roulin, A. Use of nutrient profiling to identify healthy versus unhealthy snack foods and whether they can be part of a healthy menu plan. *J. Nutr. Intermediar. Metabol.* **2017**, *9*, 1–5. [CrossRef]
3. Potter, R.; Stojceska, V.; Plunkett, A. The use of fruit powders in extruded snacks suitable for Children's diets. *LWT-Food Sci. Technol.* **2013**, *51*, 537–544. [CrossRef]
4. Roe, L.S.; Meengs, J.S.; Birch, L.L.; Rolls, B.J. Serving a variety of vegetables and fruit as a snack increased intake in preschool children. *Am. J. Clin. Nutr.* **2013**, *98*, 693–699. [CrossRef] [PubMed]
5. Beckett, S.T. *The Science of Chocolate*, 2nd ed.; The Royal Society of Chemistry: Cambridge, UK, 2008; pp. 61–79.
6. Yeganehzad, S.; Kiumarsi, M.; Nadali, N.; Ashkezary, M.R. Formulation, development and characterization of a novel functional fruit snack based on fig (*Ficus carica* L.) coated with sugar-free chocolate. *Heliyon* **2020**, *6*, e04350. [CrossRef] [PubMed]
7. Steinberg, F.M.; Bearden, M.M.; Keen, C.L. Cocoa and chocolate flavonoids: Implications for cardiovascular health. *J. Am. Diet. Assoc.* **2003**, *103*, 215–223. [CrossRef]
8. Szefer, P.; Grembecka, M. Mineral Components in Food Crops, Beverages, Luxury Food, Spices, and Dietary Food. In *Mineral Components in Foods*, 1st ed.; Szefer, P., Nriagu, J.O., Eds.; CRC Press & Taylor Francis Group: Boca Raton, UK, 2006; Chapter 7; pp. 343–435. [CrossRef]
9. Grembecka, M.; Szefer, P. Differentiation of confectionery products based on mineral composition. *Food Anal. Method* **2012**, *5*, 250–259. [CrossRef]
10. Goff, H.D. Low-temperature stability and the glassy state in frozen foods. *Food Res. Int.* **1992**, *25*, 317–325. [CrossRef]
11. Levine, H.; Slade, L. Principles of cryostabilization technology from structure/property relationships of carbohydrate/water systems: A review. *Cryo-Lett.* **1988**, *9*, 21–63.
12. Roos, Y.H.; Karel, M. Applying state diagrams to food processing and development. *Food Technol.* **1991**, *45*, 68–71.
13. Slade, L.; Levine, H. Beyond water activity: Recent advances based on an alternative approach to the assessment of food quality and safety. *Crit. Rev. Food Sci.* **1991**, *30*, 115–360. [CrossRef] [PubMed]
14. Moraga, G.; Martinez-Navarrete, N.; Chiralt, A. Water sorption isotherms and glass transition in strawberries: Influence of pretreatment. *J. Food Eng.* **2004**, *62*, 315–321. [CrossRef]
15. Roos, Y. *Time-Dependent Phenomena*, 1st ed.; Phase Transitions in Foods; Academic Press: London, UK, 1995; pp. 158–171.
16. Krokida, M.K.; Maroulis, Z.B. Effect of drying method on shrinkage and porosity. *Dry. Technol.* **1997**, *15*, 2441–2458. [CrossRef]
17. Jankovic, M. Physical properties of convectively dried and freeze-dried berrylike fruits. *Publ. Fac. Agric. Belgrade* **1993**, *38*, 129–135.
18. Mastrocola, D.; Dalla Rosa, M.; Massini, R. Freeze-dried strawberries rehydrated in sugar solutions: Mass transfers and characteristics of final products. *Food Res. Int.* **1997**, *30*, 359–363. [CrossRef]
19. Irzyniec, Z.; Klimczak, J.; Michalowski, S. Freeze-drying of the black currant juice. *Dry Technol.* **1995**, *13*, 417–424. [CrossRef]
20. Krokida, M.K.; Karathanos, V.T.; Maroulis, Z.B. Effect of freeze-drying conditions on shrinkage and porosity of dehydrated agricultural products. *J. Food Eng.* **1998**, *35*, 369–380. [CrossRef]
21. Mazza, G.; Miniati, E. *Anthocyanins in Fruits, Vegetables and Grains*, 1st ed.; CRC Press & Taylor Francis Group: Boca Raton, UK, 1993; pp. 126–248.
22. Shishehgarha, F.; Makhlouf, J.; Ratti, C. Freeze-drying Characteristics of strawberries. *Dry. Technol.* **2002**, *20*, 131–145. [CrossRef]
23. Nattress, L.A.; Ziegler, G.R.; Hollender, R.; Peterson, D.G. Influence of hazelnut paste on the sensory properties and shelf life of dark chocolate. *J. Sens. Stud.* **2004**, *19*, 133–148. [CrossRef]
24. Steele, R. *Understanding and Measuring the Shelf-Life of Food*, 1st ed.; Woodhead Publishing: Cambridge, UK, 2004; Volume 2, pp. 340–356.
25. Kita, A.; Lachowicz, S.; Filutowska, P. Effects of package type on the quality of fruits and nuts panned in chocolate during long-time storage. *LWT-Food Sci. Technol.* **2020**, *125*, 109212. [CrossRef]
26. Jakubczyk, E.; Kamińska-Dwórznicka, A.; Ostrowska Ligęza, E.; Górska, A.; Wirkowska Wojdyła, M.; Mańko-Jurkowska, D.; Górska, A.; Bryś, J. Application of different compositions of apple puree gels and drying methods to fabricate snacks of modified structure, storage stability and hygroscopicity. *Appl. Sci.* **2021**, *11*, 10286. [CrossRef]

27. Peleg, M. Assessment of a semi-empirical four parameter general model for sigmoid moisture sorption isotherms. *J. Food Eng.* **1993**, *16*, 21–27. [CrossRef]
28. Ostrowska-Ligęza, E.; Jakubczyk, E.; Górska, A.; Wirkowska, M.; Bryś, J. The use of moisture sorption isotherms and glass transition temperature to assess the stability of powdered baby formulas. *Therm. Anal. Calorim.* **2014**, *118*, 911–918. [CrossRef]
29. Ostrowska-Ligęza, E.; Dolatowska-Żebrowska, K.; Wirkowska-Wojdyła, M.; Bryś, J.; Górska, A. Comparison of thermal characteristics and fatty acids composition in raw and roasted cocoa beans from Peru (Criollo) and Ecuador (Forastero). *Appl. Sci.* **2021**, *11*, 2698. [CrossRef]
30. Gordon, M.; Taylor, J.S. Ideal copolymers and the second-order transitions of synthetic rubbers. I. Non-crystalline copolymers. *J. Appl. Chem.* **1952**, *2*, 493–500. [CrossRef]
31. Ciurzyńska, A.; Lenart, A. The influence of osmotic dehydration on chemical composition of freeze-dried strawberries. *Postęy Tech. Przetwórstwa Spożywczego* **2009**, *1*, 9–13. (In Polish)
32. de Jesús Ornelas-Paz, J.J.; Yahia, E.M.; Ramirez-Bustamante, N.; Perez-Martinez, J.; Escalante-Minakata, M.P.; Ibarra-Junquera, V.; Acosta-Muniz, C.; Guerrero-Prieto, V.; Ochoa-Reyes, E. Physical attributes and chemical composition of organic strawberry fruit (*Fragaria x ananassa* Duch, *Cv.* Albion) at six stages of ripening. *Food Chem.* **2013**, *138*, 372–381. [CrossRef]
33. Raemy, A.; Nouzille, C.; Lambelet, P.; Marabi, A. Overview of calorimetry as a tool for efficient and safe food-processing design. In *Calorimetry in Food Processing*, 1st ed.; Kaletunc, G., Ed.; IFT Press: Oxford, UK; Wiley-Blackwell: Hoboken, NJ, USA, 2009; pp. 202–236.
34. Mishra, K.; Kohler, L.; Kummer, N.; Zimmermann, S.; Ehrengruber, S.; Kämpf, F.; Dufour, D.; Nyström, G.; Fischer, P.; Windhab, E.J. Rheology of cocoa butter. *J. Food Eng.* **2021**, *305*, 110598. [CrossRef]
35. Ostrowska-Ligęza, E.; Marzec, A.; Gorska, A.; Wirkowska-Wojdyła, M.; Bryś, J.; Rejch, A.; Czarkowska, K. A comparative study of thermal and textural properties of milk, white and dark chocolates. *Thermochim. Acta* **2019**, *671*, 60–69. [CrossRef]
36. Wang, Y.; Truong, T.; Li, H.; Bhandari, B. Co-melting behaviour of sucrose, glucose & fructose. *Food Chem.* **2019**, *275*, 292–298.
37. Lee, J.W.; Thomas, L.C.; Schmidt, S.J. Can the thermodynamic melting temperature of sucrose, glucose, and fructose be measured using rapid-scanning differential scanning calorimetry (DSC)? *J. Agric. Food Chem.* **2011**, *59*, 3306–3310. [CrossRef] [PubMed]
38. Chirife, J.; Buera, M.P. Water activity, glass transition and microbial stability in concentrated/semimoist food systems. *J. Food Sci.* **1994**, *59*, 921–927. [CrossRef]
39. Sà, M.M.; Sereno, A.M. Glass transitions and state diagrams for typical natural fruits and vegetables. *Thermochim. Acta* **1994**, *246*, 285–297. [CrossRef]
40. De Santana, R.F.; de Oliveira Neto, E.R.; Santos, A.V.; Soares, C.M.F.; Lima, Á.S.; Cardoso, J.C. Water sorption isotherm and glass transition temperature of freeze-dried *Syzygium cumini* fruit (jambolan). *J. Therm. Anal. Calorim.* **2015**, *120*, 519–524. [CrossRef]
41. Kim, S.S.; Kim, S.Y.; Kim, D.W.; Shin, S.G.; Chang, K.S. Moisture sorption characteristics of composite foods filled with chocolate. *J. Food Sci.* **1999**, *64*, 300–302. [CrossRef]
42. Ciurzyńska, A.; Lenart, A. The influence of temperature on rehydration and sorption properties of freeze-dried strawberries. *Croat. J. Food Sci. Technol.* **2009**, *1*, 15–23.
43. Tsami, E.; Krokida, M.; Drouzas, A. Effect of drying method on the sorption characteristics of model fruit powders. *J. Food Eng.* **1998**, *38*, 381–392. [CrossRef]
44. Mosquera, L.H.; Moraga, G.; Martínez-Navarrete, N. Critical water activity and critical water content of freeze-dried strawberry powder as affected by maltodextrin and arabic gum. *Food Res. Int.* **2012**, *47*, 201–206. [CrossRef]
45. Ghosh, V.; Duda, J.L.; Ziegler, G.R.; Anantheswaran, R.C. Diffusion of Moisture through Chocolate-flavoured Confectionery Coatings. *Food Bioprod. Process.* **2004**, *82*, 35–43. [CrossRef]
46. Augusto, P.; Vissotto, F.; Bolini, H. Sensory impact of three different conching times on white chocolates with spray-dried and freeze-dried açai (*Euterpe oleracea*). *Food Sci. Technol. Int.* **2019**, *25*, 480–490. [CrossRef]
47. Roos, Y.H.; Drusch, S. *Phase Transitions in Foods*, 2nd ed.; Academic Press: Oxford, UK, 2015; p. 380.
48. Nightingale, L.M.; Lee, S.-Y.; Engeseth, N.J. Impact of storage on dark chocolate: Texture and polymorphic changes. *J. Food Sci.* **2011**, *76*, 142–153. [CrossRef] [PubMed]

applied
sciences

Article

Sous Vide Cooking Effects on Physicochemical, Microbiological and Sensory Characteristics of Pork Loin

Lidia Kurp [1],*, Marzena Danowska-Oziewicz [1],* and Lucyna Kłębukowska [2]

[1] Department of Human Nutrition, Faculty of Food Science, University of Warmia and Mazury in Olsztyn, 10-718 Olsztyn, Poland

[2] Department of Industrial and Food Microbiology, Faculty of Food Science, University of Warmia and Mazury in Olsztyn, 10-718 Olsztyn, Poland; lucyna.klebukowska@uwm.edu.pl

* Correspondence: lidia.kurp@uwm.edu.pl (L.K.); marzena.danowska@uwm.edu.pl (M.D.-O.); Tel.: +48-89-524-55-16 (L.K.); +48-89-523-37-60 (M.D.-O.)

Abstract: Pork loin slices were sous vide cooked at 60 °C and 65 °C for 2 h, 3 h and 4 h, and at 70 °C and 75 °C for 1 h, 1.5 h and 2 h. The cooking loss of the meat samples significantly increased with the temperature and time of heat treatment, but no correlation between cooking loss and moisture content in the samples was noted. All samples showed similar pH and water activity values. Regarding colour parameters, only yellowness showed significant differences between the samples and was affected by the temperature and time of cooking. Texture profile analysis revealed the lowest hardness of the samples cooked at 60 °C. Sensory analysis showed that cooking at 60 or 65 °C for 4 h ensured the most acceptable sensory features of the investigated samples, and tenderness and juiciness influenced the overall acceptability in the highest degree. All samples were microbiologically safe for consumption.

Keywords: pork loin; sous vide; physicochemical properties; microbiological quality; sensory quality

Citation: Kurp, L.; Danowska-Oziewicz, M.; Kłębukowska, L. Sous Vide Cooking Effects on Physicochemical, Microbiological and Sensory Characteristics of Pork Loin. *Appl. Sci.* **2022**, *12*, 2365. https://doi.org/10.3390/app12052365

Academic Editors: Michael Murkovic and Massimo Lucarini

Received: 12 January 2022
Accepted: 22 February 2022
Published: 24 February 2022

Publisher's Note: MDPI stays neutral with regard to jurisdictional claims in published maps and institutional affiliations.

1. Introduction

Sous vide cooking was adopted as a heat treatment technique in catering industries in the 1970s, and since that time, with increasing access to less expensive cooking equipment, it is gaining popularity in industrial food processing, gastronomy and home cooking [1–3]. This technique involves vacuum packaging a food item, cooking it in a water bath or a steam chamber at a relatively low temperature for a relatively long time, and finally rapid cooling, for example in an ice-water bath [4] or a blast chiller. The main feature of this technique is precise cooking temperature control which leads to reproducible culinary procedures and together with vacuum packaging facilitates the handling of ready-to-eat foods [3,5,6]. Sous vide cooking preserves nutritive compounds in food products and ensures their adequate texture, juiciness and flavour, the former due to the mild conditions of cooking and therefore limited changes in food components and the latter due to mild changes in protein structures and keeping moisture and volatile compounds within bags [7]. In addition, sous vide processing inhibits oxidative changes in products as a result of the lack of oxygen in bags [1,3,8] and allows food preparation that is microbiologically safe and prevented from cross contamination after cooking [8,9].

The advantage of sous vide cooked meat is its appealing texture and colour [10], while its flavour intensity is described as low [11]. The relatively low temperatures used in sous vide cooking were reported to positively affect the tenderness and juiciness of meat. Nevertheless, it was observed that particularly long cooking times resulted in more tender but less juicy meat [11]. The myofibrillar proteins begin to denature at 35–40 °C, causing shrinkage of muscle fibres, and this process continues to 80 °C [12]. The extent of myofibrillar protein coagulation influences the final texture of cooked meat as the shrinking of these proteins leads to an increase in meat toughness [3]. Collagen, a main protein of the connective tissue, requires long cooking and a temperature of at least 55 °C to hydrolyze, which leads to reduction in interfibre adhesion and to meat tenderness [3,12]. Christensen et al. [13] observed that meat toughness increased between 40 and 50 °C and then between 60 and 80 °C, while toughness decline occurred between 50 and 60 °C. The authors explained the reduction in toughness as a result of decreased breaking strength of the perimysial connective tissue caused by partial denaturation and shrinkage of the collagen fibres.

Colour determines the doneness of meat and therefore can affect the consumer's acceptance of a meat product [14,15]. The well-done state of meat requires reaching internal meat temperature of around 70 °C [12]. In raw meat, meat pigment myoglobin exists in three forms, i.e., oxymyoglobin, deoxymyoglobin and metmyoglobin, which show bright red, purple-red and brown colour, respectively. As a result of heat treatment, globin denaturates and precipitates with other meat proteins, and red ferrohemochrome and brown ferrihemochrome are formed. Denaturation of myoglobin begins between 55 °C and 65 °C and is almost completed by 80 °C [16,17]. Lien et al. [15] observed a 51.9% myoglobin denaturation in pork loin chops cooked to 62.8 °C end point temperature and its gradual increase to 85.3% when end point temperature reached 82.2 °C. Christensen et al. [11] suggested that when prolonged cooking time is applied, denaturation of myoglobin occurs below 60 °C.

The crucial factor in the sous vide technique is a balanced combination of temperature and cooking time [18,19]. Sánchez del Pulgar et al. [3] reported that chefs cook pork primals at 60–63 °C, while catering temperatures used for pork are around 75–80 °C. In the literature, sous vide cooking temperatures of pork for eating quality studies start at 48 °C [20] and reach 71 °C [21], while cooking times are between 45 min [21] and 32 h [11]. For a study of physical aspects, Zielbauer et al. [22] cooked pork at 45–74 °C for 10–2880 min. Most publications on sous vide cooking of pork investigate the quality of meat when temperatures between 50 and 60 °C are applied. Vaudagna et al. [23] suggested that beef sous vide cooking at 60–65 °C assures its safety and desirable yield and tenderness. In relation to pork safety, recommended end point temperatures range between 65 and 75 °C [24].

As the literature shows, the ranges of temperature and time regimes applied even to the same meat cuts are very wide, which makes the practical use of this technique quite difficult. Moreover, the emphasis on long cooking times to ensure microbiological safety of meat makes this technique time and energy consuming. In addition, combined scientific data on physicochemical and microbiological characteristics as well as sensory properties of sous vide cooked pork are scarce. Having in mind the practical application of sous vide cooking of meat in gastronomy and other catering services, the objective of the present study was to investigate the effect of different temperature-time combinations, featuring relatively shorter cooking times than usually applied in sous vide pork cooking, on cooking loss, instrumentally measured colour and texture, microbiological quality as well as sensory properties of pork loin. Correlation coefficients between selected physiochemically, instrumentally and sensorially assessed features of meat samples were also investigated.

2. Materials and Methods

2.1. Preparation of Samples

Pork loins (*M. longissimus thoracis et lumborum*) of commercial crossbred pigs PIC (5–6 month-old female pigs of around 110 kg weight) were purchased 24 h after slaughter from a local meat supplier and transported to the laboratory in chilled conditions, vacuum packed and stored at 4 °C for 4 days. After storage each muscle was trimmed and cut into 2.5 cm thick slices. The slices were individually weighed and vacuum-packed in PA/PE pouches (15 μm polyamide/60 μm polyethylene; heat resistance of −20 °C/+110 °C; Hendi, Austria) using chamber vacuum sealer (Edesa VAC-20 DT, Barcelona, Spain). Seven slices were randomly assigned to each treatment.

Sous vide cooking was performed using a water bath with an immersion circulator equipped with a temperature sensor (Diamond Z, Julabo GmbH, Seelbach, Germany). The samples were heat-treated at 60 °C and 65 °C for 2 h, 3 h and 4 h, and at 70 °C and 75 °C for 1 h, 1.5 h and 2 h, after the sample core reached the temperature set for the water bath. Cooking temperatures and times were selected on the basis of preliminary study and the literature data [5,11,12,21]. After cooking, the samples destined for physicochemical, instrumental and microbiological analyses were cooled in an ice-water bath and stored overnight at 4 °C before analysis. The samples destined for sensory analysis were served after cooking. Three independent replicate trials of the whole experiment, with the use of meat purchased on three separate occasions, were conducted.

2.2. Cooking Loss

Cooking loss was calculated on the basis of the difference in meat weight before and after heat treatment.

2.3. Moisture Content

Moisture content in raw and cooked comminuted meat samples was determined by drying to constant weight at 105 °C according to AOAC procedure 950.46 [25] using a forced draught laboratory oven (UF55; Memmert, Schwabach, Germany).

2.4. pH Measurement

Five grams of comminuted raw and cooked meat samples was homogenized with 45 mL of distilled water using the HO 4 A homogenizer (Edmund Bühler GmbH, Hechingen, Germany) for 2 min at 5000 rpm, and the measurements were registered after the equilibrium was reached using a pH meter (pH 210, Hanna Instruments, Woonsocket, RI, USA) calibrated with pH 4 and pH 7 buffers.

2.5. Water Activity

Water activity was determined at 20 °C on comminuted raw and cooked meat samples placed in measuring containers in the analyzer chamber (AWC-200, Novasina, Pfäffikon, Switzerland) calibrated with a set of Novasina humidity sources.

2.6. Colour Determination

Coordinates L^* (lightness/darkness), a^* (redness/greenness) and b^* (yellowness/blueness) of the CIE $L^*a^*b^*$ colour space were measured on the surface of raw and heat-treated samples using a CR-400 Chroma meter (Konica Minolta Sensing Inc., Osaka, Japan) equipped with standard observer 2° and illuminant D65 and calibrated with a white ceramic tile supplied by the manufacturer. Six measurements were taken for each treatment. Chroma (C^*), hue angle ($h°$) and total colour difference (ΔE^*) were calculated according to the following equations:

$$C* = \sqrt{a*^2 + b*^2} \tag{1}$$

$$h° = \mathrm{arctg}(b*/a*) \times (360°/2 \times 3.14) \tag{2}$$

$$\Delta E* = \sqrt{\Delta L*^2 + \Delta a*^2 + \Delta b*^2} \tag{3}$$

2.7. Instrumental Texture Analysis

Texture analysis of cooked meat samples was based on shear test and texture profile analysis (TPA) using TA.TXplus Texture Analyzer (Stable Micro Systems Ltd., Godalming, UK) equipped with a 50 kg load cell. Shear test was performed with a Warner-Bratzler shear blade with a v-shaped notch. The crosshead speed during the test was 250 mm/min. The TPA test was a two-cycle compression using the P/100 compression platen of 50 mm diameter, with sample deformation to 50% of its original height. The crosshead speed was 50 mm/min. For each treatment, 20 specimens cut parallel to the muscle fibres ($10 \times 10 \times 25$ mm for the shear test and 16 mm diameter, 20 mm height cores for the TPA) were analyzed.

2.8. Microbiological Analysis

The total number of mesophilic microorganisms [26], the number of coagulase-positive staphylococci [27] and the number of *Enterobacteriaceae* [28] were determined in the raw and cooked meat samples. For this purpose, 10 g of meat was homogenized with 90 mL sterile 0.1% peptone in a Stomacher (Lab Blender, Model 400, Seward Medical, London, UK) for 120 s. Appropriate dilutions were made with 0.1% peptone broth, and 1 mL was plated onto the culture media and incubated under optimal conditions: total mesophilic counts on a Plate Count Agar (PCA, Oxoid) for 72 h at 30 °C; coagulase-positive staphylococci on a Baird Parker Agar RPF (Baird Parker Rabbit Plasma Fibrinogen Agar; BPA, RPF RPF Agar; Merck) for 72 h at 37 °C and *Enterobacteriaceae* on a VRBD Agar (Violet Red Bile Dextrose; Merck) for 24 h at 37 °C. Typical colonies for each media were counted in plates from the dilution with 10–150 colonies.

The presence of pathogens *Salmonella* spp. [29] and *Listeria monocytogenes* [30] in the meat was also determined. The presence of *Salmonella* spp. was determined in 25 g of meat. The sample was homogenized in 225 mL of buffered peptone water and incubated at 37 °C for 24 h. Then, 1 mL of the culture was transferred to a GranuCult™ RVS Broth (RAPPAPORT-VASSILIADIS-Soya; Merck) and incubated at 37 °C. After 24 h of incubation, the XLD Agar (Xylose Lysine Deoxycholate Agar; Merck) and BPLS Agar (Brilliant-green Phenol-red Lactose Sucrose; Merck) media were inoculated. The presence of *Listeria monocytogenes* in 25 g of meat was determined after precultivation in Half-Fraser broth (Merck) and cultivation in Full Fraser broth (Merck) on Chromocult® Listeria Agar; ALOA (Merck) and Palcam Listeria Selective Agar (Merck).

2.9. Sensory Analysis

The evaluation was conducted in the sensory analysis laboratory of the Department of Human Nutrition. The sensory panel consisted of 15 employees of the Faculty of Food Science, trained according to [31] and experienced in sensory evaluation of food. Before evaluation two training sessions familiarized the panelists with the samples represented in the experiment. The assessment of experimental samples was repeated twice during each experiment replication. The panelists were provided water and bread to clean the palate between samples. Cooked meat samples were diagonally cut into 1 cm thick slices and served one slice per assessor. Each panelist received all treatments in random order. Samples were evaluated using a 10 cm structured graphic scale, according to [32], for overall appearance, flavour acceptability and overall acceptability (0–not acceptable, 10–very acceptable), colour uniformity (0–not uniform, 10–highly uniform), aroma intensity and meat flavour intensity (0–low, 10–very intense), tenderness (0–tough, 10–tender) and juiciness (0–low and 10–very high).

2.10. Statistical Analysis

The results of the study were presented as mean values and standard error of the mean (SEM). The experiment was conducted using randomized factorial design with temperature and treatment time as fixed effects and experiment replicates as random effects. In each experiment replicate, when not otherwise stated, the measurements were conducted in three replications. Data were analyzed using the General Linear Model procedure of Statistica 13 (TIBCO Software Inc., Tulsa, OK, USA). The results of the measurements were subjected to two-way ANOVA to identify the effects of cooking temperature, cooking time and their interaction on the investigated features of meat samples. The means were separated with the Tukey's test, and differences were considered significant if $p < 0.05$. Additionally, to examine potential relationships between selected attributes of the samples, Pearson's correlation coefficients were calculated.

3. Results and Discussion

3.1. Cooking Loss

The cooking losses of the meat samples subjected to sous vide cooking were in the range 18.16–36.66%, and they increased significantly with temperature and cooking time ($p < 0.001$; Table 1).

Table 1. Cooking loss, moisture content, water activity and pH value of sous vide cooked pork loin (mean $\pm$ SEM).

Sample		Cooking Loss (%)	Moisture Content (%)	Water Activity (-)	pH (-)
Raw pork loin		nd	72.45 $\pm$ 0.34 [i]	0.993 $\pm$ 0.001 [a]	5.78 $\pm$ 0.01 [a]
60 °C					
	2 h	18.16 $\pm$ 1.07 [a]	68.91 $\pm$ 0.28 [h]	0.995 $\pm$ 0.003 [a]	5.80 $\pm$ 0.37 [a]
	3 h	21.13 $\pm$ 1.12 [abc]	67.54 $\pm$ 0.58 [fgh]	0.993 $\pm$ 0.002 [a]	5.81 $\pm$ 0.10 [a]
	4 h	22.46 $\pm$ 1.05 [bc]	67.50 $\pm$ 0.41 [fgh]	0.991 $\pm$ 0.004 [a]	5.80 $\pm$ 0.10 [a]
65 °C					
	2 h	19.55 $\pm$ 0.46 [ab]	68.16 $\pm$ 0.20 [gh]	0.993 $\pm$ 0.002 [a]	5.81 $\pm$ 0.22 [a]
	3 h	24.41 $\pm$ 0.54 [cd]	66.90 $\pm$ 0.11 [efg]	0.993 $\pm$ 0.002 [a]	5.82 $\pm$ 0.11 [a]
	4 h	29.64 $\pm$ 0.74 [def]	66.31 $\pm$ 0.14 [defg]	0.994 $\pm$ 0.001 [a]	5.80 $\pm$ 0.07 [a]
70 °C					
	1 h	25.21 $\pm$ 1.52 [cde]	65.91 $\pm$ 0.37 [cdef]	0.993 $\pm$ 0.003 [a]	5.84 $\pm$ 0.17 [a]
	1.5 h	28.79 $\pm$ 0.46 [def]	64.00 $\pm$ 0.28 [abc]	0.994 $\pm$ 0.001 [a]	5.86 $\pm$ 0.02 [a]
	2 h	30.58 $\pm$ 0.43 [f]	65.27 $\pm$ 0.34 [cde]	0.995 $\pm$ 0.002 [a]	5.85 $\pm$ 0.14 [a]
75°C					
	1 h	29.05 $\pm$ 1.06 [ef]	64.90 $\pm$ 0.29 [bcd]	0.994 $\pm$ 0.002 [a]	5.88 $\pm$ 0.09 [a]
	1.5 h	31.61 $\pm$ 0.54 [fg]	63.24 $\pm$ 0.53 [ab]	0.994 $\pm$ 0.003 [a]	5.90 $\pm$ 0.03 [a]
	2 h	36.66 $\pm$ 0.63 [g]	62.32 $\pm$ 0.70 [a]	0.994 $\pm$ 0.001 [a]	5.91 $\pm$ 0.01 [a]
Level of significance					
Temperature		***	***	NS	NS
Time		***	***	NS	NS
Temperature × time interaction		*	NS	NS	NS

a, b, c, d, e, f, g, h, i—mean values in columns with different superscripts differ significantly at $p < 0.05$ according to the Turkey's test; * $p < 0.05$; *** $p < 0.001$; NS-not significant; nd–not determined.

The interaction of temperature and cooking time also showed a significant effect ($p < 0.05$) on cooking losses. Our results confirmed the observations of other authors [3,9,18,33]. Christensen et al. [11] noted a significant effect of temperature on cooking loss of pork and beef and a significant effect of temperature and time of processing on cooking loss of chicken. Different effects of temperature and time on the cooking loss of pork loin were observed by Hwang et al. [34]. Cooking losses tended to be higher when higher temperatures were applied, However, longer cooking time at 50 °C resulted in lower cooking loss, while at higher temperatures the cooking time effect was not significant.

Cooking loss is caused mainly by water loss during cooking, together with other meat components such as myofibrillar and sarcoplasmic proteins, collagen, lipids, vitamins, minerals and flavour compounds [35,36]. Most of the water is held in muscles within structures of myofibrillar proteins which undergo denaturation and shrinkage during heating, followed by water liberation and loss from meat [18,33,37,38]. The intensity of these processes increases with increasing temperature up to 90 °C. Collagen denaturation and subsequent shrinkage occur between 53 and 63 °C and between 60 and 70 °C, respectively. Further heating causes collagen solubilization and gelatine formation which also influence water loss from meat [7,9,37,39]. As the temperatures applied in our study were in the range 60–75 °C, the above changes can explain our observations.

3.2. Moisture Content

The moisture content in the raw meat was 72.45%, and as a result of cooking it declined significantly ($p < 0.05$) to 62.32–68.91% (Table 1). Meat cooked at 60 °C/2 h showed the highest moisture retention, and the lowest moisture retention was observed in samples heated at 75 °C/2 h. The significant effect ($p < 0.001$) of both temperature and time of cooking on the moisture content was noted, while the interaction of these parameters was not significant. Similar moisture contents in pork hams sous vide cooked at 61 and 71 °C and significantly higher moisture contents in meat cooked at lower temperatures were reported by Jeong et al. [21]. Limited differences in moisture content with increasing temperature of sous vide cooking and no significant effect of cooking time of lamb loins were noted by Roldan et al. [9]. The present results are in accordance with observations of Zielbauer et al. [22], who reported that higher temperatures and longer cooking times lead to increased water losses. The authors suggested that temperatures above 74 °C and cooking times longer than 240 min do not increase further moisture liberation from meat.

Loss of water in cooked meat is caused by a leakage of cellular juice under the influence of elevated temperature. Additionally, myofibrillar protein shrinkage, which starts at 40 °C, leads to a subsequent decline in the interfibrillar volume which in turn reduces myofibril's ability to hold water. Finally, compression of the muscle fiber bundles due to the contraction of the perimysial connective tissue at temperatures 56–62 °C contributes to water release from meat [3,36]. Low temperatures applied during sous vide cooking, and consequently low end point internal meat temperatures, apparently favor the meat's ability to retain water in its structures due to lower meat fiber shrinkage.

3.3. Water Activity

Water activity of food product is a useful indicator of its susceptibility to degradation processes caused by the growth of microorganisms and the intensity of biochemical and chemical reactions [1,40,41]. It is particularly important when the food is intended for storage. Water activity of raw meat was 0.996. No significant differences were noted between raw and cooked samples in the present study (Table 1). The values observed for heat-treated samples were between 0.991 and 0.995 and corresponded to the values obtained for a sous vide cooked turkey cutlet by Akoğlu et al. [1] and Bıyklı et al. [42]. Slightly lower value of 0.92 were reported for sous vide cooked pork loin by Díaz et al. [43]. Nevertheless, all values observed in our study were high and could not affect diversity in storage stability of investigated samples, particularly in terms of microbiological spoilage [40].

3.4. pH Value

The pH value of meat influences its water-holding capacity [14] and the reaction pathways of the Maillard reaction, and therefore affects flavour and consequently storage stability of cooked meat [41]. Analysis of pH values did not reveal significant differences between raw (5.78) and cooked meat samples (5.80–5.91; Table 1). Díaz et al. [43] reported similar pH values for pork loin sous vide cooked at 70 °C/12 h after initial roasting. Our observation does not support that of Hwang et al. [34] who noted a significant increase in pH values of sous vide cooked pork loin (5.79–6.04) compared to raw meat (5.68). The probable reason for different pattern of pH changes in both experiments could be different cooking parameters applied by Hwang et al. [34], particularly the longer cooking times, namely 12 and 24 h vs. 1–4 h investigated in our study. It suggests that at shorter cooking time, processes in protein fraction that are related to distinct pH changes are limited. A slight increase in the pH of chicken breast fillets affected by the temperature and the interaction between temperature and cooking time was noted by Haghighi et al. [44].

According to other authors, the pH increase of heated meat may be ascribed to the loss of the free acidic group associated with protein denaturation [14,38,45]; cleavage of bonds involving imidazole, sulphydryl and hydroxyl groups [36]; an exposure of basic amino residues [34] and formation of free hydrogen sulfide when cooking takes place at temperatures above 80 °C [45]. Becker et al. [14] ascribed lower pH values of pork cooked at 60 °C/2 h than samples cooked at 53 °C/20 h, 58 °C/20 h or 180 °C/50 min, to a lower protein denaturation.

3.5. Colour

Colour parameters of the raw and cooked samples are presented in Table 2. Lightness values L^* of heat-treated meats (70.05–71.63) were significantly higher ($p < 0.05$) than L^* value of the raw meat (54.72). No significant differences were noted between heated meats. However, slightly higher values were observed with increased heating temperatures. Similar observations were made by Haghighi et al. [44] regarding poultry meat and in relation to pork by Becker et al. [18], Hwang et al. [34] and Jeong et al. [21].

Table 2. CIE $L^*a^*b^*$ colour parameters of sous vide cooked pork loin (mean ± SEM).

Sample	Colour L^*	Colour a^*	Colour b^*	Chroma C^*	Hue Angle $h°$	ΔE^*
Raw pork loin	54.72 ± 0.8 [b]	8.35 ± 0.33 [b]	5.30 ± 0.22 [d]	9.89 ± 0.44 [c]	32.40 ± 0.79 [c]	nd
60 °C						
2 h	70.16 ± 1.05 [a]	7.45 ± 0.20 [a]	14.60 ± 0.12 [c]	15.59 ± 0.15 [ab]	59.71 ± 0.61 [ab]	21.20
3 h	70.05 ± 1.01 [a]	7.53 ± 0.18 [a]	14.72 ± 0.10 [c]	15.93 ± 0.11 [ab]	60.88 ± 0.58 [ab]	21.28
4 h	70.61 ± 0.71 [a]	7.44 ± 0.06 [a]	14.65 ± 0.21 [c]	15.44 ± 0.19 [a]	60.41 ± 0.38 [a]	19.44
65 °C						
2 h	70.68 ± 0.56 [a]	7.45 ± 0.06 [a]	14.59 ± 0.08 [c]	15.58 ± 0.08 [ab]	60.72 ± 0.18 [ab]	25.61
3 h	71.02 ± 1.05 [a]	7.49 ± 0.43 [a]	14.43 ± 0.36 [bc]	15.99 ± 0.47 [ab]	61.71 ± 1.07 [ab]	21.92
4 h	70.61 ± 0.33 [a]	7.49 ± 0.10 [a]	14.34 ± 0.20 [abc]	16.04 ± 0.21 [ab]	61.82 ± 0.34 [ab]	18.96
70 °C						
1 h	70.89 ± 0.82 [a]	7.57 ± 0.21 [a]	14.13 ± 0.31 [abc]	16.18 ± 0.35 [ab]	62.41 ± 0.51 [ab]	15.56
1.5 h	70.57 ± 0.08 [a]	7.58 ± 0.08 [a]	14.08 ± 0.07 [abc]	16.33 ± 0.08 [ab]	62.93 ± 0.26 [ab]	15.82
2 h	70.81 ± 0.74 [a]	7.62 ± 0.15 [a]	13.58 ± 0.15 [ab]	16.38 ± 0.17 [ab]	62.95 ± 0.48 [ab]	16.07
75 °C						
1 h	71.21 ± 0.58 [a]	7.63 ± 0.19 [a]	13.41 ± 0.20 [a]	16.44 ± 0.25 [ab]	63.00 ± 0.45 [b]	19.68
1.5 h	71.63 ± 0.94 [a]	7.76 ± 0.20 [a]	13.90 ± 0.11 [abc]	16.55 ± 0.15 [b]	62.94 ± 0.62 [ab]	19.15
2 h	71.45 ± 0.10 [a]	7.86 ± 0.04 [a]	13.46 ± 0.09 [abc]	16.41 ± 0.09 [ab]	63.03 ± 0.19 [b]	17.01
Level of significance						
Temperature	NS	NS	***	***	***	nd
Time	NS	NS	***	*	**	nd
Temperature × time interaction	NS	NS	NS	NS	NS	nd

a, b, c, d—mean values in columns with different superscripts differ significantly at $p < 0.05$ according to the Turkey's test; * $p < 0.05$; ** $p < 0.01$; *** $p < 0.001$; NS—not significant; nd—not determined.

Roldan et al. [9] pointed out an opposite pattern of L^* values for lamb, namely slightly higher values for meat cooked at the lowest temperature. Lightness of cooked meats is affected by myofibrillar protein denaturation and aggregation [20,21] as well as moisture presence on the meat surface [18]. According to Bojarska et al. [46] a deeper penetration of light into muscle tissue with greater hydration of muscular proteins results in darker colour, which might explain our observations related to L^* and moisture content.

Changes in meat redness (a^*) as a result of cooking are related to the myoglobin content and the degree of its denaturation. In the present study, no significant differences were observed between a^* values of all investigated samples, including the raw meat ($a^* = 8.35$ for the raw meat and 7.44–7.86 for the cooked meats). It was reported that the higher the cooking temperature, the lower the a^* values [18,20,23,44,47] due to more intensive pigment denaturation [44]. In the study on myoglobin denaturation in aqueous muscle extracts, Geileskey et al. [16] observed that over 50% of the myoglobin was still present in the extract at 60 °C, while at 70 and 80 °C it was almost completely precipitated. Our observations regarding redness of cooked samples probably could be explained by the longer cooking times applied at lower cooking temperatures, namely 2, 3 and 4 h at 60 and 65 °C, while

1, 1.5 and 2 h were cooking times at 70 and 75 °C, and as result, we observed a similar extent of myoglobin changes in those two groups of samples. Such an explanation seems to support the findings of Hwang et al. [34], who did not observe the differences in L^*, a^* and b^* colour values of pork loin sous vide cooked at 50, 55 and 60 °C for 12 and 24 h compared to meat heated at 75 °C for 30 min.

The yellowness (b^*) of samples increased significantly ($p < 0.005$) due to the heat treatment from 5.30 noted in Table 2 for raw meat with values from 13.41–14.72 as determined for cooked meats. The significant effect of both temperature and time ($p < 0.001$) of cooking on the b^* values was observed, but the temperature/time interaction effect was not significant. In the study of Becker et al. [18] the b^* values of pork loins increased with temperature and prolonged time of cooking. Higher b^* values of cooked meat can be ascribed to an increase in metmyoglobin level, which results in a more brownish colour [48].

Chroma (C^*, saturation) and hue angle ($h°$) are colour indicators that reflect colour intensity and tone, respectively, considering both redness and yellowness coordinates. In our study the C^* values of cooked samples were relatively low (15.44–16.55) and significantly affected by the temperature ($p < 0.001$) and time ($p < 0.05$) of cooking but not their interaction. Similar effects of cooking parameters were noted on the $h°$ values of samples ($p < 0.001$ and $p < 0.01$, resp.). Actual $h°$ values of cooked meats (59.71–63.03) indicated a lower share of redness and a higher share of yellowness in sample colour as hue angle values close to 0° indicate redder colour, while values close to 90° indicate more yellow appearance [49,50]. Total colour differences ΔE^*, calculated for the cooked samples in relation to the raw meat, showed high values (15.56–25.61), which means that regardless of cooking parameters applied, colour changes were recognizable for an unexperienced observer [51].

3.6. Instrumental Texture Analysis

3.6.1. Shear Force

Low shear force values of cooked meat are desirable as they reflect greater tenderness of ready to eat product. Tenderness is related to myofibrils and connective tissue proteins and their changes during heat treatment, mainly denaturation of myofibrillar proteins and solubilization of connective tissue [33].

The maximum force noted during a shear test was selected to characterize the texture of the samples (Table 3). The highest shear force was noted for meat cooked at 60 °C/4 h (25.61 N), and the lowest for the sample cooked at 65 °C/3 h (13.29 N). At two lower temperatures, i.e., 60 and 65 °C, the most tender meats were obtained when heating was conducted for 3 h (17.63 and 13.29 N, respectively), while at two higher temperatures, i.e., 70 and 75 °C, the lowest shear force values were observed for meats cooked for 2 h (14.66 and 14.61 N, resp.). It can be pointed out that when the same cooking time at all temperatures was taken into consideration (2 h), the shear force values decreased with increased temperature of the heat treatment. Both temperature and time of cooking had a significant effect ($p < 0.001$) on meat shear force, but their interaction effect was not significant.

Our results are partially in agreement with the experiment of Christensen et al. [20], where the shear force values of pork *Longissimus dorsi* were in the range 12.6–41.1 N and decreased with cooking temperature increases, while the effect of cooking time was not uniform and depended on the cooking temperature. In the study by Hwang et al. [34], the shear force of pork loin was affected by process temperature, but process time had significant effect on meat tenderness only at 50 °C. It was suggested that the tenderization of longer cooked (24 vs. 12 h) meat was supported by activity of intrinsic proteases which remained active at this low temperature. Thermal inactivation of proteases at temperatures higher than 55 °C was the probable reason that heat induced structural changes of meat proteins had a decisive role in meat tenderness formation. Correlation between cathepsins activity and shear force of pork has been recently reported by Dominguez-Hernandez et al. [19]. Likewise, Jeong et al. [21] observed a significant effect of cooking temperature on the shear force values of pork ham sous vide cooked at 61 °C and 71 °C, while cooking time effect was noted for meat cooked at 71 °C. Increased meat tenderness upon cooking is associated with the collagen conversion to gelatin [38,45]. Ismail et al. [52] reported increasing collagen solubility in beef cooked at 45, 65 and 75 °C, while Vasanthi et al. [45] observed an increase in collagen solubility with temperature (80–100 °C) and time of cooking (30–60 min). In our study temperatures higher than 55 °C were applied. Therefore, changes in meat texture apparently have been a result of heating and not of proteases activity.

Table 3. Texture characteristics of sous vide cooked pork loin on the basis of shear force and TPA test (mean ± SEM).

Sample	Shear Force [N]	Hardness [N]	Springiness [cm]	Cohesiveness [-]	Chewiness [N × cm]
60 °C					
2 h	23.80 ± 0.95 ef	60.26 ± 3.33 a	0.50 ± 0.01 abc	0.56 ± 0.01 de	17.51 ± 0.01 ab
3 h	17.63 ± 1.70 abcd	58.84 ± 3.67 a	0.46 ± 0.01 ab	0.51 ± 0.01 ab	16.00 ± 0.01 ab
4 h	25.61 ± 0.68 f	54.77 ± 4.19 a	0.46 ± 0.02 ab	0.51 ± 0.02 a	13.65 ± 0.01 a
65 °C					
2 h	18.24 ± 1.39 bcd	79.29 ± 4.06 bcde	0.53 ± 0.01 cd	0.57 ± 0.01 cd	24.01 ± 0.01 cde
3 h	13.29 ± 1.01 a	71.12 ± 4.23 abc	0.51 ± 0.01 abcd	0.59 ± 0.01 def	21.19 ± 0.01 bcd
4 h	19.95 ± 0.52 cde	72.91 ± 2.85 abcd	0.45 ± 0.01 a	0.52 ± 0.01 abc	17.34 ± 0.01 abc
70 °C					
1 h	22.47 ± 1.30 def	69.20 ± 2.22 ab	0.52 ± 0.01 bcd	0.60 ± 0.00 def	21.32 ± 0.01 bcd
1.5 h	18.40 ± 1.19 abcd	90.11 ± 3.46 cdef	0.49 ± 0.01 abc	0.55 ± 0.01 abcd	24.14 ± 0.01 bcde
2 h	14.66 ± 0.53 ab	90.04 ± 2.92 def	0.53 ± 0.01 cd	0.60 ± 0.01 def	28.13 ± 0.01 def
75 °C					
1 h	21.97 ± 1.20 def	95.29 ± 5.68 ef	0.57 ± 0.01 d	0.62 ± 0.01 ef	34.34 ± 0.01 f
1.5 h	17.98 ± 0.36 abcd	121.84 ± 5.42 g	0.57 ± 0.01 d	0.64 ± 0.01 f	45.22 ± 0.00 g
2 h	14.61 ± 0.44 abc	107.46 ± 3.28 fg	0.51 ± 0.01 abcd	0.56 ± 0.01 bcd	31.85 ± 0.01 ef
Level of significance					
Temperature	***	***	***	***	***
Time	***	***	***	***	***
Temperature × time interaction	NS	NS	***	***	***

a, b, c, d, e, f, g—mean values in columns with different superscripts differ significantly at $p < 0.05$ according to the Turkey's test; *** $p < 0.001$; NS—not significant.

3.6.2. Texture Profile Analysis (TPA)

TPA delivers more information on the texture of meat products than shear force which in turn is a useful indicator of initial meat tenderness [48].

Texture profile attributes are presented in Table 3. The treatment temperature that resulted in the lowest hardness was 60 °C, and the values declined with prolonged cooking time from 60.26 N to 54.77 N. Comparable hardness was observed for the samples cooked at 65 °C for 3 or 4 h and at 70 °C for 1 h. In general, the higher the cooking temperature, the higher the hardness of samples noted. Similar increases with increased temperature of process were observed in other texture attributes of samples. It is worth noting that at 60 and 65 °C, hardness and chewiness values tended to decline with prolonged cooking time, while the opposite tendency occurred at 70 and 75 °C. This observation presumably could be explained by the relatively mild myofibrillar protein denaturation at lower temperatures and increasing collagen hydrolyzation with prolonged cooking time while at higher temperatures the changes in myofibrillar proteins predominated, leading to increased meat toughness. Likewise in the case of shear force, temperature and time of treatment influenced meat hardness significantly ($p < 0.001$), but not their interaction. Other texture attributes were affected by temperature and cooking time as well as their interaction.

Jeong et al. [21] noted lower hardness and chewiness of pork ham sous vide cooked at 61 °C and 71 °C for 45 and 90 min, compared to the control samples boiled in a pot for 45 min. Sánchez del Pulgar et al. [3] did not observe any significant effect of cooking at 60 °C for 5 or 12 h on texture attributes of pork cheeks, while during treatment at 80 °C, increasing cooking time from 5 to 12 h caused a decline in these parameters. These finding were supported by the histological analyses which showed a bigger formation of completely denatured collagen fibres of the samples cooked at 80 °C than at 60 °C. Brüggemann et al. [53] observed collagen shrinkage at 57 °C and collagen denaturation between 59 °C and 61 °C. Palka et al. [54] ascribed a lower hardness, cohesiveness and chewiness of beef cooked at 60 °C than at 70 or 80 °C to the myosin and actin denaturation (at 40–60 °C and 66–73 °C, respectively) and shrinking of collagen (56–62 °C), which might as well be an explanation of observations made in the present study.

3.7. Microbiological Analyses

The microbial counts in raw and cooked meat are shown in Table 4. The samples of raw loins showed counts for mesophilic bacteria at about 3.59 log CFU/g. The number of *Enterobacterioceae* was determined in similar values (3.07 log CFU/g), whereas for coagulase-positive staphylococci it was 1.95 log CFU/g. This means that the main contamination of the raw pork loin was fecal microflora. The presence of pathogens belonging to *Salmonella* spp. and *Listeria monocytogenes* has not been detected. These results are within the standards specified in [55]. The obtained results are consistent with the values determined by Roldan et al. [9] who tested lamb loins.

After heat treatment, the total number of aerobic microorganisms was reduced by about 2 logs and ranged from 1.10 to 2.01 log CFU/g. Fecal microflora, *Enterobacteriaceae* and coagulase-positive staphylococci were determined in the population at the level <1 log CFU/g. The results obtained in the research indicate that the applied parameters of the sous vide processing were sufficient to reduce the microflora in the pork loin. The inactivation effects of Gram-negative *Enterobacteriaceae* are particularly satisfactory. Other authors also obtained a reduction of the *Enterobacteriaceae* population in pork loin to <1 log CFU/g [43]. The study of Jeong et al. [21] indicated the effectiveness of sous vide treatment in relation to fecal microflora, while only a part of the *Enterobacteriaceae* family coliforms was determined.

3.8. Sensory Analysis

The investigation of organoleptic properties of cooked meats showed that the highest scores for meat flavour intensity, flavour acceptability and overall acceptability, were given to the sample heated at 60 °C/4 h (Table 5). In respect to meat flavour intensity, this sample was significantly different from the sample heated at 60 °C/2 h. In terms of flavour acceptability, this sample was very different from the sample cooked at 75 °C/2 h. In terms of overall acceptability, this sample was superior to the samples cooked at 75 °C/1.5 h and at 75 °C/2 h. The highest perception of juiciness was noted in meat cooked at 60 °C/2 h, but a significant difference was shown, mainly compared to the samples cooked at 75 °C. The sample cooked at 65 °C/4 h received the highest scores for other sensory traits considered in the study, namely overall appearance, colour uniformity, aroma intensity and tenderness. No significant differences were observed between samples in respect to aroma intensity. The scores for juiciness, tenderness, flavour acceptability and overall acceptability slightly declined with increased treatment temperature. According to the analysis of variance, the temperature of cooking significantly affected tenderness, juiciness, flavour acceptability and overall acceptability. Time of treatment showed a significant effect on overall appearance, tenderness, juiciness, meat flavour intensity and overall acceptability. Interaction of cooking parameters influenced meat traits related to visual quality, namely overall appearance and colour uniformity.

Table 4. The microflora of raw and sous vide cooked pork loin (mean ± SEM).

Sample		Total Mesophilic Microorganisms	Coagulase-Positive Staphylococci	Enterobacteriaceae	Salmonella spp.	Listeria Monocytogenes
		[log cfu/g]			Presence in 25 g	
Raw pork loin		3.59 ± 0.21 [c]	1.95 ± 0.26 [b]	3.07 ± 0.09 [b]	NP	NP
60 °C						
	2 h	1.79 ± 0.16 [a]	<1.00 [a]	<1.00 [a]	NP	NP
	3 h	1.33 ± 0.45 [a]	<1.00 [a]	<1.00 [a]	NP	NP
	4 h	2.01 ± 0.79 [b]	<1.00 [a]	<1.00 [a]	NP	NP
65 °C						
	2 h	1.43 ± 0.31 [a]	<1.00 [a]	<1.00 [a]	NP	NP
	3 h	1.16 ± 0.22 [a]	<1.00 [a]	<1.00 [a]	NP	NP
	4 h	1.20 ± 0.16 [a]	<1.00 [a]	<1.00 [a]	NP	NP
70 °C						
	1 h	1.65 ± 0.40 [a]	<1.00 [a]	<1.00 [a]	NP	NP
	1.5 h	1.73 ± 0.04 [a]	<1.00 [a]	<1.00 [a]	NP	NP
	2 h	1.10 ± 0.14 [a]	<1.00 [a]	<1.00 [a]	NP	NP
75 °C						
	1 h	1.30 ± 0.25 [a]	<1.00 [a]	<1.00 [a]	NP	NP
	1.5 h	1.43 ± 0.31 [a]	<1.00 [a]	<1.00 [a]	NP	NP
	2 h	1.10 ± 0.14 [a]	<1.00 [a]	<1.00 [a]	NP	NP

a, b, c—mean values in columns with different superscripts differ significantly at $p < 0.05$ according to the Turkey's test; NP—not present.

Table 5. Sensory attributes of sous vide cooked pork loin (mean ± SEM).

Sample	Overall Appearance [1]	Colour Uniformity [2]	Aroma Intensity [3]	Tenderness [4]	Juiciness [5]	Meat Flavour Intensity [3]	Flavour Acceptability [1]	Overall Acceptability [1]
60 °C								
2 h	7.79 ± 0.33 [a]	8.36 ± 0.27 [abc]	6.79 ± 0.46 [a]	6.14 ± 0.38 [abc]	7.43 ± 0.25 [d]	6.86 ± 0.31 [a]	7.50 ± 0.20 [ab]	7.36 ± 0.27 [ab]
3 h	8.29 ± 0.19 [ab]	8.14 ± 0.35 [abc]	7.29 ± 0.37 [a]	6.50 ± 0.59 [abc]	6.57 ± 0.42 [cd]	7.64 ± 0.41 [ab]	8.14 ± 0.21 [b]	7.79 ± 0.15 [ab]
4 h	8.71 ± 0.24 [ab]	7.79 ± 0.46 [ab]	7.64 ± 0.60 [a]	7.00 ± 0.51 [bc]	6.43 ± 0.44 [bcd]	8.64 ± 0.27 [b]	8.43 ± 0.33 [b]	8.14 ± 0.35 [b]
65 °C								
2 h	8.23 ± 0.20 [ab]	7.77 ± 0.26 [ab]	7.15 ± 0.32 [a]	7.00 ± 0.28 [bc]	6.54 ± 0.40 [bcd]	7.23 ± 0.38 [ab]	7.23 ± 0.36 [ab]	7.08 ± 0.35 [ab]
3 h	8.00 ± 0.23 [a]	7.46 ± 0.31 [a]	7.77 ± 0.32 [a]	6.15 ± 0.62 [abc]	4.54 ± 0.57 [abc]	7.85 ± 0.30 [ab]	6.85 ± 0.48 [ab]	6.69 ± 0.43 [ab]
4 h	9.17 ± 0.21 [b]	9.33 ± 0.19 [c]	8.33 ± 0.26 [a]	8.17 ± 0.53 [c]	6.42 ± 0.53 [bcd]	8.50 ± 0.40 [ab]	7.08 ± 0.38 [ab]	7.25 ± 0.49 [ab]
70 °C								
1 h	7.81 ± 0.23 [a]	7.94 ± 0.30 [abc]	6.88 ± 0.40 [a]	6.31 ± 0.37 [abc]	5.56 ± 0.52 [abcd]	7.94 ± 0.21 [ab]	7.25 ± 0.50 [ab]	6.50 ± 0.41 [ab]
1.5 h	8.92 ± 0.23 [ab]	9.08 ± 0.26 [bc]	8.08 ± 0.26 [a]	7.00 ± 0.49 [abc]	6.17 ± 0.42 [bcd]	8.25 ± 0.35 [ab]	7.17 ± 0.34 [ab]	7.33 ± 0.31 [ab]
2 h	8.27 ± 0.14 [ab]	8.55 ± 0.25 [abc]	7.64 ± 0.31 [a]	6.00 ± 0.62 [abc]	5.36 ± 0.51 [abcd]	7.82 ± 0.35 [ab]	6.64 ± 0.43 [ab]	6.36 ± 0.53 [ab]
75 °C								
1 h	8.07 ± 0.21 [ab]	8.53 ± 0.19 [abc]	7.33 ± 0.40 [a]	4.67 ± 0.41 [a]	4.40 ± 0.46 [ab]	7.40 ± 0.40 [ab]	6.67 ± 0.54 [ab]	6.33 ± 0.47 [ab]
1.5 h	7.93 ± 0.27 [a]	8.20 ± 0.24 [abc]	7.80 ± 0.30 [a]	5.00 ± 0.52 [ab]	3.80 ± 0.43 [a]	7.93 ± 0.32 [ab]	6.80 ± 0.60 [ab]	6.20 ± 0.55 [a]
2 h	8.67 ± 0.19 [ab]	9.00 ± 0.21 [bc]	7.67 ± 0.41 [a]	5.50 ± 0.54 [ab]	4.58 ± 0.53 [abc]	8.08 ± 0.36 [ab]	6.00 ± 0.44 [a]	6.00 ± 0.49 [a]
Level of significance								
Temperature	NS	NS	NS	***	***	NS	***	***
Time	***	NS	NS	**	**	**	NS	*
Temperature x time interaction	**	***	NS	NS	NS	NS	NS	NS

a, b, c, d—mean values in columns, with different superscripts differ significantly at $p < 0.05$ according to the Turkey's test; * $p < 0.05$; ** $p < 0.01$; *** $p < 0.001$; NS—not significant; Scale 1 (0—not acceptable, 10—very acceptable); Scale 2 (0—not uniform, 10—highly uniform); Scale 3 (0—not detectable, 10—very intense); Scale 4 (0—tough, 10—tender); Scale 5 (0—low, 10—very high).

Our results regarding meat flavour intensity agree with those of Christensen et al. [11], who observed increasing intensity of meaty flavour of pork with increased temperature and cooking time using temperatures below 60 °C. Aaslyng et al. [56] pointed out that meat dish flavour is a complex phenomenon with origins from animal breed and handling before and after slaughter and then is related to meat processing, including heat treatment. The presence of precursors resulting from the type of meat cut and its origin, as well as temperature and treatment time determine the composition of flavour compounds. The Maillard reaction and a degradation of fatty acids are main sources of numerous volatile flavour compounds generated during heat treatment of meat. In addition, some importance to pork aroma may be ascribed to thiamine degradation [41]. The non-volatile compounds present in meat, such as monosodium glutamate, inosine monophosphate which originates from ATP breakdown, organic acids e.g., lactate [56] as well as derivatives of sulphur amino acids transformations [57] also contribute to meat flavour. In our study, cooking temperatures below 100 °C were applied, which significantly limited the extent of the Maillard reactions [4,41,56,58]. Anderson et al. [59] did not observe significant differences in beef lipids, based on fatty acids analysis between raw beef and beef cooked at 70 °C. Therefore, the limited extent of the above mentioned reactions at temperatures applied in our study presumably was the reason that no significant differences were observed between samples with regard to their aroma intensity. Moreover, Clausen et al. [6] did not observe a change in the free glutamate content in beef tenderloin sous vide cooked at 54 and 64 °C, which together with aroma volatile compounds delivering reactions might account for only small differences in sample flavour acceptability in the present study. On the other hand, Rotola-Pukkila et al. [60] reported lower glutamic and aspartic acid concentrations in pork loins sous vide cooked at 70 °C than at 60 and 80 °C, no effect of cooking parameters on inosine monophosphate and increased concentration of adenosine monophosphate with increased temperature and time of treatment.

The lowest juiciness of the samples cooked at 75 °C in our experiment confirms the findings of Christensen et al. [11] and agrees with the hypothesis verified by Becker et al. [14]. The hypothesis implied that transversal and longitudinal meat shrinkages, which take place at 45–60 °C and 60–90 °C, resp., decrease the water holding capacity of meat and as a result reduce its juiciness.

In our study, cooking at 75 °C produced a less tender meat than cooking at lower temperatures. These results correspond to those of TPA hardness, cohesiveness and chewiness. According to Baldwin [12], meat tenderness increases when cooked between 50 °C and 65 °C and then declines with temperature increases up to 80 °C. Clausen et al. [6] observed a slight decrease in beef tenderloin tenderness with longer cooking time both at 54 and 64 °C, and increased temperature. In the experiment of Christensen et al. [11] tenderness of pork increased both with temperature and cooking time.

3.9. Relationships between Selected Attributes of Cooked Samples

Correlation coefficients between selected quality attributes of sous vide cooked pork loin samples are presented in Table 6.

Table 6. Correlation coefficients between selected attributes of sous vide cooked pork loin.

Attribute	Moisture Content	Shear Force	Hardness	Chewiness	Springiness	Cohesiviness	Tenderness	Juiciness	Overall Acceptability
Cooking loss	−0.031	−0.157	0.369 ***	0.411 ***	0.280 **	0.316 ***	−0.172	−0.322 ***	−0.275 **
Moisture content	-	0.261 **	−0.150	−0.193	−0.216 *	−0.160	−0.153	−0.044	0.016
Shear force	-	-	−0.077	−0.064	−0.131 *	−0.212 ***	0.056	0.068	0.122
Hardness	-	-	-	0.588 ***	0.392 ***	0.336 ***	−0.156 *	−0.162 *	−0.197 *
Chewiness	-	-	-	-	0.512 ***	0.504 ***	−0.162 *	−0.219 **	−0.312 ***
Springiness	-	-	-	-	-	0.583 ***	0.008	−0.092	−0.200 *
Cohesiviness	-	-	-	-	-	-	−0.090	−0.091	−0.240 **
Tenderness	-	-	-	-	-	-	-	0.617 ***	0.561 ***
Juiciness	-	-	-	-	-	-	-	-	0.662 ***

$* \, p < 0.05$, $** \, p < 0.01$, $*** \, p < 0.001$.

Overall acceptability of cooked meat was positively and significantly correlated with sensorially assessed tenderness and juiciness (r = 0.561 and r = 0.662, resp.), while negatively and significantly, it was correlated with cooking loss (r = −0.275) and instrumentally measured texture attributes (correlation coefficients from −0.312 to −0.197). Other high correlation coefficients (r > 0.5; $p < 0.001$) were noted for hardness and chewiness (r = 0.588), chewiness and springiness (r = 0.512), chewiness and cohesiveness (r = 0.504), springiness and cohesiveness (r = 0.583) and finally for tenderness and juiciness (r = 0.617). Significant correlation coefficients (data not shown in the table) were also observed for overall acceptability and other features, such as meat flavour intensity (r = 0.236; $p < 0.01$), flavour acceptability (r = 0.370; $p < 0.001$) and chroma (r = 0.208; $p < 0.01$).

Only a few studies reported the relationships between investigated features of sous vide cooked pork. Díaz et al. [43] observed that appearance, meaty odour and meaty flavour were the main features that contributed positively and significantly to overall acceptance of pork stored for 5 or 10 weeks (r > 0.6), while the negative correlation coefficients were noted between overall acceptance and rancid, warmed-over and acidic odours and flavours. Jeong et al. [21] noted a similar tendency of changes in moisture content and shear force of experimental material. The positive relationship between shear force and moisture content in meat was also reported by Botinestean et al. [48], Roldan et al. [61] and Souza et al. [62], as well as observed in our own study (r = 0.261, $p < 0.01$). It is interesting that the correlation coefficient between cooking loss and moisture content in our study was low and insignificant. This observation could confirm that increased cooking losses might not agree with moisture reduction due to the loss of other meat components such as soluble proteins e.g., collagen [4].

4. Conclusions

The quality of cooked meat is a multidimensional feature, which is affected by meat cut characteristics and physico-chemical and sensorial properties as well as microbiological safety of final products. The research presented in the literature most often deal with selected aspects of the sous vide method: different meat cut quality, e.g., storage stability of pork loin cooked at only one temperature/time combination and assessed on the basis of physico-chemical and sensory analyses [43]; physico-chemical properties of pork cheeks cooked using different combinations of two temperatures and two times of heat treatment [3] or physico-chemical and basic microbiological quality aspects of pork ham cooked using combinations of two temperatures, two times and two vacuum degrees [21].

The present study, where we applied a wide range of cooking parameter combinations and a wide range of analyses (physico-chemical, sensorial and microbiological), revealed that cooking at 60 or 65 °C for 4 h ensured the most attractive and acceptable sensory traits of pork loin, which was also partially confirmed by the results of TPA analysis. Instrumentally measured hardness, cohesiveness and chewiness showed the lowest values for the 60 °C/4 h sample, while the lowest springiness was observed for the 65 °C/4 h sample. The sensory features that influenced the overall acceptability of sous vide cooked pork loin in the highest degree were tenderness and juiciness, as can be seen based on coefficients of correlation. That means the texture attributes were the most important for sous vide cooked pork perception. Regarding cooking loss, meat preparation at 60 °C/4 h was more beneficial than at 65 °C/4 h, however it was not reflected in moisture content and sensorially assessed juiciness. The results obtained in the research indicate that the applied parameters of the sous vide processing were sufficient to reduce the microflora in the pork loin to the level safe for consumption.

Author Contributions: Conceptualization, experiment design, writing—review and editing, funding acquisition, supervision—L.K. (Lidia Kurp), M.D.-O.; project administration—M.D.-O.; methodology, formal analysis, data analysis, writing—original draft—L.K. (Lidia Kurp), L.K. (Lucyna Kłębukowska). All authors have read and agreed to the published version of the manuscript.

Funding: Project financially supported by the Minister of Education and Science in the range of the program entitled "Regional Initiative of Excellence" for the years 2019–2022, Project No. 010/RID/2018/19, amount of funding 12.000.000 PLN.

Data Availability Statement: Not applicable.

Conflicts of Interest: The authors declare no conflict of interest.

References

1. Akoğlu, I.T.; Bıyıklı, M.; Akoğlu, A.; Kurhan, Ş. Determination of the quality and shelf life of sous vide cooked turkey cutlet stored at 4 and 12 °C. *Rev. Bras. Cienc. Avic.* **2018**, *20*, 1–8. [CrossRef]
2. Bhat, Z.F.; Morton, J.D.; Zhang, X.; Mason, S.L.; Bekhit, A.E.A. Sous-vide cooking improves the quality and in-vitro digestibility of *Semitendinosus* from culled dairy cows. *Food Res. Int.* **2020**, *127*, 108708. [CrossRef] [PubMed]
3. Sánchez del Pulgar, J.; Gázquez, A.; Ruiz-Carrascal, J. Physico-chemical, textural and structural characteristics of sous-vide cooked pork cheeks as affected by vacuum, cooking temperature, and cooking time. *Meat Sci.* **2012**, *90*, 828–835. [CrossRef] [PubMed]
4. Dominguez-Hernandez, E.; Salaseviciene, A.; Ertbjerg, P. Low-temperature long-time cooking of meat: Eating quality and underlying mechanisms. *Meat Sci.* **2018**, *143*, 104–113. [CrossRef]
5. Bryan, E.E.; Smith, B.N.; Dilger, R.N.; Dilger, A.C.; Boler, D.D. Technical note: A method for detection of differences in cook loss and tenderness of aged pork chops cooked to differing degrees of doneness using sous-vide. *J. Anim. Sci.* **2019**, *97*, 3348–3353. [CrossRef]
6. Clausen, M.P.; Christensen, M.; Djurhuus, T.H.; Duelund, L.; Mouritsen, O.G. The quest for umami: Can sous vide contribute? *Int. J. Gastron. Food Sci.* **2018**, 129–133. [CrossRef]
7. James, B.J.; Yang, S.W. Effect of cooking method on the toughness of bovine *M. Semitendinosus*. *Int. J. Food Eng.* **2012**, *8*, 1–18. [CrossRef]
8. Hong, G.E.; Kim, J.H.; Ahn, S.J.; Lee, C.H. Changes in meat quality characteristics of the sous-vide cooked chicken breast during refrigerated storage. *Korean J. Food Sci. Anim. Resor.* **2015**, *35*, 757–764. [CrossRef]
9. Roldán, M.; Antequera, T.; Martin, A.; Mayoral, A.I.; Ruiz, J. Effect of different temperature-time combinations on physicochemical, microbiological, textural and structural features of sous-vide cooked lamb loins. *Meat Sci.* **2013**, *93*, 572–578. [CrossRef]
10. Christensen, L.; Ertbjerg, P.; Løje, H.; Risbo, J.; van den Berg, F.W.J.; Christensen, M. Relationship between meat toughness and properties of connective tissue from cows and young bulls treated at low temperatures for prolonged times. *Meat Sci.* **2013**, *93*, 787–795. [CrossRef]
11. Christensen, L.; Gunvig, A.; Tørngren, M.A.; Aaslyng, M.D.; Knøchel, S.; Christensen, M. Sensory characteristics of meat cooked for prolonged times at low temperature. *Meat Sci.* **2012**, *90*, 485–489. [CrossRef] [PubMed]
12. Baldwin, D.E. Sous vide cooking: A review. *Int. J. Gastron. Food Sci.* **2012**, *1*, 15–30. [CrossRef]
13. Christensen, M.; Purslow, P.P.; Larsen, L.M. The effect of cooking temperature on mechanical properties of whole meat, single muscle fibres and perimysial connective tissue. *Meat Sci.* **2000**, *55*, 301–307. [CrossRef]
14. Becker, A.; Boulaaba, A.; Pingen, S.; Krischek, C.; Klein, G. Low temperature cooking of pork meat—Physicochemical and sensory aspects. *Meat Sci.* **2016**, *118*, 82–88. [CrossRef]
15. Lien, R.; Hunt, M.C.; Anderson, S.; Kropf, D.H.; Loughin, T.M.; Dikeman, M.E.; Velazco, J. Effects of endpoint temperature on the internal color of pork loin chops of different quality. *J. Food Sci.* **2002**, *67*, 1007–1010. [CrossRef]
16. Geileskey, A.; King, R.D.; Corte, D.; Pinto, P.; Ledward, D.A. The kinetics of cooked meat haemoprotein formation in meat and model systems. *Meat Sci.* **1998**, *48*, 189–199. [CrossRef]
17. King, N.J.; Whyte, R. Does it look cooked? A review of factors that influence cooked meat color. *J. Food Sci.* **2006**, *71*, R31–R40. [CrossRef]
18. Becker, A.; Boulaaba, A.; Pingen, S.; Röhner, A.; Klein, G. Low temperature, long time treatment of porcine *M. longissimus thoracis et lumborum* in a combi steamer under commercial conditions. *Meat Sci.* **2015**, *110*, 230–235. [CrossRef]
19. Dominguez-Hernandez, E.; Ertbjerg, P. Effect of LTLT heat treatment on cathepsin B and L activities and denaturation of myofibrillar proteins of pork. *Meat Sci.* **2021**, *175*, 108454. [CrossRef]
20. Christensen, L.; Ertbjerg, P.; Aaslyng, M.D.; Christensen, M. Effect of prolonged heat treatment from 48 °C to 63 °C on toughness, cooking loss and color of pork. *Meat Sci.* **2011**, *88*, 280–285. [CrossRef]
21. Jeong, K.O.H.; Shin, S.Y.; Kim, Y.S. Effects of sous-vide method at different temperatures, times and vacuum degrees on the quality, structural, and microbiological properties of pork ham. *Meat Sci.* **2018**, *143*, 1–7. [CrossRef] [PubMed]
22. Zielbauer, B.I.; Franz, J.; Viezens, B.; Vilgis, T.A. Physical aspects of meat cooking: Time dependent thermal protein denaturation and water loss. *Food Biophys.* **2016**, *11*, 34–42. [CrossRef]
23. Vaudagna, S.R.; Pazos, A.A.; Guidi, S.M.; Sanchez, G.; Carp, D.J.; Gonzales, C.B. Effect of salt addition on *sous vide* cooked whole beef muscles from Argentina. *Meat Sci.* **2008**, *79*, 470–482. [CrossRef] [PubMed]
24. Ángel-Rendón, S.V.; Filomena-Ambrosio, A.; Hernández-Carrión, M.; Llorca, E.; Hernando, I.; Quiles, A.; Sotelo-Díaz, I. Pork meat prepared by different cooking methods. A microstructural, sensorial and physicochemical approach. *Meat Sci.* **2020**, *163*, 108089. [CrossRef] [PubMed]

25. AOAC International. Official methods of analysis 950.46. In *Moisture in Meat*, 18th ed.; AOAC International: Gaithersburg, MD, USA, 2005.
26. *ISO 4833-1:2013*; Microbiology of the Food Chain—Horizontal Method for the Enumeration of Microorganism—Part 1: Colony Count at 30 °C by the Pour Plate Technique. 2005. Available online: https://www.iso.org/standard/59509.html (accessed on 11 January 2022).
27. *ISO/DIS 6888-2:2020*; Microbiology of the Food Chain—Horizontal Method for the Enumeration of Coagulase-Positive Staphylococci (*Staphylococcus aureus* and Other Species)—Part 2: Technique Using Rabbit Plasma Fibrinogen Agar Medium. Available online: https://webstore.ansi.org/standards/iso/isodis68882020-2405117 (accessed on 11 January 2022).
28. *ISO 21528-2:2017*; Microbiology of the Food Chain—Horizontal Method for the Detection and Enumeration of Enterobacteriaceae—Part 2: Colony—Count Technique. Available online: https://www.iso.org/standard/63504.html (accessed on 11 January 2022).
29. *ISO 6579-1:2017*; Microbiology of the Food Chain—Horizontal Method for the Detection, Enumeration and Serotyping of *Salmonella*—Part 1: Detection of *Salmonella* spp. Available online: https://www.iso.org/obp/ui/#iso:std:iso:6579:-1:ed-1:v1:en (accessed on 11 January 2022).
30. *ISO 11290-1:2017*; Microbiology of the Food Chain—Horizontal Method for the Detection, and Enumeration of *Listeria monocytogenes* and of *Listeria* spp.—Part 1: Detection Method. Available online: https://www.iso.org/standard/60313.html (accessed on 11 January 2022).
31. *PN-EN ISO 8586:2014-03*; Sensory Analysis. General Guidance for Selection, Training and Monitoring of Selected Assessors and Experts. Polish Committee for Standardization: Warsaw, Poland, 2014.
32. *ISO 4121:2003*; Sensory Analysis—Guidelines for the Use of Quantitative Response Scales. Available online: https://www.iso.org/standard/33817.html (accessed on 11 January 2022).
33. Ayub, H.; Ahmad, A. Physiochemical changes in sous-vide and conventionally cooked meat. *Int. J. Gastron. Food Sci.* **2019**, *17*, 100145. [CrossRef]
34. Hwang, S.I.; Lee, E.J.; Hong, G.P. Effects of temperature and time on the cookery properties of sous-vide processed pork loin. *Food Sci. Anim. Resour.* **2019**, *39*, 65–72. [CrossRef]
35. Gerber, N.; Scheeder, M.R.L.; Wenk, C. The influence of cooking and fat trimming on the actual nutrient intake from meat. *Meat Sci.* **2009**, *81*, 148–154. [CrossRef]
36. Oz, F.; Aksu, M.; Turan, M. The effects of different cooking methods on some quality criteria and mineral composition of beef steaks. *J. Food Process. Pres.* **2017**, *41*, e13008. [CrossRef]
37. Tornberg, E. Effect of heat on meat proteins—Implications on structure and quality of meat products. *Meat Sci.* **2005**, *70*, 493–508. [CrossRef]
38. Lawrie, R.A.; Ledward, D.A. *Lawrie's Meat Science*; Woodhead Publishing Limited: Cambridge, UK, 2006.
39. Purslov, P.P.; Oiseth, S.; Hughes, J.; Warner, R.D. The structural basis of cooking loss in beef: Variations with temperature and ageing. *Food Res. Int.* **2016**, *89*, 739–748. [CrossRef]
40. Park, Y.W. Moisture and water activity. In *Handbook of Processed Meat and Poultry Analysis*, 1st ed.; Nollet, L.M.L., Toldra, F., Eds.; CRC Press: Boca Raton, FL, USA; London, UK; New York, NY, USA, 2008; pp. 35–40.
41. Resconi, V.C.; Escudero, A.; Campo, M.M. The development of aromas in ruminant meat. *Molecules* **2013**, *18*, 6748–6781. [CrossRef] [PubMed]
42. Bıyklı, M.; Akoğlu, A.; Kurhan, Ş.; Akoğlu, I.T. Effect of different sous vide cooking temperature-time combinations on the physicochemical, microbiological, and sensory properties of turkey cutlet. *Int. J. Gastron. Food Sci.* **2020**, *20*, 100204. [CrossRef]
43. Díaz, P.; Nieto, G.; Garrido, M.D.; Bañón, S. Microbial, physical-chemical and sensory spoilage during the refrigerated storage of cooked pork loin processed by the *sous vide* method. *Meat Sci.* **2008**, *80*, 287–292. [CrossRef] [PubMed]
44. Haghighi, H.; Belmonte, A.M.; Masino, F.; Minelli, G.; Lo Fiego, D.P.; Pulvirenti, A. Effect of time and temperature on physico-chemical and microbiological properties of sous vide chicken breast fillets. *Appl. Sci.* **2021**, *11*, 3189. [CrossRef]
45. Vasanthi, C.; Venkataramanujam, V.; Dushyanthan, K. Effect of cooking temperature and time on the physico-chemical, histological and sensory properties of female carabeef (buffalo) meat. *Meat Sci.* **2007**, *76*, 274–280. [CrossRef]
46. Bojarska, U.; Batura, J.; Cierach, M. The effect of measurement site on the evaluation of tom breast muscle color. *Pol. J. Food Nutr. Sci.* **2003**, *53*, 45–49.
47. Vaudagna, S.R.; Sánchez, G.; Neira, M.S.; Insani, E.M.; Picallo, A.B.; Gallinger, M.M.; Lasta, J.A. Sous vide cooked beef muscles: Effects of low temperature–long time (LT–LT) treatments on their quality characteristics and storage stability. *Int. J. Food Sci. Tech.* **2002**, *37*, 425–441. [CrossRef]
48. Botinestean, C.; Keenan, D.F.; Kerry, J.P.; Hamill, R.M. The effect of thermal treatments including sous-vide, blast freezing and their combinations on beef tenderness of *M. Semitendinosus* steaks targeted at elderly consumers. *LWT Food Sci. Technol.* **2016**, *74*, 154–159. [CrossRef]
49. Smith, D.S. Quantifying color variation: Improved formulas for calculating hue with segment classification. *Appl. Plant Sci.* **2014**, *2*, 1300088. [CrossRef]
50. Sun, S.; Rasmussen, F.D.; Cavender, G.A.; Sullivan, G.A. Texture, color and sensory evaluation of sous-vide cooked beef steaks processed using high pressure processing as method of microbial control. *LWT Food Sci. Technol.* **2019**, *103*, 169–177. [CrossRef]
51. Mokrzycki, W.; Tatol, M. Color difference Delta E—A survey. *Mach. Graph. Vis.* **2011**, *20*, 383–411.

52. Ismail, I.; Hwang, Y.-H.; Bakhsh, A.; Joo, S.-T. The alternative approach of low temperature-long time cooking on bovine *semitendinosus* meat quality. *Asian-Australas. J. Anim. Sci.* **2019**, *32*, 282–289. [CrossRef] [PubMed]
53. Brüggemann, D.A.; Brewer, J.; Risbo, J.; Bagatolli, L. Second harmonic generation microscopy: A tool for spatially and temporally resolved studies of heat induced structural changes in meat. *Food Biophys.* **2010**, *5*, 1–8. [CrossRef]
54. Palka, K.; Daun, H. Changes in texture, cooking losses, and myofibrillar structure of bovine *M. semitendinosus* during heating. *Meat Sci.* **1999**, *51*, 237–243. [CrossRef]
55. Commission Regulation (EC) No. 2073/2005 of 15 November 2005 on Microbiological criteria for foodstuffs. *Official Journal of the European Union*, L 388, 48, 1-26. The Commission of the European Communities: Brussels, Belgium, 2005. Available online: https://eur-lex.europa.eu/legal-content/EN/TXT/PDF/?uri=OJ:L:2005:338:FULL&from=EN (accessed on 14 September 2021).
56. Aaslyng, M.D.; Meinert, L. Meat flavor in pork and beef—From animal to meal. *Meat Sci.* **2017**, *132*, 112–117. [CrossRef]
57. Przybylski, W.; Jaworska, D.; Kajak-Siemaszko, K.; Sałek, P.; Pakuła, K. Effect of heat treatment by the sous-vide method on the quality of poultry meat. *Foods* **2021**, *10*, 1610. [CrossRef]
58. Chansataporn, W.; Nopharatana, M.; Samuhasaneetoo, S.; Siriwattanayotin, S.; Tangduangdee, C. Effects of temperature on the main intermediates and products of the Maillard reaction in the chicken breast meat model system. *Int. J. Agric. Technol.* **2019**, *15*, 539–556.
59. Anderson, D.B.; Breidenstein, B.B.; Kauffan, R.G.; Cassens, R.G.; Bray, R.W. Effect of cooking on fatty acid composition of beef lipids. *Int. J. Food Sci. Technol.* **1971**, *6*, 141–152. [CrossRef]
60. Rotola-Pukkila, M.; Pihlajaviita, S.T.; Kaimainen, M.; Hopia, A. Concentration of umami compounds in pork meat and cooking juice with different cooking times and temperatures. *J. Food Sci.* **2015**, *80*, C2711–C2716. [CrossRef]
61. Roldán, M.; Antequera, T.; Armenteros, M.; Ruiz, J. Effect of different temperature-time combinations on lipid and protein oxidation of sous-vide cooked lamb loins. *Food Chem.* **2014**, *149*, 129–136. [CrossRef]
62. Souza, C.M.; Boler, D.D.; Clark, D.L.; Kutzler, L.W.; Holmer, S.F.; Summerfield, J.W.; Cannon, J.E.; Smit, N.R.; McKeith, F.K.; Killefer, J. The effects of high pressure processing on pork quality, palatability, and further processed products. *Meat Sci.* **2011**, *87*, 419–427. [CrossRef] [PubMed]

Article

Attempt to Develop an Effective Method for the Separation of Gamma-Decalactone from Biotransformation Medium

Jolanta Małajowicz *, Agata Górska, Joanna Bryś, Ewa Ostrowska-Ligęza and Magdalena Wirkowska-Wojdyła

Department of Chemistry, Institute of Food Sciences, Warsaw University of Life Sciences, 159c Nowoursynowska St., 02-776 Warsaw, Poland; agata_gorska@sggw.edu.pl (A.G.); joanna_brys@sggw.edu.pl (J.B.); ewa_ostrowska_ligeza@sggw.edu.pl (E.O.-L.); magdalena_wirkowska@sggw.edu.pl (M.W.-W.)

* Correspondence: jolanta_malajowicz@sggw.edu.pl; Tel.: +48-22-59-37-618

Abstract: Gamma-decalactone (GDL) is a fragrance compound obtained in the process of β-oxidation of ricinoleic acid, which is derived from the hydrolysis of castor oil. The biotechnological method of the synthesis of this lactone has been improved for over two decades, but the vast majority of research results have been based only on determining the concentration of the lactone by chromatographic methods without separating it from the biotransformation medium. In this study, we attempted to separate GDL from the medium in which the lactone was synthesized by *Yarrowia lipolytica* from castor oil. The effectiveness of liquid–liquid extraction, hydrodistillation, and adsorption on the porous materials (zeolite, vermiculite and resin Amberlite XAD-4) was compared. The influence of the solvent on the efficiency of GDL extraction, the influence of the acidity of the medium on the amount of GDL in the distillate, and the level of lactone adsorption on the above-mentioned adsorbents were compared by calculating the initial adsorption rate. The adsorption isotherm was determined for the most effective adsorbent. Among the five solvents tested, the most effective was diethyl ether, used at the ratio of 1:1. The extraction was characterized by higher efficiency than hydrodistillation; the difference in GDL determinations by these two methods ranged from 12.8 to 22%. The purity of the distillates was much higher than that of the extracts at 88.0 ± 3.4% compared to 53.0 ± 1.8%. The acidification of the biotransformation medium increased the concentration of the lactone in both the reaction mixture and the distillate. GDL was most efficiently adsorbed on Amberlite XAD-4 resin, for which the lactone isotherm adsorption was linear. The amount of lactone adsorbed on Amberlite XAD-4 within 1 h was approx. 80% (2.45 g), of which 1.96 g was then desorbed with ethanol. In the context of industrial applications, adsorption of GDL on Amberlite XAD-4 seems to be the most appropriate method due to material costs, the ease of the process, and low environmental burden.

Keywords: gamma-decalactone; separation; solvent extraction; hydrodistillation; adsorption; Amberlite XAD-4

Citation: Małajowicz, J.; Górska, A.; Bryś, J.; Ostrowska-Ligęza, E.; Wirkowska-Wojdyła, M. Attempt to Develop an Effective Method for the Separation of Gamma-Decalactone from Biotransformation Medium. *Appl. Sci.* **2022**, *12*, 2084. https://doi.org/10.3390/app12042084

Academic Editor: Andrea Salvo

Received: 14 January 2022
Accepted: 14 February 2022
Published: 17 February 2022

Publisher's Note: MDPI stays neutral with regard to jurisdictional claims in published maps and institutional affiliations.

1. Introduction

Lactones are one of the most important groups of substances that shape the sensory properties of food and cosmetic products. They exhibit a pleasant, fruity-floral scent and are a group of compounds that are often isolated from the natural environment. However, the possibilities of using industrial extraction from natural sources are limited, as the sensory-relevant plant components are often present in low concentration, and therefore, isolating and obtaining them in sufficiently large amounts is usually expensive. The technically troublesome and economically unprofitable isolation of cyclic esters from the natural environment, with the simultaneous willingness to meet consumer trends for a healthy lifestyle and interest in what is natural, contributed to the dissemination of the research on the biotechnological method of synthesizing these compounds [1,2].

In the last two decades, a particularly great interest has been noted in the biotechnological synthesis of gamma-decalactone (GDL), a compound with a characteristic peach fragrance used in the production of confectionery, cakes, wafers, chewing gums, and granulated and powder beverages [3]. The research conducted by the scientists from France (Waché, Dufossé, Blin-Perrin) [4–6], Belgium (Alchihab, Destain) [7,8], Portugal (Aguedo, Gomes, Braga) [9,10], and Brazil (Soares, Pereira de Andrade) [11,12] has provided extensive knowledge on the biotransformation of fatty acids to gamma-decalactone, based on β-oxidation, which takes place in yeast cell peroxisomes. Particular attention was paid to the β-oxidation process of ricinoleic acid [(*R*)-12-hydroxy-9-octadecanoic acid], which is the main component (80–90%) of castor oil obtained from castor bean seeds (*Ricinus communis*). The conversion of this acid into GDL involves cleavage from the carboxy terminus of the eight-carbon fragment as a result of four β-oxidation cycles. From the 18-carbon chain, the final 10-carbon compound (4-hydroxydecanoic acid) is obtained, which is then cyclized in an acidic environment to gamma-decalactone [13].

Many microorganisms show the capability of the biosynthesis of GDL (*Candida, Pichia, Sporobolus, Rhodotorula, Yarrowia*) and are highly active in the process of β-oxidation and reduction of the unsaturated bond in ricinoleic acid (gamma-decalactone precursor) [6,7,13–16]. Among them, *Yarrowia lipolytica* yeast is characterized by the highest production efficiency, especially when castor oil is used as a substrate in the reactions. This is due to the high extracellular lipolytic activity of this yeast induced by hydrophobic substrates [17,18].

The numerous studies conducted so far on the biotechnological synthesis of GDL have mainly focused on the improvement of reaction efficiency through the selection and optimization of individual process parameters, as well as the selection of the appropriate strain, which is often subject to additional genetic modifications [12,19–22]. However, these studies have been mostly carried out on a small scale, and the reaction efficiency, at the level of a few grams of lactone per liter of the culture medium, has been estimated on the basis of chromatographic or spectrophotometric determinations without separating a product from the mixture. The most frequently used analytical methods include GC [19,22–24], GC-MS [7,9,25,26], solid-phase microextraction in combination with GC-MS (SPEME-GC-MS) [27], and less often, a spectrophotometric method based on the ferric hydroxamate reaction [28].

So far, only a few research papers on the possibility of separating GDL from the culture medium have been published. Generally, the recovery of products in the bioprocesses that involve microorganisms is a difficult stage, especially for fragrances, due to their volatility and low solubility as well as the complexity of the biotransformation medium itself. Liquid–liquid extraction (LLE), adsorption, or pervaporation are most often used to extract fragrances [29–31]. So far, attempts to recover GDL from biotransformation media have been carried out with the use of activated carbon Acticarbone NC35 [32,33] and porous resins, e.g., Porapak Q, Chromosorb 105, SM4 Tenax-TA, Amberlite XAD-2 and XAD-4, and Macronet (MN-202, MN-102 and MN-100) [30,32–34].

In this study, we conducted research to verify the possibility of secreting GDL separation from biotransformation media, in which a lactone was synthesized from castor oil via *Yarrowia lipolytica* KKP379 yeast. The *Y. lipolytica* KKP379 yeast strain was selected for the study due to our experience in working with this strain and knowledge of some optimal conditions for its efficient biosynthesis of gamma-decalactone [35]. The level of GDL separation efficiency was compared with the use of traditional liquid–liquid extraction, hydrodistillation, and adsorption.

2. Materials and Methods

2.1. Microorganisms and Reagents

Yarrowia lipolytica KKP379, used for biosynthesis of GDL, was obtained from the Collection of Industrial Microorganisms, Prof. Wacław Dąbrowski Institute of Agricultural and Food Biotechnology, State Research Institute (Warsaw, Poland). The culture media components, yeast extract, peptone, glucose, and agar, were obtained from BTL Sp. Z o.o. (Łódź, Poland). The castor oil was from Carl Roth (Karlsruhe, Germany). The Tween 80 came from Acros Organics (Geel, Belgium). Diethyl ether, ethyl alcohol 96%, hexane, heptane, dichloromethane, chloroform, magnesium sulfate, and hydrochloric acid were provided by Avantor Performance Materials (Gliwice, Poland). Gamma-decalactone >98% and gamma-undecalactone 98% were purchased from Sigma-Aldrich (Saint Louis, MO, USA). Amberlite XAD-4, zeolite, and vermiculite were from Merck (Darmstadt, Germany). The general characteristics of the adsorbents, in accordance with the manufacturer's declarations, are presented in Table 1.

Table 1. Characteristics of the adsorbents.

Adsorbent	Composition	Specific Surface Area (m^2/g)	Average Pore Diameter (nm)
Zeolite	Porous crystalline alumino silicate, SiO_2/AlO_3	700	0.3–0.8
Amberlite XAD-4	Polystyrene resin	750	40
Vermiculite	Multilayer structure magnesium aluminium silicate inorganic	394	0.9–1.4

2.2. Cultivation of Microorganisms

Yarrowia lipolytica KKP379 was multiplied for 48 h on petri dishes with YPGA medium (10 g/L yeast extract, 20 g/L peptone, 20 g/L glucose, 20 g/L agar) at 27 °C and inoculated to a 250 mL Erlenmeyer flask containing 50 mL of YPG medium (10 g/L yeast extract, 20 g/L peptone, 20 g/L glucose). Flasks were shaken at 140 rpm and 27 °C for 24 h, until the cultures reached the late-logarithmic growth phase. This preculture was used to inoculate the biotransformation medium.

2.3. Biotransformation of Castor Oil into Gamma-Decalactone

One milliliter of the preculture was transferred into 500 mL Erlenmeyer flasks with biotransformation medium (20 g/L peptone, 100 g/L castor oil, 5 g/L Tween 80). The final biotransformation medium volume was 100 mL and the initial concentration of yeast cells was about $OD_{600} \approx 0.25$. The biotransformation was conducted at 27 °C and 140 rpm for 7 days. After finishing the reaction, to stop metabolism of yeast and to achieve the complete lactonization of 4-hydroxydecanoic acid, 1 M HCl was added to the samples. Before the separation of GDL, the pH of the mixture was adjusted to approximately 3.

2.4. Liquid–Liquid Extraction

Extraction in the liquid–liquid system was carried out using model water solutions and biotransformation mediums. The 100 mL of model solutions contained 0.3 g of GDL and 10 g of castor oil (the composition was similar to the biotransformation mediums). Solutions were emulsified before extraction using an Ultra Turrax T25 homogenizer (IKA, Staufen im Breisgau, Germany), and homogenized for 5 min at 15,000 rpm. The liquid–liquid extraction method was adapted from the methodology of Groguenin et al. [24]. For extraction, 1.5 mL or 3 mL of model solutions or biotransformation mediums were taken. They were then added to 20 µL of internal standard (gamma-undecalactone) and extracted with 1.5 mL of the either diethyl ether, hexane, heptane, chloroform, or dichloromethane as the solvent. The organic layer was separated and dehydrated by anhydrous $MgSO_4$. Finally, 1 µL of aliquots was injected into GC system.

2.5. Hydrodistillation

GDL was separated by hydrodistillation from a 100 mL sample of the biotransformation mediums. Yeast cells were centrifuged (speed of 8000 rpm for 5 min, Centrifuge MPW-223) from the medium prior to distillation. To improve the distillation process, 20 mL of distilled water was added to the samples. One hundred milliliters of distillate was collected, which was then subjected to liquid–liquid extraction (1:1, *v:v*) using diethyl ether and determined by gas chromatography.

2.6. Adsorption of Gamma-Decalactone

Absorption of GDL from biotransformation mediums was carried out with the aid of three different adsorbents: zeolite, Amberlite XAD-4, and vermiculite. Prior to use, the adsorbents were cleaned for 24 h with ethyl alcohol. Sequentially, the washed adsorbents were dried for 8 h at 80 °C to eliminate alcohol residues.

The adsorption process was carried out at a temperature of 25 °C. The adsorbents were added to the biotransformation mediums at a concentration of 30 g/L. The mediums with adsorbents were stirred at 140 rpm for 6 h, and every hour, the samples were taken for chromatographic analysis.

2.7. Recovery of Gamma-Decalactone from the Adsorbents

After the specified adsorption time, the adsorbents were separated from the culture broth by filtration on the Büchner funnel. Porous materials were washed three times with distilled water and dried at room temperature for 2 h. GDL was extracted from the adsorbents with ethyl alcohol at a ratio of 3 mL of ethanol par 1 g of adsorbent, using 5 extraction cycles. For quantification, the ethanolic fractions were collected and dried by anhydrous magnesium sulfate, and gamma-undecalactone was added as an internal standard. The ethanolic extract was analyzed by gas chromatography.

2.8. Quantification of Gamma-Decalactone

The content of GDL was analyzed using a gas chromatography instrument (YL 6100 Young Lin Instrument) equipped with a capillary column BPX (30 m × 0.25 mm) and a flame ionization detector. The N_2 was the carrier gas, and flow rate was 1.1 mL/min. The analysis conditions were as follows: a 1 μL sample was injected at 250 °C, the detector temperature was 280 °C, and the oven temperature was maintained at 165 °C for 1 min, increased to 180 °C at a rate of 3 °C/min, raised to 220 °C at a rate of 5 °C/min and held constant for 2 min.

The identification of other lactones in the broth and the purity of the samples were determined using a GC-MS system (Varian 3400 GC with a CPWAX-52-CB column, coupled to Varian Saturn II MS). The conditions of mass spectroscopy were as follows: electron ionization source was set to 80 eV, the emission current was 35.5 μA, mass spectrometry (MS) Quad was 160 °C, and the MS source was 220 °C.

2.9. Statistical Analysis

Statistical analysis was performed using Statistica 13.3 software (TIBCO Software Inc., Palo Alto, CA, USA). The results were considered statistically different at $\alpha = 0.05$. The data were expressed as mean ± relative standard deviation (±RSD).

3. Results

In the studies on the effectiveness of GDL secretion, model media were used which were similar in composition to the biotransformation or microbiological media with the addition of 10 g/dm^3 of castor oil, from which the lactone was synthesized by *Yarrowia lipolytica* yeast by β-oxidation. The effectiveness of liquid–liquid extraction (using a few selected solvents), hydrodistillation, and adsorption on natural materials and polymer resins was analyzed.

3.1. Recovery of Gamma-Decalactone by Liquid–Liquid Extraction

Solvent extraction is a very simple and rapid method of isolating fragrance compounds. Research on the separation of GDL began with analyzing the influence of the type of solvent used at the level of its recovery. For this purpose, model solutions were used with a composition that reflected the microbiological media for lactone synthesis. The model solutions contained GDL at the concentration of 3 g/L and castor oil. The concentration of the ingredients was selected based on the previous research on GDL synthesis, in which, an average of about 3 g/L of lactone was obtained from 10 g/L of castor oil by biotransformation reactions with the participation of *Yarrowia lipolytica* yeast [35].

In this study, five different organic solvents with different polarities were compared in order to obtain the best recovery of GDL. Table 2 presents a comparison of the results of the recovery test with two different solvent-to-sample ratios.

Table 2. Gamma-decalactone recovery degree from model solutions by different solvents.

Extraction Solvent Type	Solvent Volume per Sample Volume (mL/mL)	Average Percent Recovery and RSD * Values (%)
Hexane	1:2	84.3 (4.1)
	1:1	92.8 (2.1)
Heptane	1:2	72.1 (3.7)
	1:1	87.0 (2.7)
Diethyl ether	1:2	87.9 (2.1)
	1:1	99.9 (1.8)
Chloroform	1:2	85.5 (3.3)
	1:1	98.8 (1.3)
Dichloromethane	1:2	89.4 (4.4)
	1:1	101.1 (2.5)

* RSD—Relative Standard Deviation.

The data indicate that diethyl ether is one of the best solvents for GDL extraction. The recovery with its participation reached almost 100% (99.9 ± 1.8%). Slightly lower or less accurate (burdened with a bigger error) extraction efficiency was obtained with the use of chloroform or dichloromethane (with a recovery of 98.8 ± 1.3% and 101.1 ± 2.5%, respectively). The use of a larger volume of organic solvent for lactone extraction clearly improved the level of its recovery, regardless of the solvent used. With a 1:1 (*v:v*) solvent-to-sample ratio, 0.1% (for diethyl ether) to 13% (for heptane) of the GDL was not extracted. For a 1:2 (*v:v*) solvent-to-sample extraction ratio, 10.6% (for dichloromethane) to 27.9% (for heptane) of the GDL remained in the solution. The RSD values were less than 5% in all the experiments, with higher deviations for the 1:2 (*v:v*) extraction, regardless of the type of solvent. The extraction method that used equal amounts of solvent and sample for diethyl ether, chloroform, and dichloromethane showed good recoveries and high repeatability.

However, the problem with using this technique is the purity of the extracted compound. In the extracts, the GDL purity was only about 51.5 ± 1.2%–53.0 ± 1.8% (the results obtained using GC-MS analyses). The samples were contaminated mainly with ricinoleic acid derived from the hydrolysis of castor oil as well as with triglycerides of the unreacted substrate.

3.2. Recovery of Gamma-Decalactone from Biotransformation Media by Hydrodistillation

Attempts to separate GDL from biotransformation media were also carried out via hydrodistillation (HD). Castor oil biotransformation reactions (10 g/L) were performed with *Yarrowia* yeast for eight days. Each day the reaction mixture from the flask was distilled, and the GDL concentration in the distillate was compared with the values obtained from the direct liquid–liquid extraction with diethyl ether. The data presented in Figure 1 show that the use of hydrodistillation allowed the separation of GDL from the media, regardless of the level of the substrate conversion, and the use of this method is associated with a partial loss of the lactone. The concentration of the compound in the distillates on each sampling day was lower than in extracts, and the difference in determinations with the lactone content above 0.5 g/L ranged from 12.8% to 22% (from 0.25 g/L to 0.63 g/L). In the range of lower lactone concentrations, the differences in the distillates compared to the extracts were as high as 73%. However, this may be due to an error of sample determination as it is difficult to separate the lactone effectively at its low concentrations. However, taking into account the purity of the product in the distillate (about 88% ± 3.4%), this process seems to be more advantageous than liquid–liquid extraction.

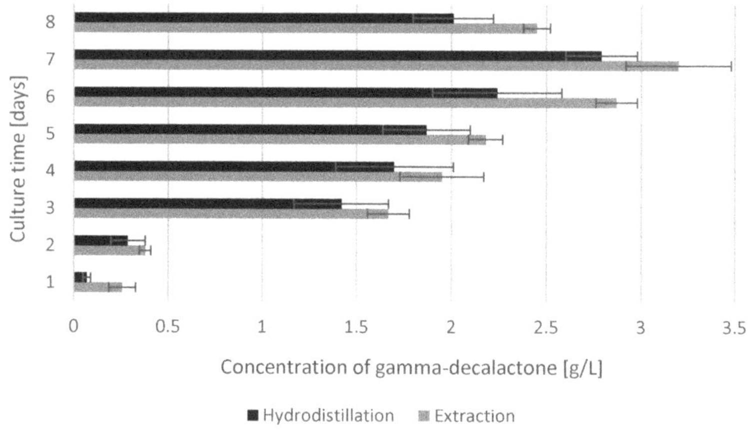

Figure 1. Comparison of gamma-decalactone concentration in biotransformation medium by extraction and hydrodistillation methods.

While conducting research on GDL hydrodistillation, the influence of the pH of the reaction mixture on the level of GDL concentration in the distillate was analyzed. On the seventh day of cultivation, when the lactone concentration was at its maximum (Figure 1) or close to it, the medium was acidified or alkalized to the pH at the levels of 1, 3, 5, or 7, and then the GDL was separated with the usage of hydrodistillation and determined chromatographically. The results of GDL concentration in the media before and after pH modifications as well as in the distillate and distillation residue are summarized in Table 3.

Table 3. Influence of acidity on gamma-decalactone levels in biotransformation medium and distillate.

	Concentration of Gamma-Decalactone and RSD Values (g/L)				
GDL concentration immediately after biotransformation reactions	2.97 (0.21)	2.76 (0.09)	2.38 (0.10)	2.76 (0.04)	2.75 (0.31)
GDL concentration after acidification/alkalization of the substrate	pH = 1	pH = 3	pH = 5	pH = 7	pH = 5.68 *
	3.25 (0.15)	3.68 (0.26)	2.52 (0.33)	1.73 (0.22)	-
GDL concentration in distillate	2.42 (0.35)	2.67 (0.01)	2.07 (0.04)	1.39 (0.05)	1.93 (0.07)
GDL concentration in distillation residue	0.73 (0.08)	0.86 (0.34)	0.63 (0.37)	0.41 (0.07)	0.92 (0.10)

* Natural pH in biotransformation medium (without modification).

The presented data clearly show that the pH of the medium has a significant influence on the concentration of GDL in the biotransformation medium, and thus, finally, in the distillate. With the decrease in the pH (acidification of the medium), the concentration of the lactone in the post-reaction mixture increased. The highest content of the lactone was recorded at the pH of 3 or 1. Acidification contributed to an increase in the concentration of the GDL in the extracted samples by approximately 33.3% and 8.8% (from 2.76 ± 0.09 to 3.68 ± 0.26 for pH 3, and from 2.97 ± 0.21 to 3.25 ± 0.15 for pH = 1). This is due to the spontaneous cyclization of 4-hydroxydecanoic acid into the neutral lactone form. However, this reaction is not total and some of the acid is not transformed into GDL.

Alkalization of the medium to pH 7, in relation to the natural acidity at the pH level of approximately 5.7, gave the opposite result. The concentration of the lactone in the medium, and thus, in the distillate, decreased by about 36% on average. With an almost identical concentration of the fragrance compound in the microbiological media after 7 days of the reaction, the differences in the concentration of the lactone between the media with pH 3 and 7 were close to 2 g/L. Therefore, the acidification of the post-reaction mixture makes sense in the context of the higher GDL concentration in the distillate. The efficiency of the distillation process, regardless of the acidity of the solutions, was at the level of 70–80%. After distillation, part of the GDL remained in the flask, which is a result of good solubility of this compound in the hydrophobic fraction (unreacted castor oil fraction).

3.3. Adsorption of Gamma-Decalactone on Porous Materials

The possibility of using adsorption was also analyzed in the research on GDL secretion. For this purpose, one of the selected adsorbents, Amberlite XAD-4, zeolite, or vermiculite, at the concentration of 30 g/L, was introduced into the biotransformation media after the finished reaction and separated from the yeast cells.

The GDL adsorption kinetics were analyzed within 6 h of the introduction of the adsorbent by monitoring the decrease in GDL concentration in the solutions (Figure 2). The quantity of the GDL adsorbed corresponded to the difference between the initial concentration (measured immediately after the finished biotransformation reaction) and the final concentration (after the adsorption).

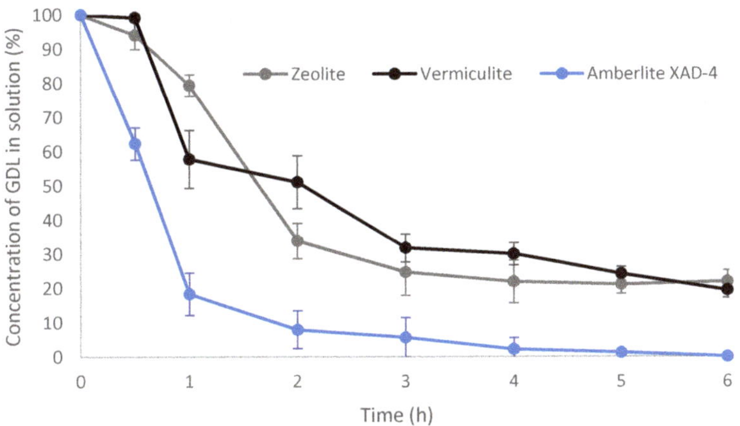

Figure 2. Kinetics of gamma-decalactone adsorption from biotransformation medium on three selected adsorbents: zeolite, vermiculite, and Amberlite XAD-4.

The data presented in Figure 2 show that at a concentration of GDL in the biotransformation media of about 3 g/L, the GDL was most effectively adsorbed on Amberlite XAD-4. After 1 h, the amount of the compound adsorbed was over 80%, which was approximately 2.45 g. In the case of vermiculite, after the same time, approximately 43% of the compound was absorbed, i.e., nearly 2 times less. The level of adsorption was the lowest with zeolite; after 1 h, there was still about 80% of the unadsorbed peach lactone in the medium. However, regardless of the type of the adsorbent used, the equilibrium state was reached after approximately 3 h of contact. At equilibrium, the adsorption capacities of zeolite and vermiculite were similar at about 70–80%, while for Amberlite XAD-4 it was over 95%.

In order to obtain a more complete picture of the adsorption kinetics from biotransformation media, the initial adsorption rate (based on the first hour of the process) was determined for each of the adsorbents used (Table 4). The data confirm that in the first hour of the process, the highest adsorption rate, at the level of 1.47×10^{-3} g GDL/(adsorbent (g) $\times$ minute), is shown by Amberlite XAD-4. This value is twice as high as compared to vermiculite and more than four times as high as compared to zeolite.

Table 4. Initial adsorption rate of gamma-decalactone on the zeolite, vermiculite, and Amberlite XAD-4 after 1 h.

	Zeolite (30 g/L)	Vermiculite (30 g/L)	Amberlite XAD-4 (30 g/L)
Initial concentration of GDL in biotransformation medium (g/L)	3.15 ± 0.23	2.97 ± 0.19	3.23 ± 0.17
Adsorbent quantity of GDL after 1 h (g/L)	0.65	1.29	2.64
Initial adsorption rate *	0.36×10^{-3}	0.72×10^{-3}	1.47×10^{-3}

* The initial adsorption rate is expressed in grams of GDL per gram of adsorbent per minute.

As the Amberlite XAD-4 resin showed the highest GDL adsorption efficiency, an adsorption isotherm was determined for it, through which the aroma compound separation was fully characterized (Figure 3). The equilibrium concentration of GDL was measured in the range of 0.1—5 g GDL/L (concentration range so far achieved in a microbial culture, depending on the initial concentration of castor oil).

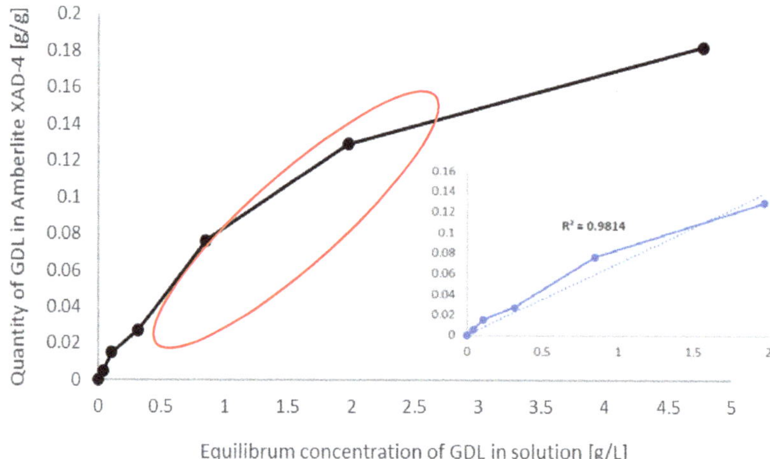

Figure 3. Adsorption isotherm of gamma-decalactone on the Amberlite XAD-4.

The curve shown in Figure 3 is a type 1 isotherm with a visible and achievable level of saturation, characteristic for sorbents in which sorption is limited by the steric effect to one or two molecular layers. Lower lactone concentrations in the medium are up to 4 g/L on average (preadsorption concentration). The adsorption isotherm has a linear relationship $R^2 = 0.9814$ (points marked with a red line, highlighted in the additional image).

Considering the possibility of using adsorption in the separation of GDL from biotransformation media, the effectiveness of removing the adsorbed lactone was also analyzed. The desorption was carried out in 5 cycles with ethyl alcohol, using the ratio of 3 mL of ethanol/g of adsorbent. Each of the adsorbents showed good desorption abilities. Figure 4 shows the amount of the desorbed GDL in 5 extraction cycles, taking into account the efficiency of each stage. The most effective desorption was with the Amberlite XAD-4 resin; about 81% of the lactone was desorbed. In the case of zeolite and vermiculite, the level of desorption was comparable and amounted to 75–76%. It is noteworthy that the ricinoleic acid was extracted from the adsorbents together with the desorbed lactone. Its concentration in the extracts after desorption ranged from 2.03 ± 0.44 g/L (in the first desorption cycle) to about 0.22 ± 0.12 g/L (in the fourth or fifth desorption cycles). With each subsequent cycle, the concentration of the desorbed lactone and ricinoleic acid decreased.

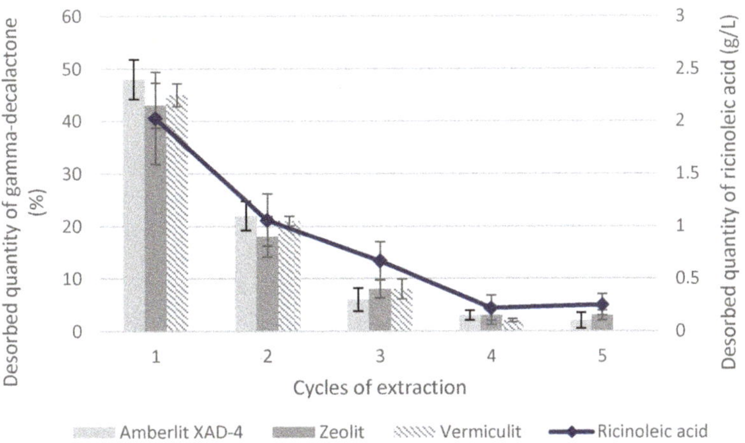

Figure 4. Desorption of the gamma-decalactone from adsorbents (zeolite, vermiculite, and Amberlite XAD-4) using five cycles of extraction with ethyl alcohol.

4. Discussion

Despite the large amount of literature in the field of the optimization of GDL biosynthesis, there are no data on the separation and purification of this compound from the post-reaction mixtures. The published results are based only on the chromatographic determinations of the samples taken directly from individual culture stages. Considering the fact that the efficiency of the extraction, purification, and isolation of this compound is a key stage which generates significant costs in relation to the entire biotechnological cycle and is decisive for the application of the methods of biotechnological synthesis of GDL in industry, research was undertaken on how to effectively isolate this compound from reaction mixtures. Our focus was on extraction, hydrodistillation, and lactone adsorption.

Liquid–liquid extraction is the most mature and commonly used method of separating fragrances compounds. Solvents commonly used in the extraction are dichloromethane, chloroform, ethyl acetate, diethyl ether, and *n*-hexane [36,37]. They were also used (except for ethyl acetate) in GDL extraction. The best efficiency obtained in this study with the use of diethyl ether (99.9%) justifies its widespread use in the extraction of lactones from microbiological media [19,24–26]. The similar efficiency and reproducibility of GDL extraction for dichloromethane and chloroform allow the conclusion that due to difficulties in extraction on a larger scale with diethyl ether (due to its high volatility and tendency to form explosive peroxides), it is possible to replace it with these solvents. The effectiveness of dichloromethane and chloroform is also confirmed in the literature. In the research on quantitative determination of *N*-acylhomoserine lactones, conducted by the team of Morin et al. [38], the best extraction results with almost the same yields were obtained for the above-mentioned solvents. Canbay [37], conducting liquid–liquid extraction of the volatile compounds from rose aromatic water, obtained complete extraction for chloroform, dichloromethane, and ethyl acetate. *n*-Hexane showed the worst extraction parameters, similar to our GDL extraction study. The effectiveness of dichloromethane in the extraction of terpenoids from aqueous solutions was also confirmed by the team of Agarwal et al. [39].

The role of the volume of the solvent used in the extraction is highlighted by Obi et al. [40], who conducted research on determination of lactones in milk fat. The authors conducted their study with the use of 0.5 to 1.5 mL of the solvent per 0.5 g of the extracted sample. The results showed that the use of a large amount of an organic solvent for the extraction of lactone clearly improves the recovery rate. This conclusion is consistent with the data obtained in this study, where solvent-to-sample ratios of 1:2 or 1:1 (*v:v*) were used for GDL extraction. The difference in the extraction efficiency for the solvents used in these two volume variants was from 9 to 17%. Obi et al. [40] showed that with a small amount of an organic solvent relative to the sample, long-chain lactones (above C_6) are captured by triacylglycerols that are present in the samples due to their hydrophobic nature. When a larger volume of solvent is used for extraction, the lactones are not affected by the triacylglycerol matrix effect. Since the biotechnological synthesis of GDL uses an oil-in-water emulsion base (addition of 10 g/L castor oil as a precursor to GDL), it is highly probable that the lactone partially dissolves in the triglycerides of the unreacted oil. This may also explain the poorer reproducibility of the extraction results for the samples extracted with a 1:2 solvent-to-sample ratio (higher standard deviations).

Although the efficiency of the performed GDL extraction is high, it is necessary to point out a disadvantage of using organic solvents in the media with the addition of lipids. In the target biotransformation media, the precursor of GDL synthesized by microorganisms is castor oil or ricinoleic acid. These are introduced into the media in excess, as they are also sources of carbon for the growth and maintenance of the vital functions of microorganisms. Not fully hydrolyzed castor oil and the ricinoleic acid not used in the β-oxidation cycle are extracted together with GDL, contaminating it significantly (unpublished data). Hence, the extraction technique itself is not sufficient and should be supported, e.g., with chromatography or fractional distillation. However, this technique is not always effective due to the small differences in the boiling points of individual compounds.

Considering the significant contamination of GDL with the lipid fraction, which is extracted regardless of the solvent used, attempts were made in further studies to separate the lactone by hydrodistillation. In the extensive literature on the subject of GDL biosynthesis, only one publication was found in which Alchihab et al. [7] used the hydrodistillation method to isolate GDL. Hydrodistillation is a traditional technique to recover essential oils and bioactive compounds from plants [41]. Despite its disadvantages, including long process time and low efficiency as well as partial degradation of unsaturated compounds due to thermal effects, it is widely used [41–43]. The use of hydrodistillation in this research to separate GDL from biotransformation media allowed for its recovery, and the efficiency of this technique was approximately 12–22% lower than in liquid–liquid extraction. It is assumed that the lower concentration of lactone in distillation may be partly due to its loss as a result of long-term heating of samples and possible hydrolysis of this compound [44]. Based on the data from Figure 1, it can be concluded that at the concentration of GDL in the biotransformation medium below 0.5 g/L, distillation exhibits low effectiveness. However, it should be emphasized that the purity of the distillates was higher than that of the extracts. In the distillates, apart from GDL, 3-hydroxy-γ-decalactone, dec-2-en-olide, and dec-3-en-olide were identified. These compounds had been previously detected during the fermentation process that was conducted by the yeast *Yarrowia lipolytica* W29 on solid media, as mentioned by Try et al. [25].

Since the microbial synthesis of GDL is based on the transformation in the β-oxidation cycle of ricinoleic acid to 4-hydroxydecanoic acid, which subsequently cyclizes to GDL, it was justified to analyze the influence of the reaction environment on the level of acid cyclization, and hence, the final concentration of GDL in the distillate. Acidity turned out to be an important factor that partially determined the concentration of GDL in the distillate. Increasing the acidity of the reaction mixture from the natural pH of about 5.68 to the pH of 3 or 1 resulted in an increase in the concentration of the lactone in the biotechnological medium, ranging from 0.25 g to 0.75 g/L (percentage range 8–33%; Table 3). The results of Feron et al. [45] confirm this research. The team carried out the bioconversion of methyl ricinoleate to gamma-decalactone using the yeast *Sporidiobolus ruinenii* and observed that acidification of the reaction medium before the extraction process increases the concentration of lactone in the aqueous phase. By changing the acidity of the post-reaction mixture from pH 6 to pH 2, an over 2.5-fold increase in the GDL concentration was observed. Romero-Guido et al. [46] also emphasizes that lactonization of 4-hydroxydecanoic acid occurred particularly easily in an acidic environment.

In the research on effective separation of GDL from biotransformation media, adsorption on porous adsorbents (Amberlite XAD-4 resin, zeolite, and vermiculite) was also tested. Porous materials are widely and willingly used due to their properties, mainly their large specific surface, high porosity, absorptive capacity, low chemical activity, high mechanical strength, and possibility of regeneration [47]. The use of porous sorbents has often been successful in increasing microorganism production and recovery of volatile compounds from fermentation broth [32–34,48]. However, it should be remembered that adsorption from biotransformation media causes certain difficulties which result from the fact that many components are present in cell culture supernatants. These ingredients come from both the growth medium and the extracellular products that are produced by the cells of microorganisms. Additionally, GDL adsorption takes place in the presence of unreacted lipid precursor (ricinoleic acid) (Figure 4) and lipid substrate (castor oil).

Among the three tested adsorbents, the highest adsorption efficiency was obtained for Amberlite XAD-4. After 6 h of adsorption on this resin, less than 1% GDL in the solution was determined. Probably, the ability of Amberlite XAD-4 resin to adsorb GDL is higher than that of zeolite or vermiculite due to the higher specific surface area ($750\ m^2/g$) compared to other materials ($700\ m^2/g$ and $394\ m^2/g$, respectively). Nongonierma et al. [47] confirms that the adsorbents with the largest surface area seem to be the most effective in the separation of organic compounds. According to Alchihab et al. [34], hydrophobic interactions, chemical structure, and high specific surface area are the main driving forces behind the adsorption of GDL to Amberlite XAD-4 resin in aqueous solutions.

In order to better understand the mechanism of GDL adsorption from biotransformation media, which is influenced by the interactions between the components of the culture medium (especially lipids) and lactone, the sorption isotherm was drawn for the most effective Amberlite XAD-4 adsorbent. The obtained linear dependence over a certain range of GDL concentrations in the biotransformation medium (up to 4 g/L) can be attributed to the low solubility of this lactone in water. This is mentioned in the research by Alchihab et al. [34], who obtained a linear relationship for each of the sorbents by carrying out the adsorption of GDL on the MN-100, MN-102, and MN-202 resins. The peach lactone adsorption method was closely related to the solubility of this compound in water, which is about 0.6 g/L. The same conclusions were made by Souchon et al. [33] while testing activated carbon and three porous polystyrene-type polymers (Porapak Q, Chromosorb 105 and Resin SM4).

Desorption of GDL from individual adsorbents, carried out with ethyl alcohol, presented relatively good efficiency; 75–81% of adsorbed lactone. The chromatographic analysis of the extracts after desorption from Amberlite XAD-4 revealed the content of unreacted ricinoleic acid (Figure 4). Its amount decreased during individual cycles of the process, and the purity of the GDL after desorption was approximately $77 \pm 2\%$. The phenomenon of ricinoleic acid adsorption on the surface of adsorbents has been reported by Guan et al. [49].

5. Conclusions

The development of an effective method of GDL separation from biotransformation media seems to be important in the context of the attempts to apply biotechnological synthesis of peach lactone in industry. Due to the existing gap in the research on GDL separation, an attempt was made to compare the effectiveness of three techniques: liquid–liquid extraction, hydrodistillation, and adsorption, in fragrance recovery. The relatively low concentration of GDL in the fermentation broth (approx. 3 g/L) and the presence of an unreacted lipid substrate (castor oil or ricinoleic acid), in which the lactone is partially dissolved, make its recovery difficult.

Bearing in mind the selectivity of the process, its efficiency, and the speed of execution, among the three techniques used, adsorption on Amberlite XAD-4 is recommended. The six-hour adsorption on this resin resulted in less than 1% of lactone remaining in the biotransformation medium. GDL adsorption on resin is based, inter alia, on the use of Van der

Waals forces, which enable quick adsorption and easy desorption of the fragrance. Moreover, in the context of industrial GDL bioproduction and its separation from the medium, Amberlite XAD-4 has several advantages: it is inexpensive, does not chemically react with the fragrance, is easily regenerated, and can be easily adapted on an industrial scale. In addition, considering the aspect of environmental protection, it must be remembered that the desorption of GDL from Amberlite XAD-4 resin requires only small amounts of ethanol and takes place with an efficiency of approximately 80%.

In the context of the purity of the released GDL, the hydrodistillation process seemed to be more advantageous, as the purity of the distillate was about 10% higher than that of the ethanol extract after lactone desorption from the resin (88 $\pm$ 3.4% and 77 $\pm$ 2%, respectively). However, by choosing a separation method, one should consider the resultant of many factors.

This article only outlines the direction of further research that is necessary to optimize the GDL adsorption process, both in economic terms (studies on higher adsorption efficiency by selecting, e.g., the amount of adsorbent in relation to the adsorbed compound) and ecological terms (minimizing the amounts of reagents used).

Author Contributions: Conceptualization, J.M.; methodology, J.M. and J.B.; formal analysis, J.M., M.W.-W. and E.O.-L.; investigation, J.M. and J.B.; writing—original draft preparation, J.M.; writing—review and editing, J.M. and A.G.; visualization, J.M., E.O.-L. and M.W.-W.; supervision, J.M. and A.G.; funding acquisition, A.G. All authors have read and agreed to the published version of the manuscript.

Funding: The study was financially supported by the Ministry of Education and Science within funds of the Institute of Food Sciences of Warsaw University of Life Sciences (WULS), for scientific research.

Institutional Review Board Statement: Not applicable.

Informed Consent Statement: Not applicable.

Data Availability Statement: Not applicable.

Conflicts of Interest: The authors declare no conflict of interest.

References

1. Longo, A.M.; Sanromán, M.A. Production of Food Aroma Compounds: Microbial and Enzymatic Methodologies. *Food Technol. Biotechnol.* **2006**, *44*, 335–353.
2. Braga, A.; Guerreiro, C.; Belo, I. Generation of flavors and fragrances through biotransformation and de novo synthesis. *Food Bioproc. Tech.* **2018**, *11*, 2217–2228. [CrossRef]
3. Gopinath, M.; Vijayakumar, L.; Dhanasekar, R.; Viruthagiri, T. Microbial biosynthesis of γ-decalactone and its applications—A review. *Glob. J. Biotechnol. Biochem.* **2008**, *3*, 60–68.
4. Blin-Perrin, C.; Molle, D.; Dufossé, L.; Le-Quere, J.L.; Viel, C.; Mauvais, G.; Feron, G. Metabolism of ricinoleic acid into gamma-decalactone: Beta-oxidation and long chain acyl intermediates of ricinoleic acid in the genus *Sporidiobolus* sp. *FEMS Microbiol. Lett.* **2000**, *188*, 69–74. [CrossRef]
5. Dufossé, L.; Feron, G.; Mauvais, G.; Bonnarme, P.; Durand, A.; Spinnler, H.E. Production of γ-decalactone and 4-hydroxy-decanoic acid in the genus *Sporidiobolus*. *J. Ferment. Bioeng.* **1998**, *86*, 169–173. [CrossRef]
6. Waché, Y.; Aguedo, M.; Choquet, A.; Gatfield, I.L.; Nicaud, J.M.; Belin, J.M. Role of β-oxidation enzymes in γ-decalactone production by the yeast *Yarrowia lipolytica*. *Appl. Environ. Microbiol.* **2001**, *67*, 5700–5704. [CrossRef] [PubMed]
7. Alchihab, M.; Destain, J.; Aguedo, M.; Majad, L.; Ghalfi, H.; Wathelet, J.P.; Thonart, P. Production of gamma-decalactone by a psychrophilic and a mesophilic strain of the yeast *Rhodotorula aurantiaca*. *Appl. Biochem. Biotechnol.* **2009**, *158*, 41–50. [CrossRef] [PubMed]
8. Alchihab, M.; Destain, J.; Aguedo, M.; Wathelet, J.P.; Thonart, P. The utilization of gum tragacanth to improve the growth of *Rhodotorula aurantiaca* and the production of gamma-decalactone in large scale. *Appl. Biochem. Biotechnol.* **2010**, *162*, 233–241. [CrossRef]
9. Aguedo, M.; Gomes, N.; Escamilla Garcia, E.; Waché, Y.; Mota, M.; Teixeira, J.A.; Belo, I. Decalactone production by *Yarrowia lipolytica* under increased O_2 transfer rates. *Biotechnol. Lett.* **2005**, *27*, 1617–1621. [CrossRef]
10. Gomes, N.; Aguedo, M.; Teixeira, J.; Belo, I. Oxygen mass transfer in a biphasic medium: Influence on the biotransformation of methyl ricinoleate into γ-decalactone by the yeast *Yarrowia lipolytica*. *Biochem. Eng. J.* **2007**, *35*, 380–386. [CrossRef]
11. Soares, G.P.A.; Souza, K.S.T.; Vilela, L.F.; Schwan, R.F.; Dias, D.R. γ-decalactone production by *Yarrowia lipolytica* and *Lindnera saturnus* in crude glycerol. *Prep. Biochem. Biotechnol.* **2017**, *47*, 633–637. [CrossRef] [PubMed]

12. Pereira de Andrade, D.; Carvalho, B.F.; Schwan, R.F.; Dias, D.R. Production of γ-decalactone by yeast strains under different conditions. *Food Technol. Biotechnol.* **2017**, *55*, 225–230. [CrossRef] [PubMed]
13. Waché, Y.; Aguedo, M.; LeDall, M.T.; Nicaud, J.M.; Belin, J.M. Optimization of *Yarrowia lipolytica's* β-oxidation pathway for γ-decalactone production. *J. Mol. Catal. B Enzym.* **2002**, *19–20*, 347–351. [CrossRef]
14. Okui, S.; Uchiyama, M.; Mizugaki, M. Metabolism of hydroxy fatty acids. II. intermediates of the oxidative breakdown of ricinoleic acid by genus *Candida*. *J Biochem.* **1963**, *54*, 536–540. [CrossRef]
15. Endrizzi, A.; Awadé, A.C.; Belin, J.M. Presumptive involvement of methyl ricinoleate β-oxidation in the production of γ-decalactone by the yeast *Pichia guilliermondii*. *FEMS Microbiol Lett.* **1993**, *114*, 153–159. [CrossRef]
16. Lee, S.L.; Cheng, H.Y.; Chen, W.C.; Chou, C.C. Effect of physical factors on the production of γ-decalactone by immobilized cells of *Sporidiobolus salmonicolor*. *Process Biochem.* **1999**, *34*, 845–850. [CrossRef]
17. Najjar, A.; Robert, S.; Guérin, C.; Violet-Asther, M.; Carrière, F. Quantitative study of lipase secretion, extracellular lipolysis, and lipid storage in the yeast *Yarrowia lipolytica* grown in the presence of olive oil: Analogies with lipolysis in humans. *Appl. Microbiol. Biotechnol.* **2011**, *89*, 1947–1962. [CrossRef]
18. Spagnuolo, M.; Hussain, M.S.; Gambill, L.; Blenner, M. Alternative substrate metabolism in *Yarrowia lipolytica*. *Front. Microbiol.* **2018**, *9*, 1077. [CrossRef]
19. Braga, A.; Belo, I. Production of γ-decalactone by *Yarrowia lipolytica*: Insights into experimental conditions and operating mode optimization. *J. Chem. Technol. Biotechnol.* **2015**, *90*, 559–565. [CrossRef]
20. Małajowicz, J.; Nowak, D.; Fabiszewska, A.; Iuliano, A. Comparison of gamma-decalactone biosynthesis by yeast *Yarrowia lipolytica* MTLY40-2p and W29 in batch-cultures. *Biotechnol. Biotechnol. Equip.* **2020**, *34*, 330–340. [CrossRef]
21. Guo, Y.; Song, H.; Wang, Z.; Ding, Y. Expression of POX2 gene and disruption of POX3 genes in the industrial *Yarrowia lipolytica* on the γ-decalactone production. *Microbiol. Res.* **2012**, *167*, 246–252. [CrossRef] [PubMed]
22. Guo, Y.; Feng, C.; Song, H.; Wang, Z.; Ren, Q.; Wang, R. Effect of POX3 gene disruption using self-cloning CRF1 cassette in *Yarrowia lipolytica* on the γ-decalactone production. *World J. Microbiol. Biotechnol.* **2011**, *27*, 2807–2812. [CrossRef]
23. Gomes, N.; Braga, A.; Teixeira, J.A.; Belo, I. Impact of lipase-mediated hydrolysis of castor oil on γ-decalactone production by *Yarrowia lipolytica*. *J. Am. Oil Chem. Soc.* **2013**, *90*, 1131–1137. [CrossRef]
24. Groguenin, A.; Waché, Y.; Escamilla Garciaa, E.; Aguedo, M.; Husson, F.; LeDall, M.T.; Nicaud, J.M.; Belin, J.M. Genetic engineering of the β-oxidation pathway in the yeast *Yarrowia lipolytica* to increase the production of aroma compounds. *J. Mol. Catal. B Enzym.* **2004**, *28*, 75–79. [CrossRef]
25. Try, S.; De-Coninck, J.; Voilley, A.; Chunhieng, T.; Waché, Y. Solid state fermentation for the production of γ-decalactones by *Yarrowia lipolytica*. *Process Biochem.* **2018**, *64*, 9–15. [CrossRef]
26. Moradi, H.; Asadollahi, M.A.; Nahvi, I. Improved γ-decalactone production from castor oil by fed-batch cultivation of *Yarrowia lipolytica*. *Biocatal. Agric. Biotechnol.* **2013**, *2*, 64–68. [CrossRef]
27. Pérez-Olivero, S.J.; Pérez-Pont, M.L.; Conde, J.E.; Pérez-Trujillo, J.P. Determination of lactones in wines by headspace solid-phase microextraction and gas chromatography coupled with mass spectrometry. *J. Anal. Methods Chem.* **2014**, *863019*. [CrossRef]
28. Zhao, Y.; Mu, X.; Nie, Y.; Xu, Y. A new rapid spectrophotometric quantitative determination method for γ-decalactone and application in high-throughput screening for γ-decalactone producing strains. *Food Sci. Biotechnol.* **2014**, *23*, 1935–1940. [CrossRef]
29. Rong, S.; Guan, X.; Li, Q.; Guan, S.; Cai, B.; Zhang, S. Biotransformation of 12-hydroxystearic acid to gamma-decalactone: Comparison of two separation systems. *J. Microbiol. Methods* **2020**, *178*, 106041. [CrossRef]
30. Medeiros, A.B.P.; Pandey, A.; Vandenberghe, L.P.S.; Pastore, G.M.; Soccol, C.R. Production and recovery of aroma compounds produced by solid-state fermentation using different adsorbents. *Food Technol. Biotechnol.* **2006**, *44*, 47–51.
31. Bluemke, W.; Schrader, J. Integrated bioprocess for enhanced production of natural flavors and fragrances by *Ceratocystis moniliformis*. *Biomol. Eng.* **2001**, *17*, 137–142. [CrossRef]
32. Souchon, I.; Spinnler, H.E.; Dufosse, L.; Voilley, A. Trapping of γ-decalactone by adsorption on hydrophobic sorbents: Application to the bioconversion of methyl ricinoleate by the yeast *Sporidiobolus salmonicolor*. *Biotechnol. Technol.* **1998**, *12*, 109–113. [CrossRef]
33. Souchon, I.; Rojas, J.A.; Voilley, A.; Grevillot, G. Trapping of aromatic compounds by adsorption on hydrophobic sorbents. *Sep. Sci. Technol.* **1996**, *31*, 2473–2491. [CrossRef]
34. Alchihab, M.; Aldric, J.M.; Aguedo, M.; Destain, J.; Wathelet, J.P.; Thonart, P. The use of Macronet resins to recover γ-decalactone produced by *Rhodotorula aurantiaca* from the culture broth. *J. Ind. Microbiol. Biotechnol.* **2010**, *37*, 167–172. [CrossRef]
35. Małajowicz, J.; Kozłowska, M. Factors affecting the yield in formation of fat-derived fragrance compounds by *Yarrowia lipolytica* yeast. *Appl. Sci.* **2021**, *11*, 9843. [CrossRef]
36. Wang, J.; Quan, C.; Wang, X.; Zhao, P.; Fan, S. Extraction, purification and identification of bacterial signal molecules based on *N*-acyl homoserine lactones. *Microb. Biotechnol.* **2011**, *4*, 479–490. [CrossRef]
37. Canbay, H.S. Effectiveness of liquid-liquid extraction, solid phase extraction, and headspace technique for determination of some volatile water-soluble compounds of rose aromatic water. *Int. J. Anal. Chem.* **2017**, *4870671*. [CrossRef]
38. Morin, D.; Grasland, B.; Vallee-Rehel, K.; Dufau, C.; Haras, D. On-line high-performance liquid chromatography-mass spectrometric detection and quantification of *N*-acylhomoserine lactones, quorum sensing signal molecules, in the presence of biological matrices. *J. Chromatogr. A* **2003**, *1002*, 79–92. [CrossRef]
39. Agarwal, S.G.; Gupta, A.; Kapahi, B.K.; Baleshwar, A.; Thappa, R.K.; Suri, O.P. Chemical composition of rose water volatiles. *J. Essent. Oil Res.* **2005**, *17*, 265–267. [CrossRef]

40. Obi, J.; Yoshinaga, K.; Tago, A.; Nagai, T.; Yoshida, A.; Beppu, F.; Gotoh, N. Simple quantification of lactones in milk fat by solvent extraction using gas chromatography–mass spectrometry. *J. Oleo Sci.* **2018**, *67*, 941–948. [CrossRef]
41. Costa, P.; Grosso, C.; Gonçalves, S.; Andrade, P.B.; Valentão, P.; Bernardo-Gil, M.G.; Romano, A. Supercritical fluid extraction and hydrodistillation for the recovery of bioactive compounds from *Lavandula viridis* L'Hér. *Food Chem.* **2012**, *135*, 112–121. [CrossRef]
42. Schaneberg, B.T.; Khan, I.A. Comparison of extraction methods for marker compounds in the essential oil of lemon grass by GC. *J. Agric. Food Chem.* **2002**, *50*, 1345–1349. [CrossRef] [PubMed]
43. Yamini, Y.; Khajeh, M.; Ghasemi, E.; Mirza, M.; Javidnia, K. Comparison of essential oil compositions of *Salvia mirzayanii* obtained by supercritical carbon dioxide extraction and hydrodistillation methods. *Food Chem.* **2008**, *108*, 341–346. [CrossRef]
44. Memarzadeh, S.M.; Ghasemi Pirbalouti, A.; AdibNejad, M. Chemical composition and yield of essential oils from *Bakhtiari savory (Satureja bachtiarica* Bunge. *) under different extraction methods.* Ind. Crops Prod. **2015**, *76*, 809–816. [CrossRef]
45. Feron, G.; Dufosse, L.; Pierard, E.; Bonnarme, P.; Quere, J.L.; Spinnler, H. Production, identification, and toxicity of (gamma)-decalactone and 4-hydroxydecanoic acid from *Sporidiobolus* spp. *Appl. Environ. Microbiol.* **1996**, *62*, 2826–2831. [CrossRef] [PubMed]
46. Romero-Guido, C.; Belo, I.; Ngoc Ta, T.; Cao-Hoang, L.; Alchihab, M.; Gomes, N.; Thonart, P.; Teixeira, J.A.; Destain, J.; Waché, Y. Biochemistry of lactone formation in yeast and fungi and its utilisation for the production of flavour and fragrance compounds. *Appl. Microbiol. Biotechnol.* **2011**, *89*, 535–547. [CrossRef]
47. Nongonierma, A.; Voilley, A.; Cayot, P.; Le Quéré, J.L.; Springett, M. Mechanisms of extraction of aroma compounds from foods, using adsorbents. Effect of various Parameters. *Food Rev. Inter.* **2006**, *22*, 51–94. [CrossRef]
48. Holst, O.; Mattiasson, B. Solid sorbent used in extractive bioconversion. In *Extractive Bioconversion*; Mattiasson, B., Holst, O., Eds.; Marcel Dekker Inc.: New York, NY, USA, 1991; pp. 189–205.
49. Guan, S.; Rong, S.; Wang, M.; Cai, B.; Li, Q.; Zhang, S. Enhanced biotransformation productivity of gamma-decalactone from ricinoleic acid based on the expanded vermiculite delivery system. *J. Microbiol. Biotechnol.* **2019**, *29*, 1071–1077. [CrossRef]

applied
sciences

MDPI

Article

Phytochemical Profile of Eight Categories of Functional Edible Oils: A Metabolomic Approach Based on Chromatography Coupled with Mass Spectrometry

Carmen Socaciu [1,*], Francisc Dulf [2], Sonia Socaci [1], Floricuta Ranga [1], Andrea Bunea [3], Florinela Fetea [1] and Adela Pintea [4]

[1] Department of Food Science, Faculty of Food Science and Technology, University of Agricultural Sciences and Veterinary Medicine Cluj-Napoca, 400372 Cluj-Napoca, Romania; sonia.socaci@usamvcluj.ro (S.S.); floricuta.ranga@usamvcluj.ro (F.R.); florinelafetea@usamvcluj.ro (F.F.)

[2] Faculty of Agriculture, University of Agricultural Sciences and Veterinary Medicine Cluj-Napoca, 400372 Cluj-Napoca, Romania; francisc.dulf@usamvcluj.ro

[3] Faculty of Animal Science and Biotechnology, University of Agricultural Sciences and Veterinary Medicine Cluj-Napoca, 400372 Cluj-Napoca, Romania; andrea.bunea@usamvcluj.ro

[4] Faculty of Veterinary Medicine, University of Agricultural Sciences and Veterinary Medicine Cluj-Napoca, 400372 Cluj-Napoca, Romania; apintea@usamvcluj.ro

* Correspondence: carmen.socaciu@usamvcluj.ro

Featured Application: The data presented here recommend a systematic analytical flow based on a metabolomic approach to be applied for different foods, food bioresources or byproducts valorization, identifying the profile of valuable components, including phytochemicals, as authenticity biomarkers and health-promoting molecules.

Citation: Socaciu, C.; Dulf, F.; Socaci, S.; Ranga, F.; Bunea, A.; Fetea, F.; Pintea, A. Phytochemical Profile of Eight Categories of Functional Edible Oils: A Metabolomic Approach Based on Chromatography Coupled with Mass Spectrometry. *Appl. Sci.* **2022**, *12*, 1933. https://doi.org/10.3390/app12041933

Academic Editor: Agata Górska

Received: 11 January 2022
Accepted: 10 February 2022
Published: 12 February 2022

Publisher's Note: MDPI stays neutral with regard to jurisdictional claims in published maps and institutional affiliations.

Abstract: Functional vegetable oils are highly considered not only for their nutritional value, but also for their health benefits. The profile of phytochemicals responsible for their quality is useful also for the identification of possible mislabeling or adulteration. The comparative composition of eight categories (sunflower, pumpkin, hempseed, linseed, soybean, walnut, sea buckthorn and olive) of commercial vs. authentic oils was determined. Fatty acids, volatiles, carotenoids, tocopherols, and phenolic components were analyzed by gas- and liquid chromatography-based techniques coupled with diode array, mass spectrometry, or fluorescence detection. Classification models, commonly used in metabolomics, e.g., principal component analysis, partial least squares discriminant analysis, hierarchical clusters and heatmaps have been applied to discriminate each category and individual samples. Carotenoids, tocopherols, and phenolics contributed mostly, qualitatively, and quantitatively to the discrimination between the eight categories of oils, as well as between the authentic and the commercial ones. This metabolomic approach can be easily implemented and the heatmaps can be considered as "identity" cards of each oil category and the quality of commercial oils, comparative to the authentic ones of the same botanical and geographical origin.

Keywords: edible functional oils; food identity; phytochemicals' profile; gas- and liquid chromatography; mass spectrometry; chemometrics; metabolomics

1. Introduction

Vegetable oils are essential food products of high nutritional value and a major agri-food commodity subjected to adulteration before commercialization. Consumers are appreciating oils' nutritional potential but are concerned about their quality and safety, while the food producers are careful to their quality, influenced by multiple environmental and technological factors affecting mostly the fatty acids, such as oxidation and higher levels of trans derivatives [1–3]. Over the years, the Codex standards for fats and oils have been modified to enhance the supervision of vegetable oils quality and authenticity, and available databases and international standards are evolving (https://www.fao.org/3/t4660t/t4660t0e.htm accessed 15 January 2021). The EU legislation by Regulation (EEC) 2568/91 established methods for olive oil analysis, but not for other types of vegetable oils. Meanwhile, in the context of the fight against food fraud, the publication of EU regulation 2017/625 focused on legal regulations related to edible oils. Authenticity is a very important quality criterion, considering the significant differences of composition, safety, and price between the cold-pressed (extra virgin, virgin) and refined oils, as well the mislabeling of commercial products. Moreover, along with olive oils, other edible oils are considered functional or specialty (e.g., pumpkin, walnut, soybean, hemp, linseed, sea buckthorn), since, beside the nutritional function, they contain bioactive compounds with health benefits and reduced risk of some diseases, mainly cardiovascular and neurovegetative, due to their high amounts of essential polyunsaturated lipids rich in linoleic acid and linolenic fatty acids, phospholipids, phytosterols, and bioactive phenolics [4].

Edible vegetable oils contain mainly triacylglycerols (up to 98%) and different phytochemicals (micronutrients) as tocopherols, phytosterols and carotenoids, essential components for our diet, having high nutritional value. The most consumed edible oils in Central and Eastern Europe are sunflower, soybean, rapeseed oils, but also increasingly olive, pumpkin, grapeseed and palm oils [5,6]. Most publications related to the identity (quality and authenticity) of oils are dominated by studies on different olive oil varieties and commercial quality. The health benefits of olive oil are well documented, being mainly attributed to their high content in monounsaturated fatty acids (MUFAs), e.g., oleic and palmitoleic acids, but also to sterols, carotenoids, tocopherols and polyphenolic compounds (tyrosol derivatives and oleuropein), all being known for their antioxidant activity [7–13]. These natural antioxidants are found in many other functional oils, but fewer data were available regarding their composition and identity, dependent on their botanical phenotype and geographical distribution. Metabolomics is an advanced technology which offers a comprehensive characterization of small, bioactive molecules (phytochemicals) which play a central role in oils' identity. Analytical platforms such as gas chromatography and liquid chromatography coupled with diode array, fluorescence or mass spectrometry-detection techniques (GC–MS and LC–DAD-RF-MS) and nuclear magnetic resonance (NMR) spectroscopy are essential prerequisites used to ensure a systematic study about the quality and authenticity biomarkers of vegetable oils. Mainly, chromatography is applied for the separation and mass spectrometry for identification. The GC-MS is a standardized method to identify and quantify fatty acids, volatile components, and sterols, while LC-DAD-RF-MS or fluorescence (HPLC-RF) are useful for the separation, identification and quantification of carotenoids, tocopherols and polyphenols ([14–22]. Meanwhile, infrared spectrometry also brings added value information, if subsequent multivariate classification methods are applied, e.g., by principal component analysis (PCA), partial least squares-discriminant analysis (PLS-DA) or cluster analysis (CA) [23–25]. Actually, any analysis needs an adequate chemometric evaluation by multivariate analysis, for improving the classification performance, particularly for challenging classification tasks such as the discrimination of edible oils of different biological or geographical origins, and genuine vs. adulterated commercial products [26–28]. The data fusion based on these complementary analytical techniques brings new knowledge and allows the identification of specific oil biomarkers (e.g., fatty acids, sterols and tocopherols) to ensure correct classification. Metabolomics concepts are very useful for real progress in the multivariate classifications, by different sophisticated statistical or machine learning methods [8,10,12,29]. The objective of this study was to compare the phytochemicals' profile of 30 edible oils from eight categories, based on their botanical origins (sunflower, pumpkin, hempseed, linseed, soybean, sea buckthorn and olive oil), to discriminate their specific identity biomarkers. There were four advanced techniques applied—GC-MS, HPLC-DAD, HPLC-RF and LC-ESI+MS—to separate and identify five different components (fatty acids, volatiles, tocopherols, carotenoids, and phenolic derivatives) from each vegetable oil (genuine vs. commercial) and to provide the value of qualitative and quantitative biomarkers for their recognition and authenticity.

The chemometric analysis included a variety of classification models, including principal component analysis (PCA), partial least square discriminant analysis and hierarchical clustering by heatmaps. The differences among oils have been described and the specific recognition biomarkers were identified, creating an "identity card" for each category of oil.

2. Materials and Methods

2.1. Oil Samples

Eight oil categories, including 30 edible oils produced and commercialized in Romania (excepting olive oils) were investigated. From each category, one sample was authentic, process-controlled, and carefully supervised, as provided from trusted manufacturers. The commercial samples were purchased from retailers being labelled as "cold-pressed" "extra virgin", or "virgin" for olive oils. Table 1 contains data about their botanical origin, the declared/labelled quality grade, and sample codes which were used in the classification studies. The last column shows the codes of authentic samples.

Table 1. The botanical origin, the labelled quality, and codes used in all investigations and statistical classifications of commercial and authentic oils.

Botanical Origin/Group nr.	Commercial Samples: Labelled Quality/Sample Code	Authentic Samples Code
Sunflower (I)	Cold-pressed/SFO1, SFO2, SFO3	aSFO4
Pumpkin (II)	Cold-pressed/PO1, PO2, PO3	aPO4
Linseed (III)	Cold-pressed/LSO1, LSO2	aLSO3
Hempseed (IV)	Cold-pressed/HO1, HO2, HO3	aHO4
Soybean (V)	Cold-pressed/SO2, SO3	aSO1
Walnut (VI)	Cold-pressed/WO1, WO2, WO3	aWO4
Sea buckthorn (VII)	Cold-pressed/SBO2, SBO3, SBO4	aSBO1
Olive (VIII)	Extra virgin Italy (EVOO1), virgin Spain (EVOO2), Extra virgin oil Greece (EVOO3), virgin Greece (EVOO4)	-

Three aliquots of 50 mL from each oil were used for analysis and the measurements were done in triplicate from each aliquot.

2.2. UV-VIS Spectrometry

As a preliminary investigation, a volume of 10 mL from each oil was mixed with 10 mL hexane, vortexed for 3 min, centrifuged and filtered through 0.25 microns membrane. The UV-VIS spectra were recorded using a Jasco V 530 double beam spectrometer, in the region 200–700 nm, the maximum absorption being registered for each sample.

2.3. Fatty Acid Analysis by GC-MS

The GC-MS analysis of fatty acids was done by the standard transesterification procedure followed by separation of ethyl esters (FAMEs) using a PerkinElmer Clarus 600 T GC-MS (PerkinElmer, Inc., Shelton, CT, DOOR) with a column Supelcowax 10 (60 m × 0.25 mm i.d., 0.25 μm thickness (Supelco Inc., Bellefonte, PA, USA). Temperature program: initial temperature 140 °C, then increase to 220 °C by 7 °C/min, the final temperature remained constant for 23 min up to the end. Injector temperature: 210 °C, split 1:24. Injected volume: 0.5 μL. Eluent: He, 0.8 mL/min. MS conditions: ionization energy 70 eV, trap current 100 μA, source temperature 150 °C. Mass scanning range m/z: 22–395, with 0.14 scan/s. The identification of fatty acids was based on retention times, by comparison with retention times of fatty acids from a standards mixture (37 component FAME Mix, SUPELCO# 47885-U) and by comparing the mass spectra with those from the MS database (NIST MS Search 2.0). Individual fatty acid levels were expressed in % of area of the total identified fatty acid peak areas.

2.4. Volatiles' Profile by ITEX/GC-MS

The extraction followed by separation of volatile compounds was performed using the "in-tube extraction" ITEX/GC-MS technique applied on a GC-MS Shimadzu QP-2010 (Shimadzu Scientific Instruments, Kyoto, Japan) equipped with AOC-5000 Combi-PAL autosampler (CTC Analytics, Zwingen, Switzerland) and ZB-5ms capillary column, 30 m × 0.25 mm id × 0.25 μm (film thickness) (Phenomenex, Torrance USA). A volume of 0.1 mL oil was placed in the headspace vial, sealed, and incubated at 60 °C for 20 min with continuous stirring. The volatile compounds accumulated in the headspace of the ampoule were adsorbed, using a syringe, into a Tenax fiber (ITEX-2TRAPTXTA, (G23)-Siliconert 2000, Tenax 80/100 mesh, Switzerland) and then thermally desorbed in the GC injector. After each analysis, the fibre was cleaned by passing a warm stream of Nitrogen. The temperature program for the chromatographic column for the separation of volatile compounds was as follows: maintenance 2 min at 50 °C, followed by an increase at 160 °C by 4 °C/min and then an increase at 255 °C by 15 °C/min and maintenance at this temperature for 10 min. The carrier gas was helium, at a constant flow rate of 1 mL/min. The injector temperature, ion source and interface have been set to 250 °C. The MS detector used the electronic impact ionization (EI) mode in a scanning range of 40–350 m/z. The split ratio was 1:20. Separated volatiles were identified by comparing the mass spectra with those from NIST27 and NIST147 mass spectrum libraries and verified by comparison with retention indices extracted from www.pherobase.com, accession 12 December 2021) and www.flavornet.org, accessed 10 December 2021) (for single-phase columns stationary like ZB-5ms). The results are expressed as a percentage of the total peak area.

2.5. Carotenoid Profile and Quantification by HPLC-DAD

Aliquots of 5 g from each oil were transferred to brown round bottom flasks and dissolved in 20 mL diethyl ether. Saponification was performed by stirring with 25 mL solution of KOH (30%) for 5 h, in the dark, at room temperature, under inert gas. The reaction mixture was transferred into a separation funnel containing diethyl ether and water. The organic phase was washed repeatedly with NaCl 5% solution, until complete removal of alkali (neutral pH of the water phase). The organic phase was filtered and dried over sodium sulphate and the solvent was removed completely using a rotatory evaporator, at 35 °C under reduced pressure. The residues containing carotenoids were dissolved in ethyl acetate and filtered through 0.2 μm PTFE filters prior to injection.

HPLC separation of carotenoids was performed using a Shimadzu LC20 AT HPLC system (Shimadzu Corporation, Kyoto, Japan) with a SPDM20A diode array detector and a YMC C30 reversed phase column (250 × 4.6 mm i.d., 5 μm particle size). The mobile phase consisted of solvent A, methanol/methyl tert-butyl ether/water (83:15:2, $v/v/v$), and solvent B, methyl tert-butyl ether/methanol/water (80:7:2, $v/v/v$). A gradient system was used, as follows: 0 min 0% solvent B, 20 min 0% B, 130 min 82% B, 132 min 0% B, followed by equilibration of column for 10 min. The flow rate was 0.8 mL/min and the chromatograms were recorded at 450 nm. The DAD operated in the range of 300–600 nm for the acquisition of UV-VIS spectra. Carotenoid standards β-carotene, lutein and zeaxanthin (purity ≥ 98%, ≥95% and ≥98%, respectively) were purchased from Extrasynthese (Lyon, France). The identification of individual carotenoids was based on the comparison of their retention time, elution order on C30 column and UV-VIS spectra (λmax, spectral fine structure (%III/II)) with those of the available standards and with literature data. The quantification was performed using external ten-points calibration curves constructed in the range 1–100 μg/mL for the three major carotenoids. The correlation coefficients were R^2 = 0.9912 (β-carotene), R^2 = 0.9991 (lutein) and R^2 = 0.9996 (zeaxanthin) [30].

2.6. HPLC-RF Separation and Quantification of Tocopherols

The oil samples (0.5 g) were dissolved in different volumes of hexane (15–25 mL), so that the tocopherol concentrations fell within the ranges of the calibration curves, filtered through 0.2 μm Nylon filters and injected directly into the HPLC system (Shimadzu

LC20 equipped with a Lichrosorb Si60 column (250 × 4.6 mm, 5 µm) and a fluorescence detector RF20A (excitation λ = 290 nm and emission λ = 330 nm). The mobile phase was hexane: 2-propanol (99.5: 0.5, v/v), isocratically with a flow rate of 1 mL/min. Quantitative analysis was performed by external calibration with authentic standards (α-tocopherol, β-tocopherol, γ-tocopherol, and δ-tocopherol) (Extrasynthese, Genay, France) in the concentration range of 0.25–50 µg/mL, having correlation coefficients R^2 of 0.9980.

2.7. Extraction and Quantification of Total Phenolics

To a volume of 3 mL oil, there was added first 3 mL hexane, and after vortex mixing 1 min, 5 mL of methanol: water (3:2) mixture was added and sonicated for 15 min. Then, samples were centrifuged for 10 min at 3000 rpm. The upper phase was removed, and to the remaining methanol: water phase, 3 mL hexane was added. Finally, three washes with hexane were performed in order to discard the nonpolar lipids. The clean methanolic phase was evaporated on a rotary evaporator until dryness and taken up in 1 mL methanol 95% for total polyphenols (TPs) determination using the Folin–Ciocalteu (FC) method [31]. Shortly, to 0.05 mL methanol extract, 2.35 mL water, 0.150 mL FC reagent and 0.450 mL natrium bicarbonate was added. In parallel for blank, the sample was replaced by methanol 95%. After two hours in the dark, the samples' absorption at 750 nm was determined using a BioTek multidetection spectrometer (BioTek Instruments Inc., US). The calibration was done using gallic acid solutions from 0.0625 to 1 mg/mL, the curve obtained was y = 0.9443x + 0.0608. The R^2 value was 0.9945. The concentration of total phenolics (TP-FC) was expressed in gallic acid equivalents (GAE), namely in mg GAE/100 mL oil.

2.8. LC-ESI+-MS Identification and HPLC-DAD Quantification of Individual Phenolics

Volumes of 0.3 mL from each methanolic extract, prepared as described above, were injected in an HPLC Agilent 1200 system equipped with a photodiode array detector (DAD) and coupled with a mass spectrometer (MS) single quadrupole Agilent model 6110 (Agilent Technologies, Santa Clara, CA, USA) was used for phenolics qualitative and quantitative analysis. The phenolics' separation was done on an Eclipse XDB C18 column 5 µm (4.6 × 150 mm), (Agilent Technologies, Santa Clara, CA, USA) and a two mobile-phase gradient was applied for 30 min, 25 °C, at a flow of 0.5 mL/min. Solvent A was 0.1% Ac acetic in water:ACN (99:1) and solvent B was 0.1% Ac acetic in acetonitrile. The gradient: 0 min, 5% B, 0–2 min, 5% B, 2–18 min, 5%–40% B, 18–20 min, 40%–90% B, 20–24 min, 90% B, 24–25 min, 90%–5% B, 25–30 min, 5% B. For the accurate identification of each molecule, the MS used the ESI+ ionization, with a capillary voltage of 300 V, at 300 °C, nitrogen flow 7 L/min, for a m/z range from 100 to 1200 Da. The data acquisition and processing were done using the ChemStation Agilent software. For quantitative determinations, different calibration curves were done (Figure S1) using the LC-DAD system and different pure standards: gallic acid and vanillin (for hydroxybenzoic acid derivatives, recorded at 280 nm), chlorogenic and ferulic acids (for hydroxycinnamic derivatives, recorded at 280 nm), oleuropein (for tyrosols, recorded at 280 nm), luteolin (for flavones, recorded at 340 nm), quercetin (for flavonoids, recorded at 340 nm) and pinoresinol (for lignans, recorded at 280 nm). The total polyphenols (TP-LC) were calculated afterwards in each case from the sum of each phenolic concentration, considering the calibrations and each peak area in the LC-DAD.

2.9. Multivariate Analysis

All data were statistically processed using the MetaboAnalyst 5.0 online software (29). The multivariate analysis included principal component analysis (PCA), partial least square discriminant analysis (PLSDA), heatmaps and random forest analysis. Significant differences ($p \leq 0.05$) between mean values were evaluated by one-way ANOVA. The one-way ANOVA statistics also included the post-hoc Tukey HSD test (made by MetaboAnalyst 5.0 algorithm) for a good statistical evaluation of differences between oil samples for each type of phytochemical (see Supplementary Files).

3. Results

3.1. UV-VIS Spectral Fingerprinting Related to Botanical Origin

The individual UV-VIS spectra of all oils were recorded and analyzed as a first criteria of oil recognition based on the spectral fingerprint and absorbance intensity (data not shown). The sunflower (SFO) and walnut (WO) oils showed absorbances mainly in the UV region (220–280, 325 nm and 220–280 nm, respectively). The pumpkin oil (PO) group was recognized by VIS intense peaks at 450, 526, 568, 590 and 620 nm, indicating their carotenoid, acylated anthocyanin and chlorophyll content. The spectra of linseed (LSO) and soybean (SO) oils showed higher absorbance at 448 and 443 nm, respectively, as indicators of quinones content. The hempseed (HO) and olive (EVOO) oil groups were characterized by strong absorbance at 668 nm (indicator of chlorophylls). The sea buckthorn (SBO) oils had a unique fingerprint with highest absorbance at 445–470 nm, indicating a rich carotenoid content. These spectra offered a first preliminary information about the specific fingerprints of each oil category and a rough indicator of authenticity [25].

3.2. GC-MS Profile, Univariate Analysis and Multivariate Classification of Oils Based on Fatty Acids Composition

Based on the GC-MS analysis (typical chromatograms of each oil category are presented in Figure S2), the percentages of individual fatty acids (FAs) in all oils were determined, then the average values and standard deviations were calculated for commercial oils and compared to the authentic ones. Table S1 includes the individual fatty acid levels expressed in % of area of the total identified fatty acid areas in the GC-MS chromatograms of oils under investigation. These raw data were processed by one-way ANOVA including the Tukey HSD post-hoc test (Table S2) to evaluate the significance of differences between the oil groups, considering each of the 10 fatty acids separated and identified by GC-MS. The graphic representations of the Tukey HSD statistical differentiations (made by the MetaboAnalyst 5.0 algorithm) are also included in Table S2. The specific unique composition of SBO is clearly observed, significantly different from other oils due to fatty acids C16:1n-7, C18:1n-7, C16:0, and C18:2n-6. The linseed oil (LSO) was significantly different due to the presence of C18:3n-3, the hempseed oil (HO) was differentiated due to its higher content in C18:3n-6 and C20:0, while sunflower oils (SFO) had a significant, higher content of C22:0. Olive oils were differentiated by their higher content of C18:1n-9. Stearic acid (C18:0) was present in all studied oils at different levels and cannot be considered specific, although soybean showed the highest content.

The multivariate analysis was applied for complementary information and included PCA (data not shown) and PLSDA. Figure 1 included the PLSDA scores plot and the heatmap graph which evidentiates the specific fingerprints of these oil category (Figure 1).

The PLSDA score plot (Figure 1a) shows a co-variance of 45.6% and indicates a good discrimination between groups, and especially between SBO vs. SFO and EVOO. Based on PLSDA loadings, the FAs with higher VIP scores were C20:0 and C18:3n-6 (γ-linolenic acid) as markers for HO, C18:1n-9 for EVOO, C18:2n-6 for WO, C22:0 for SFO, C18:3n-3 for LSO oils. Figure 1b illustrates the heatmap of specific fatty acids characterizing each individual sample. There were identified increased C18:0 levels for commercial SO3 and PO2 samples and C22:0 for SFO1 and SFO2 samples. These levels can be considered potential indicators of adulteration, but such data need further confirmation.

Table 2 presents comparatively the mean values and standard deviations (x ± SD) registered for the main classes of fatty acids, namely saturated (SFA), monounsaturated (MUFA) and polyunsaturated (PUFA) fatty acids identified in all eight types of oils. Samples SFO, PO, HO, SO and WO showed the highest percentages of C18:2n-6 (58 to 81% of total FAs), while LSO had maximum values of 52–67% of C18:3n-3. In SBO, palmitic acid was a major FA (C16:0 at 35.8–36.2%), while in the EVOO group of oils, oleic acid (C18:1n-9) had values from 76.5% to 82.9%. Differences were noticed also between the percentages of the other mono- or polyunsaturated acids.

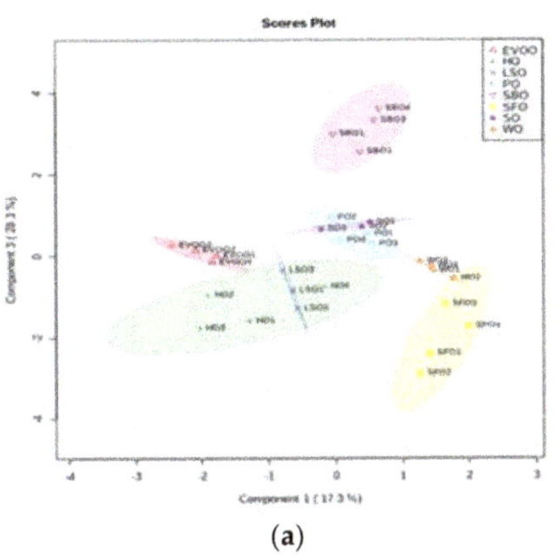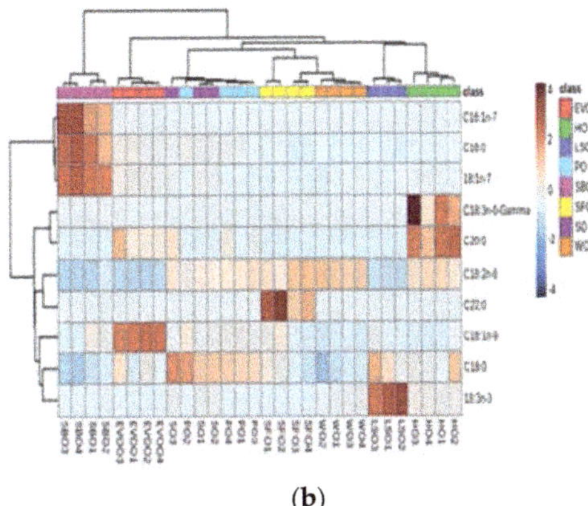

(a)   **(b)**

Figure 1. (a) The PLSDA score plot for the discriminatory analysis of the eight classes of oils. (b) The heatmap illustrating the individual fatty acids to be considered potential biomarkers of each oil group.

Table 2. Mean values and standard deviations (x ± SD) representing the percentages of saturated (SFA), monounsaturated (MUFA), polyunsaturated (PUFA) fatty acids identified in each type of oil. The ratios PUFA/MUFA and MUFA + PUFA/SFA (unsaturation index) are presented in columns 5 and 6, respectively. For oil abbreviations, see Table 1.

Oil Type	SFA (x ± SD)	MUFA (x ± SD)	PUFA (x ± SD)	PUFA/MUFA	MUFA + PUFA/SFA
SFO	5.621 ± 1.855	20.396 ± 11.349	73.982 ± 13.175	3.627	16.791
PO	11.042 ± 2.125	27.659 ± 6.014	61.297 ± 8.052	2.216	13.569
LSO	6.556 ± 1.657	17.510 ± 4.691	75.933 ± 6.247	4.337	14.253
HO	7.614 ± 1.277	13.833 ± 4.175	78.551 ± 5.409	5.678	12.133
SO	11.865 ± 0.754	21.891 ± 2.232	66.243 ± 1.863	3.026	7.428
WO	4.405 ± 0.075	9.949 ± 3.311	85.645 ± 3.360	8.608	21.700
SBO	36.783 ± 7.364	52.171 ± 4.251	11.045 ± 1.559	0.212	1.719
EVOO1,3	13.251 ± 2.183	78.306 ± 2.284	8.443 ± 1.053	0.107	6.546
EVOO2,4	10.166 ± 1.568	85.879 ± 3.068	3.955 ± 0.869	0.046	8.836

The ratios PUFA/MUFA and MUFA + PUFA/SFA (considered as an unsaturation index) of each category can discriminate roughly between highly unsaturated ones (WO) and medium unsaturated oils (SFO, HO, LSO, PO, SO, EVOO). The SBO group of oils was represented mainly by palmitic C16:0 and palmitoleic C16:1 acid and showed the lowest unsaturation index, the PUFA content being mainly represented by C18:2n-6 (5.2–11.4%). The SO and EVOO groups were similar in unsaturation index, but SO had a higher percentage of PUFAs due to C18:2n-6 and C18:1n-9, compared to EVOO. The highest ratio PUFA/MUFA and unsaturation was noticed for WO, represented mainly by C18:2n-6 and C18:1n-9 (Figure 1b). The HO and LSO oils contained distinctly C18:3n-3 at 52.5–67.1% and 9.9–10.1% respectively, the first one having a superior unsaturation. The unsaturation of SFO and PO (represented mainly by C18:2n-6 and C18:1n-9 at percentages higher than 85%) was found to be higher compared to SO and EVOO.

No significant differences were noticed between the authentic samples and the commercial ones, when fatty acids were considered as potential biomarkers of authenticity. These data are in good agreement with recent publications which referred mainly to olive oils and sunflower, hemp, linseed (flaxseed), walnut and sea buckthorn oils [32–35]. For sunflower oils, recent data showed a dominant level of 64% linoleic acid C18:2n-6, while in our study the percentage reached an average of 70%; for flaxseed oils linolenic acid represented 46.5%, while in our case it was around 65%, with similar percentages being noticed by other authors [36,37]. The levels of the fatty acids were useful for the detection of frauds of olive oil (rich in C18:1) with sunflower, soybean and walnut, even at levels of adulteration (below 5%), as demonstrated previously [38]. Sunflower, soybean, linseed and olive oils were found to have different percentages of linolenic/linoleic/oleic or palmitic acid, dominated by palmitoleic acid in sea buckthorn oil, by oleic acid in olive oil, linoleic acid in sunflower and soybean [37], and linolenic in pumpkinseed oil (46.40%), while in our samples the percentage reached 60%. The characterization of walnut oils was recently published [39,40] and showed levels of 62–74% C18:2 and 10% C18:3. In our study, the level of C18:2n-6 reached 80% and 5% C18:3n-3. Fewer data were available for hempseed oils, e.g., 55% C18:2, 16% C18:3 and 11% C18:1 [41]; our data showed mean values of 70% C18:2n-6, 12% C18:1, 10% C18:3n-3, and also 1% C18:3n-6. A special FA composition was noticed for sea buckthorn oils, in our study this oil being characterized by approx. 35% C16:0, followed by approx. 25% C18:1–9, 24% C16:1n-7, 8% C18:2–6, and specifically by 5% C18:1–7, a good marker of recognition. Similar composition was noticed by other authors [42], e.g., 30–33% C16:0, 30–35% C16:1,14–18% C18:1, 5–7% C18:2, and up to 30% C18:3.

3.3. ITEX/GC-MS Profile of Volatiles, Univariate and Multivariate Analysis

Table S3 presents the individual composition of volatile oils' components expressed in % of peak area of the total volatiles, as determined by ITEX/GC-MS (mean values of duplicated separations). For oil abbreviations, see Table 1.

The univariate one-way ANOVA including Tukey's HSD post-hoc test was done (Table S4 includes the graphic representations of significant differences) to evaluate the significance of differences between the oil groups, considering the volatile molecules separated and identified by ITEX/GC-MS. The ethyl ester of hexanoic acid, heptanal, beta-pinene and beta-cis ocimene, 3 methyl butanoic acid and the ethyl ester of octanoic acid evidentiate as specific biomarkers of SBO, significantly different from other oils. Meanwhile, alpha-ocimene and acetophenone can be considered biomarkers of SFO, while olive oils were specifically recognized by their higher levels of hexyl acetate, 2,7 dimethyl 1-octanol and 3-hexenol acetate. Benzoic acid was a common component of many oils, with SFO being the richest source of it.

Figure 2a represents the multivariate data analysis, including the PLSDA score plot with a good discrimination between the different individual oils and oil groups (co-variance of 44.4%), considering all volatile molecules separated and identified. Figure 2b includes the heatmap which illustrates the differences between individual samples and oil categories.

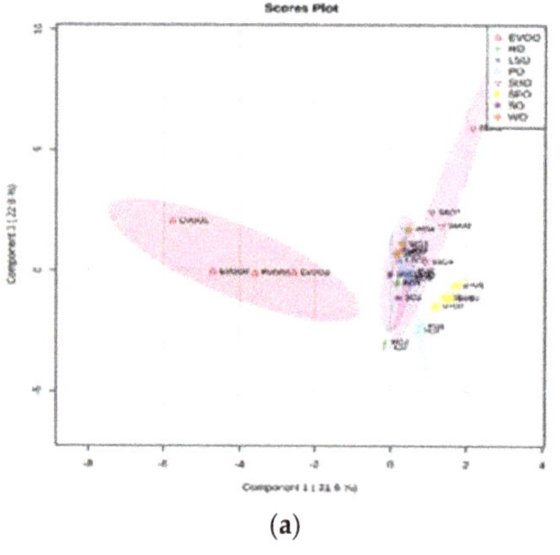

(a)

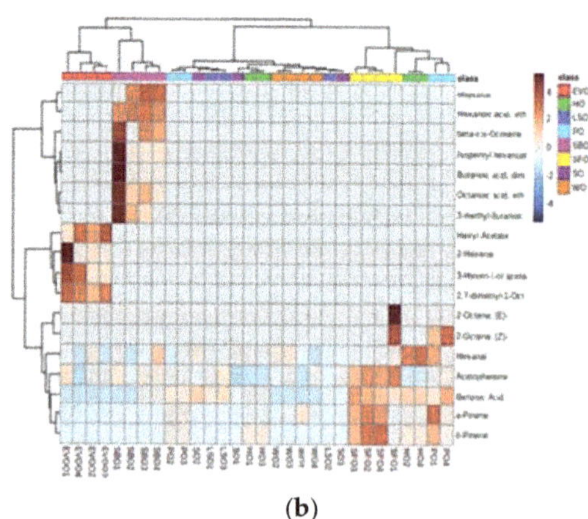

(b)

Figure 2. (**a**) PLSDA score plot for the discriminatory analysis of the eight classes of oils, based on ITEX/GC-MS analysis of volatiles. (**b**) Heatmap illustrating the volatile biomarkers of each individual oil.

The VIP scores calculated from PLSDA analysis classify the most relevant 15 molecules which may explain the discrimination between samples (data not shown). The highest values were recorded for dimethyl octane, hexyl acetate, hexenol and hexenal, all in the EVOO group. Other distinct groups of molecules (hexanoic acid, heptanal, ocimene, octanoic acid and methyl butane, isopentyl hexane) were specifically representing the SBO. We identified specific volatiles in EVOO samples, mainly hexenal, hexenol, hexyl acetate, and 2,7 dimethyl octanol, while for PO, LSO, HO and WO, mainly a-pinene and hexanal were found at different proportions, as shown in Figure 3b. The other molecules (pinenes, hexanal, 2-octene) were less specific but in different levels in the oils' groups. Comparing the authentic samples with the commercial ones, distinct differences were identified for sample EVOO1 in the EVOO group, for SBO1 (the genuine sample) comparing with the other samples in the SBO group, and for SFO1 samples in the SFO group. Such differences are possible indicators of adulteration. Different profiles were noticed for SFO, represented by a- and b-pinene, 2-octene and acetophenone, and for SBO recognized by the presence of butanoic acid derivatives, isopentyl hexanoate and b-cis-ocimene, responsible for the specific flavor of this oil. Few data were reported about volatiles responsible for the sensorial characteristics of these oils, but generally they were similar, mainly for olive oils [43–45], but also for linseed [36] and walnut [39,40].

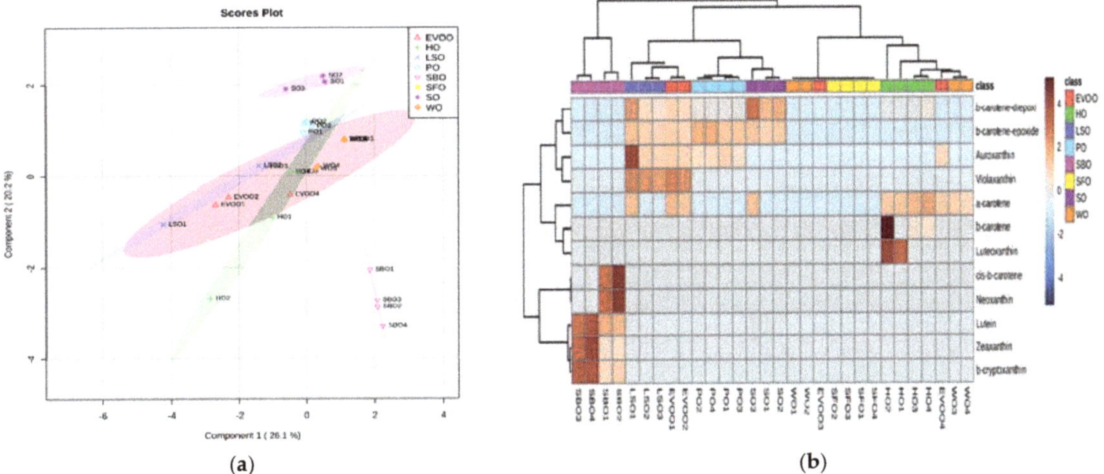

(a)

(b)

Figure 3. (**a**) PLSDA score plot for the discriminatory analysis of the eight classes of oils, considering their carotenoid contents. (**b**) Heatmap representing the Euclidian distance using Ward algorithm to identify the similarities between individual oils.

3.4. HPLC-DAD Separation, Identification and Quantitation of Carotenoids

Table 3 represents the mean concentrations of carotenoids found in commercial SFO, PO, LSO, HO, SO, WO, EVOO and SBO oils, comparative to authentic oils from each category. The retention time and the absorption spectra determined by diode array detection (DAD) were useful for the identification of individual molecules, while the peak intensity values recorded from the UV-VIS spectra were used to calculate the concentration.

Concerning the total carotenoid content, a large range of concentrations were noticed, from mean values of 0.092 mg/100 g in SFO, to 0.15 mg/100 g in WO, 0.452 and 0.257 mg/100 g in LSO and PO, respectively, 0.366–0.395 mg/100 g in EVOO, 0.471–0.906 mg/100 g in SO, and 49.470–129.751 mg/100 g in SBO. SBO was, by far, the richest source of these pigments. The univariate one-way ANOVA including Tukey's HSD post-hoc test (Table S5 includes the graphic representations of significant differences) reflected the significance of differences between the oil groups, considering carotenoids; SBO was, by far, the richest source of β-carotene, lutein, zeaxanthin and β-cryptoxanthin, while PO was the richest source of β-carotene epoxide, and SO a source of high β-carotene diepoxide. Luteoxanthin was found especially in hempseed and olive oils, while violaxanthin and auroxanthin were in LSO.

Table 3. Mean values (mg/100 g) of individual carotenoids separated and identified in commercial and authentic oils, respectively. For oil abbreviations, see Table 1.

Mean Values ± SD	LUT	ZEA	α-CAR	β-CAR	β-CAR-Diepoxide	β-CAR-Epoxide	β-CRP	ARO	NEO	VLX	LUX
SFO1-3	0.038	0.036	0	0.018	0	0	0	0	0	0	0
SD	0.001	0.001	0	0.001							
aSFO4	0.039	0.038	0	0.019	0	0	0	0	0	0	0
PO1-3	0.156	0.042	0	0.027	0.023	0.022	0	0.032	0	0	0
SD	0.053	0.001		0.003	0.002	0.002		0.005			
aPO4	0.134	0.043	0	0.026	0.022	0.027	0	0.02	0	0	0
LSO1-2	0.245	0.055	0.021	0.024	0.055	0.020	0	0.069	0.059	0.038	0
SD	0.239	0.021	0.001	0.005	0.022	0.002	0	0.044	0.030	0.002	0
aLSO3	0.065	0.040	0	0.020	0.033	0.018	0	0.036	0.034	0.035	0
HO1-3	0.196	0.040	0.020	0.041	0.019	0	0	0	0	0	0.098
SD	0.080	0.001	0.002	0.001	0.001						
aHO4	0.168	0.031	0.025	0063	0.028	0	0	0	0	0	0.0077
SO2-3	0.798	0.064	0.010	0.028	0.076	0.021	0.016	0	0	0	0
SD	0.358	0.016	0.014	0.006	0.018	0.001	0.022				
aSO1	0.396	0.043	0.000	0.022	0.057	0.019	0.000	0	0	0	0
WO1-3	0.046	0.045	0.018	0.022	0	0	0.034	0	0	0	0
SD	0.002	0.008	0.001	0.001			0.001				
aWO4	0.044	0.037	0.017	0.022	0	0	0.035	0	0	0	0
SBO2-4	16.000	94.640	0	5.077	0	0	10.218	0	3.816	0	0
SD	5.411	58.011		1.354			4.984		0.987		
aSBO1	10.903	25.422	0	7.037	0	0	3.431	0	2.677		0
EVOO2,4	0.093	0.041	0.011	7.037	0.044	0.019	0.031	0.038	0.037	0.041	0.085
SD	0.082	0.009	0.015	0.020	0	0	0	0	0	0	0
EVOO1,3	0.130	0.042	0.019	0.027	0.040	0.018	0.030	0.036	0.054	0.038	0.076
SD	0.009	0.001	0.001	0.001	0	0	0	0	0.027	0	0.002

ZEA, zeaxanthin; BC, β-carotene; β-CRP, β-cryptoxanthin; ARO, auroxanthin; NEO, neoxanthin; LUX, luteoxanthin; CAR, total carotenoids; TOC, total tocopherols.

Significant differences between the authentic and commercial samples were noticed for LSO, SO, SBO and between extravirgin (EVOO1,3) and virgin (EVOO2,4) olive oil subgroups, in all cases lutein and zeaxanthin being responsible for such differences.

Figure 3a represents the PLSDA score plot for the discriminatory analysis considering their carotenoid contents. The heatmap illustrates the similarities between oil samples inside a group and among groups (Figure 3b).

The univariate and multivariate analysis confirm the specificity of carotenoid composition for each oil category, as mentioned above. The PLSDA score plot shows in this case a co-variance of 46.3% for the first two components, with good discrimination between SBO and SO oils; the other oil groups showing less differentiation. The VIP scores shows the most significant 15 molecules to differentiate oils: SBO due to β-carotene, lutein, zeaxanthin, β-cryptoxanthin, while luteoxanthin, β-carotene and α-carotene were specifically representing HO. Meanwhile violaxanthin and auroxanthin were specific to LSO. Comparing the authentic samples with the commercial ones (Figure 3b), the heatmap shows identified distinct differences for sample EVOO3 in the EVOO group, for HO1 and HO2 comparative to the authentic sample HO4. Different concentrations of carotenoids, highly dependent on the botanical and geographical origin of vegetables, were reported in some recent publications [41,46] and are in agreement with our data.

3.5. HPLC-RF Separation, Identification and Quantitation of Tocopherol Isomers

For tocopherols' quantitation, the HPLC separation with fluorescence (RF) detection allowed the registration of the peak intensities of each type of the four isomers (α, β, γ, δ), comparative to pure standards, as presented in Table 4.

Table 4. Mean values (mg/100 g) of individual tocopherol isomers separated and identified in commercial and authentic oils. For oil abbreviations, see Table 1.

Mean Values ± SD	α-Tocopherol	β-Tocopherol	γ-Tocopherol	δ-Tocopherol
SFO1-3	50.955	2.257	4.887	0.000
SD	0.290	2.285	3.779	0.000
aSFO4	55.340	2.560	1.670	0.000
PO1-3	6.833	6.540	32.203	2.393
SD	4.923	10.375	8.394	1.315
aPO4	11.670	3.450	32.340	2.930
LSO1-2	0.835	6.375	12.270	0.385
SD	0.672	1.025	4.059	0.021
aLSO3	1.970	6.650	13.660	0.980
HO1-3	4.213	2.507	35.877	1.783
SD	1.592	1.020	6.855	0.524
aHO4	4.430	2.890	36.440	1.980
SO2-3	7.725	0.000	57.955	36.390
SD	1.633	0.000	2.496	9.433
aSO1	13.410	0.000	60.740	19.910
WO1-3	3.700	1.287	18.343	4.220
SD	1.854	1.118	4.148	0.944
aWO4	3.220	1.660	17.550	3.980
SBO2-4	73.943	29.207	2.887	0.000
SD	33.263	18.390	1.636	0.000
aSBO1	23.400	42.250	42.690	0.000
EVOO1,3	82.885	3.210	6.935	2.060
SD	105.366	1.089	7.969	1.923
EVOO2,4	154.135	10.500	13.800	0.000
SD	17.487	3.422	3.437	0.000

The total concentration of tocopherols (TOC) ranged from 12.52 mg/100 g (commercial PO1-3) to 178.435 mg/100 g in EVOO2,4. Besides EVOO, the SBO, SO and SFO oils proved to be oils rich in tocopherols. Concomitantly, variations of the four tocopherol isomer levels were noticed in some groups. The SFO oils were characterized by high α-tocopherols, while SO was rich in γ and δ-tocopherols. In the SBO group, the authentic sample had a 1:1 ratio of β- and γ-tocopherols, as major isomers, while the commercial samples showed decreased concentrations.

The univariate one-way ANOVA including Tukey's HSD post-hoc test (Table S6 including the graphic representations of significant differences) reflected the significance of differences between the oil groups, considering the different tocopherol isomers: α-tocopherol was dominant in olive oils EVOO and to a lesser extent in SBO, β-tocopherol had the highest level in SBO, and δ-tocopherol was dominant in SO samples. Meanwhile in all oils β-tocopherol was commonly found in different concentrations. The PLSDA score plot shows in this case a co-variance of 61.1% for the first two components, with good discrimination between EVOO, SBO and SO oils.

Figure 4 shows the PLSDA score plot for the discriminatory analysis considering their tocopherol contents, and also the heatmap illustrating the specificity of individual oils and oil categories.

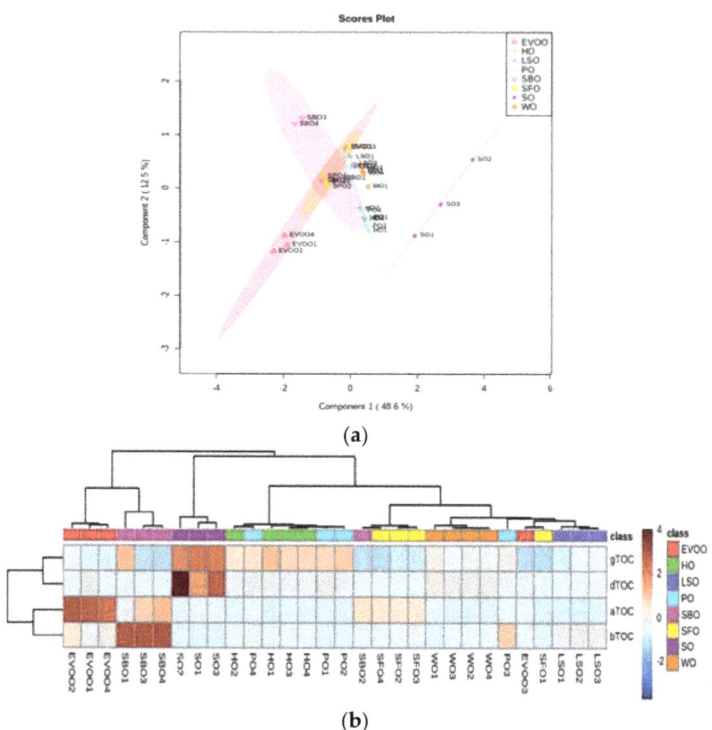

(a)

(b)

Figure 4. (**a**) PLSDA score plot for the discriminatory analysis of the eight classes of oils, considering their tocopherol contents. (**b**) Heatmap representing the Euclidian distance using the Ward algorithm to identify the similarities between individual oils.

3.6. LC-ESI⁺-MS, HPLC-DAD and Spectrometric Evaluation of Individual and Total Phenolic Derivatives

For phenolic derivatives, by the LC-ESI+-MS analysis, the separation and identification of polyphenols was done, based on the retention times, peak areas and the m/z values for parental ions and fragments. In parallel, based on the calibration curves (Figure S1) of pure standards using the HPLC-DAD system, the retention times and peak areas were considered, and the concentration of individual compounds was determined. Concomitantly, the fast spectrophotometric FC method was applied for the quantitative evaluation and compared with the data obtained by HPLC-DAD. Table S7 includes the phenolic composition of individual oils, expressed in mg/100 mL oil according to the calibrations made for each category of phenolics, as presented in Figure S1 (Supplementary File). These data were obtained by HPLC-DAD (TP-LC) and compared with the values obtained by the spectrometric method.

A large variety of phenolic derivatives were separated and identified in all samples, specific to each oil category, e.g., hydroxy benzoic and hydroxycinnamic acids (gallic, protocatechuic, vanilic, elagic, syringic and cafeic, coumaric, synapic, dicaffeoyl quinic, chlorogenic acids, respectively), flavonoids (quercetin, isorhamnetin, luteolin, gallocatechin, daidzein, genistein), as well as lignans (pino- and matairesinols), tyrosols, oleuropein and juglone.

The EVOO, SBO and SO were classified as "high-phenolic" oils compared to SFO, LSO, PO, HO, and WO. The TP values determined by HPLC-DAD were around three times superior to the values obtained by spectrometry (TP-FC), a result which is obvious considering that the FC method is based on the reducing capacity of free hydroxyl groups of phenolics' and is not an accurate determination of all phenolic derivatives. As noticed also by other authors [47], we found high statistical correlation ($p < 0.005$) between TP-HPLC and TP-FC values. Meanwhile, one should be considered that lignans and flavonoids do not show any linear relationship with FC data, suggesting caution about interpretation of FC results for oils having very different phenolic profiles.

Based on LC-ESI+-MS analysis and individual peak areas for each phenolic derivative, the multivariate analysis by PLSDA score plot shows a co-variance of 33.1% for the first two components, with good discrimination between EVOO, SBO and HO oils, the other oil groups showing less differentiation (Figure 5a). Figure 5b represents the heatmap which illustrates the specific composition of each oil category, according to their phenolic composition.

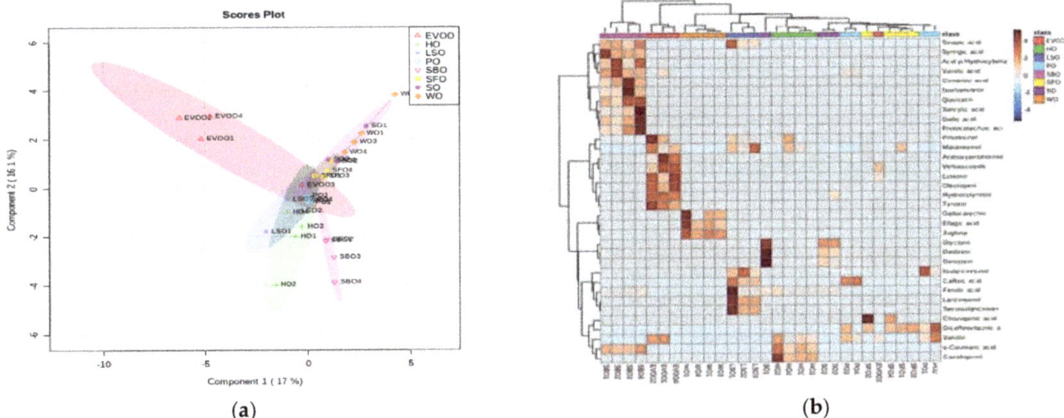

Figure 5. (a) PLSDA score plot for the discriminatory analysis of the eight categories of oils, considering their content in phenolic derivatives. (b) Heatmap representing the Euclidian distance, and by the Ward algorithm, the similarities between individual oil samples inside a group and among groups.

Based on PLSDA analysis, the VIP scores showed the most significant 15 molecules to differentiate these oils. Therefore, SFO was specifically characterized by chlorogenic and dicaffeoylquinic acids, PO by vanillin and caffeoylquinic acid, LSO by matairesinol and pinoresinol, HO by canabigenol, SO by glycitein, daidzein and genistein, WO by juglone, ellagic acid and gallocatechin, EVOO by oleuropein, tyrosol and hydroxytyrosol, luteolin, pino-, matai- and acetoxipinoresinols, and SBO by many phenolic acids from hydroxybenzoic and hydroxycinnamic subclasses. Intuitive images are given by heatmaps of individual oils (Figure 5b), phenolics being good biomarkers to discriminate qualitatively and quantitatively the individual oil samples, as well the different categories of oils.

4. Discussion

The large variability regarding the composition of edible vegetable oils is reflected by many reports and publications, as presented above. Therefore, it is very difficult to compare data released from different European regions, even for the same type of oils, since the botanical variety of the oil raw sources (seeds or fruits), the geographic location and environment conditions, the fractions used to produce the oil (seeds or pulp or whole fruit) and the technological parameters are strongly influencing their final composition.

The data presented here reflected mainly the composition of oils manufactured in Romania, at low scale (the authentic ones) or industrial scale. Their composition did not differ significantly, although the commercial samples had some quantitative differences for the five phytochemical categories.

The data obtained by HPLC-DAD-RF-MS and GC-MS, complemented with statistical multivariate analysis, brings added value for the discrimination, classification, and authentication of these edible oils. These results are complementary and in good agreement with our previous studies using the ATR-FTIR spectrometry coupled with chemometrics [25].

The univariate analysis using one-way ANOVA and Tukey's HSD post-hoc test combined with multivariate analysis showed significant discriminations among oils used in this study, either their botanical authentication, or their individual variations due to processing parameters.

The profiles of fatty acids (Figure 1a) show similarities between SO and PO oils, represented mainly by C18:0. LSO is represented mainly by C18:3n-3, while WO by C18:2n-6 and EVOOs by C18:1n-9. Specifically, SFO was identified by C22:0 and C18:2n-6, while HO by C18:3n-6 (γ-linoleic acid) and C20:0, and finally, SBO by C16:0, C16:1n-7 and C18:1n-9. Our data were compared to other publications, as presented in Section 3.2, the average levels of fatty acids being similar, in spite of their different origins.

For volatiles, the heatmap shows a good discrimination between the profiles of the eight oil groups, especially between EVOO, SFO, SBO. Stronger discriminations were observed between SFO (rich in 2-octene, acetophenone, α- and β-pinenes), for EVOOs rich in hexyl derivatives (hexenol, hexenal, hexyl acetate), and for SBO, rich in volatile acids (butanoic, hexanoic, butanoic), heptanal and ocimen. WO, LSO and SO showed less significant biomarkers, while HO was represented by a higher content of hexanal and PO by octene and pinenes. Few data were reported about volatiles responsible for the sensorial characteristics of these oils, but generally they were similar, mainly for olive oils [43–45], but also for linseeds [36] and walnut [39,40].

Beside the relevance of volatiles for the oil origin and identity, one can also consider the influence of the thermal processing on volatiles' elaboration, their different sensorial characteristics being due not only to the raw material, but also to technological parameters (temperature-time and storage conditions).

Concerning carotenoid pigments, HO was the richest source of β- and α-carotene and luteoxanthin, while SO was the richest in β-carotene epoxides. The LSO was identified easily by a high content in auroxanthin and violaxanthin, beside β-carotene epoxides. The samples EVOO had lower levels of auroxanthin, violaxanthin and α-carotene' By far, the richest content of carotenoids, especially zeaxanthin, lutein, β-carotene and β-cryptoxanthin was reported in SBO, followed by WO, EVOO, and in lower amounts in SO. Different concentrations of carotenoids, highly dependent on the botanical and geographical origin of vegetables were reported in some recent publications [41,46] and are in agreement with our previous and present data.

Significant differences were observed in relation to the different tocopherol isomers. While α-tocopherol was found as major molecule in EVOO, SBO and SFO, β-tocopherol was mostly representative for SBO, γ- and δ-tocopherol for SO. This intuitive map also helps to make a ranking of these oils according to the relative content of carotenoids and tocopherols, useful for their recognition, e.g., SBO, HO and SO with specific, different compositions, as well EVOO, PO and LSO groups having distinct groups of molecules which may be considered as biomarkers. Relevant recent articles found similar or different levels for tocopherols in similar vegetable oils [35,37,40,41,48–52].

The most significant discrimination between the oils was obtained considering the phenolic derivatives, as a mirror of their identity. The richest oils in phenolics were EVOO and SBO, represented specifically by tyrosols, oleuropein, verbascoside, luteolin and pinoresinols, and by quercetin and isorhamnetin together with many phenolic acids (gallic, syringic, vanilic, protocatecuic, cinnamic, salycilic), respectively. Then, LSO is recognized specifically by lignans (pino-, matairesinols larici, isolarici and secoisolarici resinols), WO by juglone, elagic acid and gallocatechin, and SO by daidzein, glycitein, and genistein. HO has its typical phenolics, canabigenol, o-coumaric acid and vanillin, while PO is characterized by dicaffeoyl quinic acid and vanillin. Finally, SFO is the richest source of chlorogenic acid and dicaffeoyl quinic acid. Similar phenolics, having different ranges, were also mentioned by different publications, especially the ones related to olive oils [45,53,54], but also for linseed [37], walnut [40], or sea buckthorn [48] oils.

5. Conclusions

The data presented here recommend a systematic analytical flow based on a metabolomic approach to be applied for edible oils, but also for other plant-based foods, or byproducts, identifying the profile of valuable components, including phytochemicals, as authenticity biomarkers and health-promoting molecules.

The results of our previous and current investigations demonstrated that a succession of instrumental analysis combined with chemometric models, such as PLSDA and heatmap clustering, have a good predictive ability to detect the oils' botanical origin, as well as to identify qualitative and quantitative differences between individual samples from the same phenotype and the specific features of each category of oils. Therefore, these complementary data are useful not only to identify the oil category but also the quality of commercial oils (which may be mislabeled or adulterated), comparative to the authentic ones of the same botanical and geographical origin.

Finally, we are recommending the data fusion of such complementary investigations to have a more clear and accurate evaluation and identity of commercial edible oils, and especially of functional oils (cold pressed, virgin, extravirgin) richer in phytochemicals as bioactive molecules.

This study needs validation for an extended number of samples (authentic and commercial) to have a more accurate image of the phytochemical biomarkers to be considered in an "identity" card of each type of edible oil, including their quality determined by specific processing.

Supplementary Materials: The following supporting information can be downloaded at: https://www.mdpi.com/article/10.3390/app12041933/s1. Figure S1: Calibration curves made by HPLC-DAD, with different pure standards of phenolic derivatives: gallic acid and vanillin (to quantify hydroxybenzoic acid derivatives), clorogenic and ferulic acid (for hydroxicinnamic derivatives), quercetin (for flavonoids), luteolin (for flavones), oleuropein (for tyrosols), and pinoresinol (for lignans). Figure S2: Typical GC-MS chromatograms of the different oils under investigation. For oils' abbreviations, see Table 1. Table S1: Individual fatty acid levels expressed in % of area of the total identified fatty acid areas in the GC-MS chromatograms of oils under investigation (mean values of duplicated separations). For oils' abbreviations, see Table 1. Table S2: Results of one-way ANOVA and Tukey's HSD post-hoc test to evaluate the significance of differences between the oil groups, considering each of the 10 fatty acids separated and identified by GC-MS. The graphic representations of the Tukey HSD (made by MetaboAnalyst 5.0 algorithm) are presented below. Table S3: Individual composition of volatile oils' components expressed in % of peak area of the total volatiles, as determined by ITEX/GC-MS (mean values of duplicated separations). For oils' abbreviations, see Table 1. Table S4: Results of one-way ANOVA and Tukey's HSD post-hoc test to evaluate the significance of differences between the oil groups, considering all volatiles separated and identified by ITEX GC-MS. The graphic representations of the Tukey HSD (made by MetaboAnalyst 5.0 algorithm) are presented below. Table S5: Results of one-way ANOVA and Tukey's HSD post-hoc test to evaluate the significance of differences between the oil groups, considering carotenoid molecules separated and identified by HPLC-DAD. The graphic representations of the Tukey HSD are attached below. Table S6: Results of one-way ANOVA and Tukey's HSD post-hoc test to evaluate the significance of differences between the oil groups, considering the tocopherol isomers identified by HPLC-FD. The graphic representations of the Tukey HSD are attached below. Table S7: The phenolic composition of all oil samples: mean values expressed in mg/100 mL oil and standard deviations (SD), as determined by HPLC-DAD. For details, see Section 2.

Author Contributions: Conceptualization, C.S.; methodology, F.D., S.S., F.F. and F.R.; software, C.S.; validation, A.B., S.S. and F.D.; formal analysis, F.D., S.S., F.F. and F.R.; writing—original draft preparation, C.S., F.R. and F.F.; writing—review and editing, C.S. and A.P.; supervision, project administration and funding acquisition, C.S. and A.P. All authors have read and agreed to the published version of the manuscript.

Funding: This research was funded by the Executive Unit for Financing Higher Education, Research, Development, and Innovation (UEFISCDI) within the framework of project PN-III-P1-1.2-PCCDI-2017-0046/No. 1/2018, within PNCDI III and from the National Research Development Projects to finance excellence (PFE)-14/2022-2024 granted by the Romanian Ministry of Research and Innovation.

Institutional Review Board Statement: Not applicable.

Informed Consent Statement: Not applicable.

Data Availability Statement: Not applicable.

Acknowledgments: The authors acknowledge the donations of authentic oil samples from different manufacturers.

Conflicts of Interest: The authors declare no conflict of interest.

References

1. Gunstone, F. (Ed.) *Vegetable Oils in Food Technology: Composition, Properties and Uses*, 2nd ed.; Wiley Blackwell: Hoboken, NJ, USA, 2011; 376p, ISBN 978-1-444-33991-8.
2. Salah, W.A.; Nofal, M. Review of some adulteration detection techniques of edible oils. *J. Sci. Food Agric.* **2020**, *101*, 811–819. [CrossRef] [PubMed]
3. Dubois, V.; Breton, S.; Linder, M.; Fanni, J.; Parmentier, M. Fatty acid profiles of 80 vegetable oils with regard to their nutritional potential. *Eur. J. Lipid Sci. Technol.* **2007**, *109*, 710–732. [CrossRef]
4. Hernandez, E.M. Specialty Oils: Functional and Nutraceutical Properties. In *Functional Dietary Lipids, Food Formulation, Consumer Issues and Innovation for Health*; Woodhead Publisher for Series in Food Science, Technology and Nutrition; Woodhead Publishing: London, UK, 2016; pp. 69–101.
5. Socaciu, C.; Dulf, F.; Ranga, F.; Fetea, F.; Bele, C.; Echim, C. Selected biomarkers and methods to fingerprint vegetable functional oils originating from Romania. In Proceedings of the Conference Euro Food Chem XV—Food for the Future, Copenhagen, Denmark, 5–8 July 2009; pp. 162–165.
6. Socaciu, C.; Ranga, F.; Fetea, F.; Leopold, D.; Dulf, F.; Parlog, R. Complementary advanced techniques applied for plant and food Authentication. *Czech Food Sci.* **2009**, *27*, S70–S75. [CrossRef]
7. Bajoub, A.; Bendini, A.; Fernández-Gutiérrez, F.; Carrasco-Pancorbo, A. Olive oil authentication: A comparative analysis of regulatory frameworks with especial emphasis on quality and authenticity indices, and recent analytical techniques developed for their assessment A review. *Crit. Rev. Food Sci. Nutr.* **2018**, *58*, 832–857. [CrossRef] [PubMed]
8. Lioupi, A.; Nenadis, N.; Theodoridis, G. Virgin olive oil metabolomics: A review. *J. Chromatogr. B* **2020**, *1150*, 122–161. [CrossRef] [PubMed]
9. Messai, H.; Farman, M.; Sarraj-Laabidi, A.; Hammami-Semmar, A.; Semmar, N. Chemometrics Methods for Specificity, Authenticity and Traceability Analysis of Olive Oils: Principles, Classifications and Applications. *Foods* **2016**, *5*, 77. [CrossRef] [PubMed]
10. Da Ros, A.; Masuero, D.; Riccadonna, S.; Brkić Bubola, K.; Mulinacci, N.; Mattivi, F.; Lukić, I.; Vrhovsek, U. Complementary Untargeted and Targeted Metabolomics for Differentiation of Extra Virgin Olive Oils of Different Origin of Purchase Based on Volatile and Phenolic Composition and Sensory Quality. *Molecules* **2019**, *24*, 2896. [CrossRef]
11. Servili, M.; Montedoro, G. Contribution of phenolic compounds to virgin olive oil quality. *Eur. J. Lipid Sci. Technol.* **2002**, *104*, 602–613. [CrossRef]
12. Olmo-García, L.; Carrasco-Pancorbo, A. Chromatography-MS based metabolomics applied to the study of virgin olive oil bioactive compounds: Characterization studies, agro-technological investigations and assessment of healthy properties. *TrAC Trends Anal. Chem.* **2021**, *135*, 116153. [CrossRef]
13. Boskou, D. Phenolic Compounds in Olives and Olive Oil. In *Olive Oil: Minor Constituents and Health*, 1st ed.; Boskou, D., Ed.; CRC Press: Boca Raton, FL, USA, 2009; Chapter 3; pp. 11–44.
14. Lerma Garcia, M.J. *Characterization and Authentication of Olive and Other Vegetable Oils*; Springer: Berlin/Heidelberg, Germany, 2012.
15. Wang, L.; Chen, Y.; Ye, Z.; Hellmann, B.; Xu, X.; Jin, Z.; Ma, Q.; Yang, N.; Wu, F.; Jin, Y. Screening of Phenolic Antioxidants in Edible Oils by HPTLC-DPPH Assay and MS Confirmation. *Food Anal. Methods* **2018**, *11*, 3170–3178. [CrossRef]
16. Li, C.; Yao, Y.; Zhao, G.; Cheng, W.; Liu, H.; Liu, C.; Shi, Z.; Chen, Y.; Wang, S. Comparison and analysis of fatty acids, sterols, and tocopherols in eight vegetable oils. *J. Agric. Food Chem.* **2011**, *59*, 12493–12498. [CrossRef] [PubMed]
17. Zhang, L.; Li, P.; Sun, X.; Wang, X.; Xu, B.; Wang, X.; Ma, F.; Zhang, Q.; Ding, X. Classification and adulteration detection of vegetable oils based on fatty acid profiles. *J. Agric. Food Chem.* **2014**, *62*, 8745–8751. [CrossRef] [PubMed]
18. Shen, M.; Zhao, S.; Zhang, F.; Huang, M.; Xie, J. Characterization and authentication of olive, camellia and other vegetable oils by combination of chromatographic and chemometric techniques: Role of fatty acids, tocopherols, sterols and squalene. *Eur. Food Res. Technol.* **2021**, *247*, 411–426. [CrossRef]
19. Mota, M.F.S.; Waktola, H.D.; Nolvachai, Y.; Marriott, P.J. Gas chromatography-mass spectrometry for characterization, assessment of quality and authentication of seed and vegetable oils. *TrAC Trends Anal. Chem.* **2021**, *138*, 116238. [CrossRef]

20. Bagur-González, M.G.; Pérez-Castaño, E.; Sánchez-Viñas, M.; Gázquez-Evangelista, D. Using the liquid chromatographic fingerprint of sterols fraction to discriminate virgin olive from other edible oils. *J. Chromatogr. A* **2015**, *1380*, 64–70. [CrossRef] [PubMed]
21. Yuan, C.; Xie, Y.; Jin, R.; Ren, L.; Zhou, L.; Zhu, M.; Ju, Y. Simultaneous analysis of tocopherols, phytosterols, and squalene in vegetable oils by high-performance liquid chromatography. *Food Anal. Methods* **2017**, *10*, 3716–3722. [CrossRef]
22. Esteki, M.; Simal-Gandara, J.; Shahsavari, Z.; Zandbaaf, S.; Dashtaki, E.; Heyden, Y.V. A review on the application of chromatographic methods, coupled to chemometrics, for food authentication. *Food Control* **2018**, *93*, 165–182. [CrossRef]
23. Li, X.; Kong, W.; Shi, W.; Shen, Q.A. A combination of chemometric methods and GC–MS for the classification of edible vegetable oils. *Chemometr. Intell. Lab. Syst.* **2016**, *155*, 145–150. [CrossRef]
24. Biancolillo, A.; Marini, F.; Ruckebusch, C.; Vitale, R. Chemometric Strategies for Spectroscopy-Based Food Authentication. *Appl. Sci.* **2020**, *10*, 6544. [CrossRef]
25. Socaciu, C.; Fetea, F.; Ranga, F.; Bunea, A.; Dulf, F.; Socaci, S.; Pintea, A. Attenuated Total Reflectance-Fourier Transform Infrared Spectroscopy ATR-FTIR. Coupled with Chemometrics, to Control the Botanical Authenticity and Quality of Cold-Pressed Functional Oils Commercialized in Romania. *Appl. Sci.* **2020**, *10*, 8695. [CrossRef]
26. Giacomelli, L.M.; Mattea, M.; Ceballos, C.D. Analysis and characterization of edible oils by chemometric methods. *J. Am. Oil Chem. Soc.* **2006**, *83*, 303–308. [CrossRef]
27. Gómez-Caravaca, A.M.; Maggio, R.M.; Cerretani, L. Chemometric Applications to Assess Quality and Critical Parameters of Virgin and Extra-Virgin Olive Oil A Review. *Anal. Chim. Acta* **2016**, *913*, 1–21. [CrossRef] [PubMed]
28. Godoy, A.C.; Dos Santos, P.D.S.; Nakano, A.Y.; Bini, R.A.; Siepmann, D.A.B.; Schneider, R.; Gaspar, P.A.; Pfrimer, P.W.D.; da Paz, R.F.; Santos, O.O. Analysis of Vegetable Oil from Different Suppliers by Chemometric Techniques to Ensure Correct Classification of Oil Sources to Deal with Counterfeiting. *Food Anal. Methods* **2020**, *13*, 1138–1147. [CrossRef]
29. Chong, J.; Wishart, D.S.; Xia, J. Using MetaboAnalyst 40 for Comprehensive and Integrative Metabolomics Data Analysis. *Curr. Protoc. Bioinform.* **2019**, *68*, e86. [CrossRef] [PubMed]
30. Tudor, C.; Gherasim, E.C.; Dulf, F.V.; Pintea, A. In vitro bioaccessibility of macular xanthophylls from commercial microalgal powders of *Arthrospira platensis* and *Chlorella pyrenoidosa*. *Food Sci. Nutr.* **2021**, *9*, 1896–1906. [CrossRef]
31. Singleton, V.L.; Orthofer, R.; Lamuela-Raventos, R.M. Analysis of total phenols and other oxidation substrates and antioxidants by means of the Folin-Ciocalteu reagent. *Methods Enzymol.* **1999**, *299*, 152–178.
32. Alves, A.Q.; da Silva, V.A.; Góes, A.J.S.; Silva, M.S.; de Oliveira, G.G.; Bastos, U.V.G.A.; de Castro Neto, A.G.; Alves, A.J. The Fatty Acid Composition of Vegetable Oils and Their Potential Use in Wound Care. *Adv. Skin Wound Care* **2019**, *32*, 1–8. [CrossRef]
33. Frančáková, H.; Ivanišová, E.; Dráb, S.; Krajčovič, T.; Tokár, M.; Mareček, J.; Musilová, J. Composition of Fatty Acids in Selected Vegetable Oils. *Potravinarstvo* **2015**, *9*, 538–542. [CrossRef]
34. Zielińska, A.; Nowak, I. Fatty acids in vegetable oils and their importance in cosmetic industry. *Chem. Aust.* **2014**, *68*, 103–110.
35. Kamal-Eldin, A.; Andersson, R. A multivariate study of the correlation between tocopherol content and fatty acid composition in vegetable oils. *J. Am. Oil Chem. Soc.* **1997**, *74*, 375–380. [CrossRef]
36. Krist, S.; Stuebiger, G.; Bail, S.; Unterweger, H. Analysis of volatile compounds and triacylglycerol composition of fatty seed oil gained from flax and false flax. *Eur. Lipid Sci. Technol. J.* **2006**, *108*, 48–60. [CrossRef]
37. Choo, W.S.; Birch, J.; Dufour, J.P. Physicochemical and quality characteristics of cold-pressed flaxseed oils. *J. Food Comp. Anal.* **2006**, *20*, 202–211. [CrossRef]
38. Christopoulou, E.; Lazaraki, M.; Komaitis, M.; Kaselimis, K. Effectiveness of determination of fatty acids and triglycerides for the detection of adulteration of olive oils with vegetable oils. *Food Chem.* **2004**, *84*, 463–474. [CrossRef]
39. Tsamouris, G.; Hatziantoniou, S.; Demetzos, C. Lipid analysis of Greek Walnut oil *Juglans regia* L. *Z. Naturforsch.* **2002**, *57c*, 51–56. [CrossRef]
40. Ojeda-Amador, R.M.; Desamparados Salvador, M.; Gómez-Alonso, S.; Fregapane, G. Characterization of virgin walnut oils and their residual cakes produced from different varieties. *Food Res. Int.* **2018**, *108*, 396–404. [CrossRef]
41. Montserrat-de la Paz, S.; Marín-Aguilar, F.; Garcia Gimenez, M.D.; Fernandez-Arche, M.A. Hemp (*Cannabis sativa* L.) seed oil: Analytical and phytochemical characterization of unsaponifiable fraction. *J. Agric. Food Chem.* **2014**, *62*, 1105–1110. [CrossRef]
42. Yang, B.; Kallio, H.P. Fatty acid composition of lipids in sea buckthorn (*Hippophaë rhamnoides* L.) berries of different origins. *J. Agric. Food Chem.* **2011**, *49*, 1939–1947. [CrossRef]
43. Diez-Simon, C.; Mumm, R.; Hall, R.D. Mass spectrometry-based metabolomics of volatiles as a new tool for understanding aroma and flavour chemistry in processed food products. *Metabolomics* **2019**, *15*, 41. [CrossRef]
44. Campestre, C.; Angelini, G.; Gasbarri, C.; Angerosa, F. The compounds responsible for the sensory profile in monovarietal virgin olive oils. *Molecules* **2017**, *22*, 1833. [CrossRef]
45. Lukić, I.; Lukić, M.; Žanetić, M.; Krapac, M.; Godena, S.; Brkić Bubola, K. Inter-Varietal Diversity of Typical Volatile and Phenolic Profiles of Croatian Extra Virgin Olive Oils as Revealed by GC-IT-MS and UPLC-DAD Analysis. *Foods* **2019**, *8*, 565. [CrossRef]
46. Tudor, C.; Bohn, T.; Iddir, M.; Dulf, F.V.; Focşan, M.; Rugină, D.O.; Pintea, A. Sea Buckthorn Oil as a Valuable Source of Bioaccessible Xanthophylls. *Nutrients* **2019**, *12*, 76. [CrossRef] [PubMed]
47. Alessandri, S.; Ieri, F.; Romani, A. Minor polar compounds in extra virgin olive oil: Correlation between HPLC-DAD-MS and the Folin-Ciocalteu spectrophotometric method. *J. Agric. Food Chem.* **2014**, *62*, 826–835. [CrossRef] [PubMed]

48. Franke, S.; Frohlich, K.; Werner, S.; Bohm, V.; Schone, F. Analysis of carotenoids and vitamin E in selected oilseeds, press cakes and oils. *Eur. J. Lipid Sci. Technol.* **2010**, *112*, 1122–1129. [CrossRef]

49. Zielińska, A.; Nowak, I. Abundance of active ingredients in sea-buckthorn oil. *Lipids Health Dis.* **2017**, *16*, 95. [CrossRef]

50. Tyskiewicz, K.; Gieysztor, R.; Maziarczyk, I.; Hodurek, P.; Rój, E.; Skalicka-Wozniak, K. Supercritical Fluid Chromatography with Photodiode Array Detection in the Determination of Fat-Soluble Vitamins in Hemp Seed Oil and Waste Fish Oil. *Molecules* **2018**, *23*, 1131. [CrossRef]

51. Kallio, H.; Yang, B.; Peippo, P. Effects of Different Origins and Harvesting Time on Vitamin C, Tocopherols, and Tocotrienols in Sea Buckthorn (Hippophae rhamnoides. Berries. *J. Agric. Food Chem.* **2002**, *50*, 6136–6142. [CrossRef]

52. Schwartz, H.; Ollilainen, V.; Piironen, V.; Lampi, A.-M. Tocopherol, tocotrienol and plant sterol contents of vegetable oils and industrial fats. *J. Food Comp. Anal.* **2008**, *21*, 152–161. [CrossRef]

53. García, A.; Brenes, M.; García, P.; Romero, C.; Garrido, A. Phenolic content of commercial olive oils. *Eur. Food Res. Technol.* **2003**, *216*, 520–525. [CrossRef]

54. Klikarová, J.; Česlová, L.; Kalendová, P.; Dugo, P.; Mondello, L.; Cacciola, F. Evaluation of Italian extra virgin olive oils based on the phenolic compounds' composition using multivariate statistical methods. *Eur. Food Res. Technol.* **2020**, *246*, 1241–1249. [CrossRef]

Article

The Effect of Essential Oils on the Survival of *Bifidobacterium* in In Vitro Conditions and in Fermented Cream

Mariola Kozłowska [1,*], Małgorzata Ziarno [2,*], Magdalena Rudzińska [3], Małgorzata Majcher [3], Jolanta Małajowicz [1] and Karolina Michewicz [2]

[1] Department of Chemistry, Institute of Food Science, Warsaw University of Life Sciences-SGGW, 02-776 Warsaw, Poland; jolanta_malajowicz@sggw.edu.pl

[2] Department of Food Technology and Assessment, Institute of Food Science, Warsaw University of Life Sciences-SGGW, 02-776 Warsaw, Poland; kbmichewicz@gmail.com

[3] Faculty of Food Science and Nutrition, Poznań University of Life Sciences, 60-637 Poznan, Poland; magdalena.rudzinska@up.poznan.pl (M.R.); malgorzata.majcher@up.poznan.pl (M.M.)

* Correspondence: mariola_kozlowska@sggw.edu.pl (M.K.); malgorzata_ziarno@sggw.edu.pl (M.Z.); Tel.: +48-22-593-76-22 (M.K.); +48-22-593-76-66 (M.Z.)

Citation: Kozłowska, M.; Ziarno, M.; Rudzińska, M.; Majcher, M.; Małajowicz, J.; Michewicz, K. The Effect of Essential Oils on the Survival of *Bifidobacterium* in In Vitro Conditions and in Fermented Cream. *Appl. Sci.* **2022**, *12*, 1067. https://doi.org/10.3390/app12031067

Academic Editor: Catarina Guerreiro Pereira

Received: 6 January 2022
Accepted: 18 January 2022
Published: 20 January 2022

Publisher's Note: MDPI stays neutral with regard to jurisdictional claims in published maps and institutional affiliations.

Abstract: Essential oils derived from plant materials are a mixture of compounds that exhibit antibacterial properties. Due to their distinct aroma, they also serve as a desirable natural additive for various food products, including dairy products. In this study, the essential oils of lemon peels, clove buds, and juniper berries were obtained by steam distillation and characterized using gas chromatography–mass spectrometry to determine their chemical compositions and effects on the viability of seven *Bifidobacterium* strains. Furthermore, the effect of essential oils on the viability of *Bifidobacterium animalis* subsp. *lactis* Bb-12 was investigated in cream samples during fermentation and after storage for 21 days at 6 °C. The fatty acid composition of fat extracted from essential oils containing sour cream samples and the volatile aroma compound profile of the sour cream samples were also determined chromatographically. Among the 120 compounds identified, monoterpene hydrocarbons were dominant in the essential oils of lemon peels (limonene and γ-terpinene) and juniper berries (sabinene and β-myrcene), while eugenol and eugenol acetate were abundant in the essential oil of clove buds. In addition to these compounds, butanoic and acetic acids were found in the tested sour cream samples. In turn, fat extracted from these samples was rich in saturated fatty acids, mainly palmitic acid. Among the tested strains of the genus *Bifidobacterium*, *B. animalis* subsp. *lactis* Bb-12 was the most sensitive to the essential oils of clove and juniper, as indicated by the larger growth inhibition zones. However, both the concentration and type of essential oils used had no effect on the number of cells of this strain present in the cream samples immediately after fermentation and after its 21-day storage, which suggests that the tested essential oils could be a natural additive to dairy products.

Keywords: clove buds; juniper berries; lemon peels; fatty acid composition; GC–MS; GC–TOF–MS; cream; fermentation

1. Introduction

Bifidobacteria are known for their beneficial effects on human health, including their role in the proper functioning of the digestive tract [1]. In the large intestine, they hydrolyze sugars and produce lactic and acetic acids, which are metabolic end products. They reduce the pH of the contents in the large intestine and, thus, the levels of various harmful substances, including ammonia [2]. Bifidobacteria can also utilize ammonia as a source of nitrogen. Due to their low pH, they exert an antibacterial effect, inhibiting the growth of pathogenic microorganisms such as *Shigella*, *Salmonella*, and *Staphylococcus*, as well as enteropathogenic *Escherichia coli* strains. *Bifidobacterium* can reduce blood cholesterol and produce vitamins (mainly the B group), food enzymes, casein phosphatase, and lysozyme [3]. Bifidobacteria-containing formulations used to treat constipation promote the production of organic acids by these bacteria, which stimulate normal intestinal peristalsis [4]. Studies also confirm that some of the bifidobacteria strains have excellent probiotic potential and, hence, can be included in formulas used for infants who cannot be fed with their mother's milk [5].

Bifidobacterium and *Lactobacillus acidophilus*, which have the ability to colonize and grow in the intestines, are the most commonly used microorganisms for the production of probiotic foods. Yogurts and yogurt drinks (containing live bacteria) are becoming increasing popular in the dairy market. An important aspect in the production of such products is ensuring that they have a high nutritional value, the right proportion of nutrients, and the benefits of additional vitamins, minerals, and other substances. Consumer awareness of nutrition has been growing in the past few years, posing challenges to both science and the food industry. Recent studies have been focusing on testing the possibility of replacing preservatives with natural substitutes, especially the plant-derived ones [6–9]. Essential oils extracted from plant materials are an ideal option. They are made up of organic compounds, such as esters, ethers, aldehydes, and ketones, which are characterized by a distinct aroma. Their biological activity results from the effects of individual dominant components or the synergistic action of a group of compounds. Some essential oils exhibit antimicrobial, antioxidant, and anticancer activities, while others have sedative, choleretic, and stimulatory effects on gastric function and secretion of digestive juices [10–19].

Besides enhancing the organoleptic characteristics, essential oils added in food products can promote their microbiological stabilization. They are often used as a flavoring agent for meat or meat-based products. The available literature describes the biological properties of essential oils extracted from clove buds, citrus peels, and juniper berries [10,12–15,17,18,20]. However, there are no reports on the effects of essential oils obtained from the above substances on *Bifidobacterium* bacteria, which is used as a component in probiotic fermented dairy products.

Therefore, this study aimed to investigate the effect of the essential oils extracted from clove buds, lemon peels, and juniper berries on the viability of selected *Bifidobacterium* strains in in vitro conditions. In addition, the effect of the concentration of these essential oils on the viability of the most sensitive *Bifidobacterium* strain was investigated in cream samples during their fermentation, and after storage, at 6 °C for 21 days. If there was no influence of the tested essential oils and their concentrations on the viability of the most sensitive strain (*B. animalis* subsp. *lactis* Bb-12) in the cream, we assumed that the other tested strains would not show any sensitivity to the tested essential oils added to the fermented cream. The chemical composition of the essential oils used was analyzed using gas chromatography–mass spectrometry (GC–MS). The fatty acid composition of the fat extracted from the sour cream samples was also determined chromatographically after its storage for 21 days. The volatile aroma compound profiles of sour cream samples enriched with essential oils were determined by the solid-phase microextraction (SPME)-GC–time-of-flight (TOF)-MS method.

2. Materials and Methods

2.1. Materials and Chemicals

Dried juniper berries (Species: *Juniperus communis* L.; Genus: *Juniperus* L.; Family: Cupressaceae; Order: Pinales; Class: Pinopsida; Division: Coniferophyta), clove buds (Species: *Syzygium aromaticum* L.; Genus: *Syzgium*; Family: Myrtaceae; Order: Myrtales; Class: Magnoliopsida; Division: Magnoliophyta), and fresh lemons (Species: *Citrus limon* L.; Genus: *Citrus* L.; Family: Rutaceae; Order: Sapindales; Class: Magnoliopsida; Division: Magnoliophyta) used in the study were purchased from a local market (Warsaw, Poland) in December 2017. All the solvents (methanol, dichloromethane, chloroform, and dimethyl sulfoxide (DMSO), *n*-hexane) and reagents (potassium chloride, anhydrous magnesium sulfate, sodium hydroxide, and boron trifluoride) used were of analytical grade and purchased from Avantor Performance Materials Poland (Gliwice, Poland). The fatty acid methyl esters

(FAME) reference standard (certified) mixture (consisting of 37 fatty acids ranging from C4 to C24) was obtained from Supelco (Bellefonte, PA, USA).

The *Bifidobacterium* strains used in this study (*B. animalis* subsp. *lactis* Bb-12, *B. bifidum* 89, *B. infantis* 45/03, *B. lactis* AD 600, *B. lactis* HN 019, *B. longum* AD 50, and *B. longum* KNA 1/08) were supplied by the Museum of Clean Cultures of the Division of Milk Technology of Warsaw University of Life Sciences (Poland).

2.2. Essential Oil Preparation

The lemons were washed carefully using tap water, and peels were removed and cut into pieces using a knife. Then, the pieces (40 g) were steam-distilled in a 1000 mL round-bottomed flask for 6 h. The distillate was cooled to the ambient temperature ($20 \pm 2\ °C$) and dried using anhydrous magnesium sulfate to remove the remaining water. The obtained essential oil was stored in a dark glass bottle at $-21\ °C$ until further analyses. The essential oils of juniper berries and clove buds were isolated using the same procedure, but in this case the dry plant material was ground in a blender before steam distillation (20 g of plant material and 500 mL of distilled water). The yield of essential oils was calculated using the formula: yield oil (%) = weight of oil (g)/weight of sample (g) $\times$ 100%.

2.3. The GC–MS Analysis of Essential Oils

The chemical components of lemon peel, clove bud, and juniper berry oils were determined using a Hewlett-Packard HP 7890A gas chromatograph instrument coupled with a 5975C mass spectrometer (Agilent Technologies, Santa Clara, CA, USA) with an SLB-5MS (25 m $\times$ 0.2 mm $\times$ 0.33 µm) column. The samples of essential oils were directly injected onto a gas chromatograph. Helium gas was used as a carrier at a constant flow rate of 0.8 mL/min. The initial oven temperature was set at 40 °C for 1 min, and then the temperature was raised to 180 °C at a rate of 6 °C/min and further to 280 °C at 20 °C/min. The mass spectra were recorded in the electron impact mode (70 eV) at a scan range of m/z 33–350. The identification of volatile compounds was carried out based on the comparison of their retention indices (RI) and mass spectra with appropriate standards. If standards were not available, tentative identification was performed by comparing the mass spectra of the compounds with the spectral data from NIST 05, and their RI with the literature data. The RI of each compound was calculated using a homologous series of C6–C16 *n*-alkanes. Obtained results were calculated as a percentage composition of identified compounds.

2.4. The Effects of Essential Oils on the Viability of Bifidobacterium in In Vitro Conditions

The bacterial strains were cultured overnight in a BSM broth (Merck KGaA, Darmstadt, Germany) before their inoculation. The antibacterial activity of the obtained essential oils was evaluated using the well diffusion method [21]. Briefly, the bacterial suspension was distributed into the BSM agar medium at a density of 6–7 log(CFU/mL) and then poured into a Petri dish. After the medium had solidified, 5-mm wells were cut into it and 20 µL of each essential oil was applied at a concentration of 100%, 75%, 50%, 25%, and 0%. Each of the tested essential oils were dissolved in a mixture of three organic solvents in a 4:1:1 volume ratio (chloroform:methanol:DMSO). The solvent mixture was used as the control sample (0%). After adding the essential oils, the Petri dishes with wells were incubated at 37 °C for 72 h under anaerobic conditions (Anaerocult A, Merck, Germany) in an anaerobic jar (Merck, Germany) and then the zone of growth inhibition was measured. The experiments were repeated twice, and two parallel results were obtained each time.

2.5. The Effect of the Essential Oil Concentration on the Viability of B. animalis Subsp. lactis Bb-12 in Cream during Fermentation

Ultraheat-treated cream "Łowicka tortowa" (OSM Łowicz, Łowicz, Poland), with a fat content of 36%, was used in the experiments. Each tested essential oil, at a concentration of 100.0, 50.0, 25.0, 12.5, and 0.0 µg, was added to an 100-mL portion of cream followed by a culture of *B. animalis* subsp. *lactis* Bb-12 that was added at a density of 8.2 ± 0.4 log(CFU/mL) overnight. The samples were fermented at 37 °C for 5 h and were then stored at 6 °C for 21 days. Before and after incubation, as well as after the 21 days of storage, the bifidobacteria population was determined using the deep plate method with the BSM agar as a culture medium. The Petri dishes were incubated at 37 °C for 72 h under anaerobic conditions (Anaerocult A, Merck, Germany) in an anaerobic jar. After incubation, all characteristic bacterial colonies grown were counted and expressed as log(CFU/mL). The experiments were repeated twice, and two parallel results were obtained each time.

2.6. Fat Extraction

Sour cream samples (9 g) were weighed into a 50-mL Eppendorf conical tube and a mixture of two solvents, chloroform and methanol (2:1; v/v; 30 mL), were added. The samples were slowly mixed in a vortex apparatus for 5 s and were then centrifuged at $6170 \times g$ (Biofuge Stratos, Thermo Fisher Scientific, Waltham, MA, USA) for 5 min. After centrifugation, the bottom layer was transferred into a separatory funnel using a Pasteur pipette and a 0.9% KCl solution (30 mL) was added. The resulting mixture was shaken gently and left aside for separation into two phases. The lower phase was collected and dried by adding a pinch of anhydrous magnesium sulfate. After the drying agent was filtered off through a Whatman No. 1 paper filter, the remaining solvent was evaporated under reduced pressure in a rotary evaporator (Rotavapor R-215, Büchi Labortechnik, Switzerland) at 40 °C. The extracted fat samples were weighed, flushed with nitrogen, and stored at −21 °C until further use. The percentage of total extractable fat was calculated.

2.7. Fatty Acid Analyses

The fatty acid compositions of the fat extracted from the sour cream samples were analyzed using a Trace 1300 gas chromatograph equipped with an SP-2560 capillary column (100 m × 0.25 mm × 0.2 μm; Supelco, Darmstadt, Germany) and a flame ionization detector after its derivatization to FAME, according to the AOCS Official Method Ce 1k-07 [22]. Hydrogen was used as the carrier gas at a flow rate of 1.5 mL/min. Initially, the oven temperature was set at 160 °C and was then increased at 12 °C/min to 220 °C, which was maintained for 20 min. The temperatures of the injector and detector were set at 240 °C. Fatty acids were identified by comparing their retention times with those of certified FAME. Their content was estimated in a percentage using the AOCS Official Method Ce 1h-05 (2005) [23].

2.8. Analyses of Volatiles Extracted from Cream and Sour Cream Samples

Volatiles were extracted by SPME using a carboxene/polydimethylsiloxane fiber (Supelco, Bellefonte, PA, USA) mounted in a CTC CombiPal autosampler (Agilent Technologies, Santa Clara, CA, USA). For the analysis of each volatile, 5 g of a sample was placed into 20-mL vials, spiked with 1 μg of an internal standard (naphthalene d8), and sealed with polytetrafluoroethylene/silicon septa caps. To avoid the influence of temperature on the formation of volatiles, the compounds were extracted without additional heating from the headspace at a stable room temperature of 21 °C for 30 min. The identification of the compounds was performed using a GCxGC–TOF-MS system with a ZOEX cryogenic (N2) modulator (Pegasus IV, LECO, St. Joseph, IL, USA) which was equipped with an SLB-5 MS (30 m × 0.25 mm × 0.5 μm) column (Supelco Bellefonte, PA, USA) coupled to a Supelcowax-10 (1.1 m × 0.2 mm × 0.2 μm) column (Supelco Bellefonte, PA, USA) and operated in a multidimensional chromatography mode. The temperature of the injector was set to 250 °C. During the injection, the fiber was exposed for 5 min in the splitless mode (1-min purge time). The conditions used for chromatography were as follows: helium flow, 0.8 mL/min; initial oven temperature, 40 °C (for 1 min) which was then increased to 180 °C at 6 °C/min and then further to 250 °C at 20 °C/min. For the two-dimensional analysis, the modulation time was optimized and set to 4 s and mass spectra were collected at a rate of 150 spectra/s. The transfer line was kept at 280 °C and the ion source was heated to 250 °C. Volatiles were identified by comparing the RI and the mass spectra of the eluting compounds with those of the NIST 05 library match. A mixture of *n*-alkanes (C6–C20), dissolved in pentane which was obtained from Supelco (Bellefonte, PA, USA), was used for determining the RI of the volatiles. Calculations were carried out using Chroma TOF software (version 3.34). The semiquantification of compounds, which is represented by the average relative peak area of three replicates (arbitrary units presented as the mean value of three determinations), was achieved using the characteristic ions listed in Table 4. The results do not correspond to the absolute amount of a compound present in the samples, but they were calculated and used only for evaluating the differences between the analyzed samples.

2.9. Statistical Analyses

All the results were reported as an arithmetic average and the standard deviation was calculated from the replicates. Statistical analyses were performed using the Statgraphics Plus 4.0 package (Statgraphics Technologies, Inc., The Plains, VA, USA). The differences between the mean scores were determined using the analysis of variance (ANOVA). The differences between the means of multiple groups were analyzed using a two- or three-way ANOVA with Tukey's multiple range tests ($\alpha = 0.05$, $p < 0.05$).

3. Results and Discussion

3.1. Essential Oil Composition

The chemical compositions of the essential oils obtained from lemon peels, clove buds, and juniper berries by steam distillation are presented in Table 1. The GC–MS analysis indicated that the essential oils obtained were a mixture of various compounds, which mainly included mono- and sesquiterpene hydrocarbons and their oxygenated derivatives. Among the 120 identified components representing 96.59–99.54% of the total oils, 34, 47, and 72 belonged to the etheric oil obtained from clove buds, lemon peels, and juniper berries, respectively. Monoterpene hydrocarbons (68.53%) and oxygenated monoterpenes (24.78%) were the dominant group in the oil samples from lemon peels. However, the content of sesquiterpene hydrocarbons in these samples was relatively low (4.09%). The essential oil from juniper berries also had abundant monoterpenes (hydrocarbons and oxygenated monoterpenes) which accounted for 54.66% of the total. Sesquiterpenes (38.25%) and several nonterpene components, such as 2,3-hexanedione and hydroxymethyl 2-hydroxy-2-methylpropionate, were also identified in the studied oils. The main components present in the monoterpene hydrocarbon fraction of the essential oil from juniper berries were β-myrcene (6.45%), sabinene (6.19%), α-pinene (5.71%), limonene (4.78%), p-cymene (3.98%), and β-pinene (3.59%), while α-thujene (3.12%), α-terpinene (1.61%), and γ-terpinene (1.81%) were found in lower amounts. Among the oxygen-containing monoterpenes identified in the essential oil from juniper berries, linalool, α-campholenal, (E)-pinocarveol, α-phellandren-8-ol, p-cymen-8-ol, and verbenone were dominant. In turn, the sesquiterpene hydrocarbon fraction found in this essential oil consisted of α-cubebene (1.28%), β-elemene (2.01%), caryophyllene (1.54%), α-farnesene (1.67%), humulene (3.68%), α-muurolene (1.26%), germacerene D (1.54%), β-cubebene (2.88%), γ-amorphene (1.17%), γ-cadinene (1.74%), δ-cadinene (3.50%), and germacrene B (1.76%). A study showed that a total of 70 compounds were identified in the essential oils obtained from fresh juniper berries harvested from May 2010 to January 2011 [24]. Depending on the period of harvest of juniper fruits, monoterpene hydrocarbons accounted for 18.91–45.26% of the total volatiles of the obtained essential oils. The juniper berries collected in May had a similar content of monoterpene hydrocarbons as determined in the presented study, but they differed in the amount of the main representatives of this class of compounds. Compared to the content determined in the present study, α-pinene was at a higher level but sabinene and β-myrcene were identified in lower amounts. Moreover, monoterpene hydrocarbons were predominant in the essential oils isolated by steam distillation from ripe juniper berries collected from both the southwestern and central north regions of the Republic of Macedonia [25]. The main components of the monoterpene hydrocarbon fraction were α-

and β-pinene, β-myrcene, sabinene, and limonene. The content of compounds also differed the most in the oils obtained from the Macedonian juniper berries. Two Macedonian samples contained similar levels of β-pinene and sabinene, as determined in our study, whereas the amount of α-pinene was higher and limonene was slightly lower. The differences in the chemical composition of juniper oils, as well as in the amount of individual components, may be related to the geographical origin of the plant, the degree of ripeness of the berries, and the age of the berries used in the steam distillation [26]. The essential oils obtained from ripe and unripe wild juniper fruits in the Molise region were rich in α-pinene [24]. However, it has been reported that the amount of this compound may decrease as the fruit ripens.

The chemical compositions of the essential oils obtained from citrus peels, especially the content of monoterpene hydrocarbons and oxygenated monoterpenes, may also change during fruit ripening. Among the four types of citrus (bitter orange, orange maltaise, mandarin, and lemon), the highest level of limonene, which is one of the most dominant monoterpene hydrocarbons, was already reached at the immature fruit stage [27]. These observations could be useful to obtain essential oils with a high yield and with a high content of limonene. Our results also showed that limonene (31.51%) was the major volatile compound in yellow lemon peels. Furthermore, limonene (37.94%) and β-pinene (25.44%) were also dominant components in Tunisian essential oils isolated from lemon samples [28]. These active ingredients are often used for preparing perfumes, medicines, and flavoring agents, and their presence can have an impact on the antioxidant and antibacterial properties of essential oils. The essential oil of the Meyer lemon was found to exhibit the highest activity against *Bacillus cereus*, and that of the Interdonato lemon showed the highest activity against *E. coli* (the diameter of the inhibition zone: 15 mm) [29]. Other identified volatile components in the monoterpene hydrocarbon fraction included γ-terpinene (12.53%), α-pinene (2.68%), sabinene (3.06%), β-pinene (7.14%), β-myrcene (4.49%), and p-cymene (4.06%). The oxygenated monoterpene fraction included terpinolene (1.06%), linalool (2.76%), terpinene-4-ol (2.06%), α-terpineol (3.04%), neral (2.74%), geranial (4.05%), geraniol (3.38%), neryl acetate (1.81%), and geranyl acetate (1.84%). Sesquiterpenes were the third group of chemical compounds identified in the essential oil of lemon peels. Only six compounds of this group were identified and among them α-bergamotene and bisabolene were present at a concentration above 1%. These two sesquiterpenes were also identified in the volatile profile of essential oils obtained from five selected varieties of *C. limon* [30].

Similar to the lemon essential oil, oil obtained from clove buds is used as a food flavoring agent [31], in the preparations for gums and teeth, and in aromatherapy. It has the potential to be used as a natural preservative due to its antioxidant, antimicrobial, and antifungal activities [32,33]. As shown in Table 1, a total of 34 compounds, mainly belonging to the class of sesquiterpene hydrocarbons (10.71%) and others (83.91%), were identified in the clove essential oil. The major constituents of this oil were eugenol (62.74%), eugenol acetate (15.97%), caryophyllene (6.24%), humulene (3.54%), and chavicol (1.36%), while the content of the remaining components did not exceed 1%. Eugenol, eugenol acetate, caryophyllene, and humulene were also identified as the main components in the essential oil of cloves cultivated in the Mediterranean region of Turkey [34] and clove oil from Toli-Toli and Bali [35]. Our results were comparable to those reported for clove oil obtained from Toli-Toli, especially in terms of the content of eugenol and eugenol acetate, while the Turkish clove bud oil was richer in eugenol (87%) and contained less eugenol acetate compared to the content of these components determined in this study. In turn, the clove bud oil from Bangladesh was shown to consist of 31 compounds, of which eugenol was dominant. In comparison to our research, the percentage of eugenol was lower and amounted to 49.71% [36]. However, the content of this compound was about 15% higher in the oil obtained from clove leaves. Other studies showed that this bioactive compound was the most abundant at the flowering stage in mature trees [37]. On the other hand, the oil obtained from the stem of the clove plant in Amboina Island was characterized by a higher percentage of eugenol compared to the oil obtained from the buds and leaves of the clove plant in the same region but a lower percentage of eugenol acetate and caryophyllene [38]. Amelia et al. [39] indicated that the chemical composition of the clove essential oil may be influenced by the geographic origin of plant population and the growing conditions. The authors revealed the differences between the content of major and minor constituents in Java and Manado clove oils. Clove oil from Java (55.60%) contained a lower amount of eugenol compared to Manado oil (74.64%) but a higher amount of eugenol acetate (20.54%). According to the results of our study, the content of these two compounds was between the lowest and the highest values reported for clove oils from Java and Manado. The presence of eugenol acetate may contribute more to the sweet aroma of clove oil, while eugenol and chavicol contribute to its spicy odor.

Table 1 presents the data on the yield of the three studied essential oils obtained by steam distillation. Clove oil was obtained with the highest yield (3.65%), while the yield of lemon peel oil was the lowest (0.63%). In turn, the yield of juniper berry oil was 0.95%. The chemical composition of the studied etheric oils and their yields could be influenced by seasonal variation, environmental conditions, the age of the plant, the latitude and altitude of growing site, the degree of maturity of the raw material used for distillation, and the part of the plant and method used to isolate the essential oil [37,39,40].

3.2. The Effect of Essential Oils on the Viability of Bifidobacterium in In Vitro Conditions

Each of the surveyed essential oils was tested for antibacterial action. Some of the cited studies have indicated that Gram-positive bacteria are more sensitive to these essential oils [7,11]. Thus, it could be assumed that the studied 7ifidobacterial strains, belonging to the Gram-positive bacterial group, should be inhibited by these oils. Table 2 presents the sizes of the inhibition zones observed for the studied bifidobacteria strains under the influence of the tested essential oils.

Two of the tested bifidobacteria strains (strains 45/03 and 89) were the most influenced by the essential oil of the lemon peel. Their growth inhibition zones differed from that of the remaining tested strains, as revealed by significant differences in the statistical analyses. Strain 89 showed the largest average inhibition zone, and the inhibition zone of strain 45/03 was only slightly smaller (Table 2) The average sizes of the zones of these strains were 0.98 and 0.96 mm, respectively. The other strains did not show a sensitivity to the lemon peel essential oil and their mean growth inhibition zones, observed with all the used concentrations of essential oils, were 0 mm. The obtained results indicate that the size of the bacterial growth zone is significantly determined by the bacterial strain. The maximum size of the zone of growth inhibition observed with this essential oil was, at the same time, the smallest when compared with the largest zones of growth inhibition observed with the other tested oils.

Table 1. Chemical compositions of essential oils from lemon peel (*C. limon* L. Burm.), clove bud (*S. aromaticum* L.), and juniper berry (*J. communis* L.) determined by GC–MS.

No.	Compound	KI [1]	Content (%) [2]			No.	Compound	KI	Content (%)		
			Lemon Peel	Clove Bud	Juniper Berry				Lemon Peel	Clove Bud	Juniper Berry
1.	2,3-Hexanedione	793	- [3]	-	0.42	61.	(E,E)-2,4-Decadienal	1314	-	-	0.22
2.	2-Heptanone	889	-	0.62	-	62.	Methyl geranate	1323	-	-	0.49
3.	Heptanal	896	0.05	-	-	63.	Chavicol acetate	1329	-	0.13	-
4.	α-Tricyclene	926	-	-	0.31	64.	Citronellol acetate	1350	0.25	-	
5.	α-Thujene	926	1.34	-	3.12	65.	α-Terpinyl acetate	1351	-	-	0.39
6.	α-Pinene	934	2.68	-	5.71	66.	α-Cubebene	1351	-	0.04	1.28
7.	Camphene	945	0.17	-	0.63	67.	Neryl acetate	1354	1.81	-	-
8.	Sabinene	969	3.06	-	6.19	68.	Eugenol	1356	-	62.74	-
9.	β-Pinene	972	7.14	-	3.59	69.	Geranyl acetate	1373	1.84	-	-
10.	β-Myrcene	985	4.49	-	6.45	70.	α-Copaene	1376	-	0.35	-
11.	Decane	996	-	0.08	-	71.	Tetradecene	1390	0.02	-	-
12.	Ethyl hexanoate	998	-	0.05	-	72.	Tetradecane	1398	0.08	-	-
13.	Octanal	999	0.66	-	-	73.	β-Elemene	1391	-	-	2.01
14.	α-Phellandrene	1002	0.28	-	0.67	74.	Dodecanal	1399	0.14	-	-
15.	δ-3-Carene	1007	0.02	-	0.73	75.	α-Gurjunene	1407	-	-	0.49
16.	α-Terpinene	1019	0.99	-	1.61	76.	Longifolene	1408			0.13
17.	p-Cymene	1026	4.06	-	3.98	77.	Caryophyllene	1408	0.45	6.24	1.54
18.	Eucalyptol	1031	-	0.02	-	78.	β-Copaene	1416	-	-	0.27
19.	Limonene	1033	31.51	0.04	4.78	79.	α-Bergamotene	1430	1.21	-	-
20.	2-Heptanol acetate	1044	-	0.71	-	80.	Aromadendrene	1439	-	-	0.13
21.	Ocimene	1047	0.26	0.11	-	81.	Cadina-3,5-diene	1437	-	-	0.14
22.	γ-Terpinene	1059	12.53	-	1.81	82.	α-Farnesene	1508	0.23	-	1.67

Table 1. *Cont.*

No.	Compound	KI [1]	Content (%) [2]			No.	Compound	KI	Content (%)		
			Lemon Peel	Clove Bud	Juniper Berry				Lemon Peel	Clove Bud	Juniper Berry
23.	1-Octanol	1062	0.05	-	-	83.	Humulene	1454	-	3.54	3.68
24.	cis-Sabinene hydrate	1069	0.08	-	-	84.	cis-Muurola-4(15),5-diene	1460	-	-	0.34
25.	cis-Linalool Oxide	1071	0.08	-	-	85.	Cadina-1(6),4-diene	1485	-	-	0.45
26.	Terpinolene	1085	1.06	-	-	86.	α-Muurolene	1499	-	0.13	1.26
27.	2-Nonanone	1091	-	0.45	-	87.	Germacrene D	1480	-	-	1.54
28.	Methyl benzoate	1091	-	0.24	-	88.	β-Cubebene	1420	-	-	2.88
29.	Linalool	1095	2.76	-	1.03	89.	Pentadecane	1450	0.14	0.13	-
30.	Nonanal	1110	0.73	-	-	90.	β-Selinene	1485	-	-	0.41
31.	Citronellal	1149	0.74			91.	Valencene	1489	0.19	-	-
32.	Acetic acid, phenylmethyl ester	1170	-	0.26		92.	Bicyclosesquiphellandrene	1482	-	-	1.61
33.	(Z)-p-Menth-2-en-1-ol	1188	-	-	0.76	93.	γ-Amorphene	1483	-	0.04	1.17
34.	α-Campholenal	1195	-	-	1.31	94.	α-Selinene	1494	-	0.26	-
35.	(E)-Pinocarveol	1139	-	-	0.75	95.	β-Elemene	1391	-	-	0.42
36.	Camphor	1143	-	-	0.59	96.	Bisabolene	1499	1.97	-	-
37.	α-Phellandren-8-ol	1166	-	-	2.78	97.	γ-Cadinene	1512	-	0.17	1.74
38.	Pinocarvone	1162	-	-	0.23	98.	δ-Cadinene	1510	0.04	-	3.50
39.	Terpinene-4-ol	1177	2.06	0.06	-	99.	Eugenol acetate	1524	-	15.97	-
40.	p-Cymen-8-ol	1180	-	-	2.34	100.	Epizonarene	1497	-	-	0.55
41.	α-Terpineol	1186	3.04	-	-	101.	Cadine-1,4-diene	1532	-	-	1.29
42.	Hydroxymethyl 2-hydroxy-2-methylpropionate	1189	-	-	2.29	102.	α-Amorphene	1506	-	-	0.56
43.	Methyl salicylate	1190	-	0.92	-	103.	α-Calacorene	1542	-	0.11	0.87

Table 1. *Cont.*

No.	Compound	KI [1]	Content (%) [2]			No.	Compound	KI	Content (%)		
			Lemon Peel	Clove Bud	Juniper Berry				Lemon Peel	Clove Bud	Juniper Berry
44.	Myrtenal	1193	-	-	0.67	104.	Germacrene B	1556	-	-	1.76
45.	Decanal	1203	0.13	-	-	105.	Spathulenol	1576	-	-	0.67
46.	Octyl acetate	1210	0.04	-	-	106.	Caryophyllene oxide	1581	-	0.96	0.68
47.	Verbenone	1214	-	-	1.42	107.	Humulene epoxide	1606	-	0.09	0.51
48.	Chrysanthenyl acetate	1232	-	-	0.12	108.	Di-epi-1,10-cubenol	1614	-	-	0.70
49.	Cumin aldehyde	1238	-	-	0.28	109.	α-Cadinol	1653	-	-	2.83
50.	Neral (E-Citral)	1239	2.74	-	-	110.	Cadalene	1674	-	-	0.39
51.	Carvone	1239	0.46	0.13	0.52	111.	Oplopanone	1733	-	-	1.02
52.	Chavicol	1253	-	1.36	-	112.	Benzyl benzoate	1762	-	0.34	-
53.	Geraniol	1253	3.38	-	-	113.	α-Copaene-11-ol	1891	-	-	0.15
54.	Methyl citronellate	1261	-	-	0.62	114.	Biformene	1979	-	-	0.16
55.	Geranial (Z-Citral)	1266	4.05	-	-	115.	m-Camphorene	1944	-	-	0.39
56.	Perilla aldehyde	1268	0.31	-	-	116.	p-Camphorene	1977	-	-	0.19
57.	Bornyl acetate	1280	0.07	-	0.78	117.	13-Epimanool	2056	-	-	0.65
58.	Thymol	1290	0.05	-	-	118.	Squalene	2790	-	0.08	-
59.	2-Undecanone	1291	-	0.09	0.85	119.	Heptacosane	2705	-	0.09	-
60.	Undecanal	1309	0.10	-	-	120.	Octacosane	2804	-	0.21	-
			Monoterpene hydrocarbons (%)						68.53	0.26	39.58
			Oxygenated monoterpenes (%)						24.78	0.66	15.08
			Sesquiterpene hydrocarbons (%)						4.09	10.71	31.69
			Oxygenated sesquiterpene (%)						-	1.05	6.56
			Other (%)						2.14	83.91	5.56
			Total identified (%)						99.54	96.59	98.47
			Oil yield (%)						0.63	3.65	0.95

[1] Kovats' Retention Index on DB-5MS column; [2] relative content expressed as percentage of the total oil composition obtained by GC/MS; [3] not identified.

Table 2. Size of the growth inhibition zones (mm) of *Bifidobacterium* strains observed with selected essential oils (n = 4).

Bifidobacteria Strain	The Concentration of the Essential Oil [%]	Lemon Peel Mean	±SD (mm)	Clove Bud Mean	±SD (mm)	Juniper Berry Mean	±SD (mm)
HN 019	100	0.0	±0.0 [a]	5.6	±0.4 [i,j]	0.0	±0.0 [a]
	75	0.0	±0.0 [a]	3.0	±0.2 [e,f]	0.0	±0.0 [a]
	50	0.0	±0.0 [a]	3.0	±0.1 [e,f]	0.0	±0.0 [a]
	25	0.0	±0.0 [a]	2.6	±0.7 [e]	0.0	±0.0 [a]
	0	0.0	±0.0 [a]	0.0	±0.0 [a]	0.0	±0.0 [a]
AD 50	100	0.0	±0.0 [a]	5.6	±0.5 [i,j]	1.8	±0.1 [c,d]
	75	0.0	±0.0 [a]	5.0	±1.2 [h,i]	0.0	±0.0 [a]
	50	0.0	±0.0 [a]	4.0	±0.0 [f,g]	0.8	±0.9 [b]
	25	0.0	±0.0 [a]	4.5	±1.4 [g]	0.6	±0.8 [a,b]
	0	0.0	±0.0 [a]	0.0	±0.0 [a]	0.0	±0.0 [a]
45/03	100	1.8	±1.0 [c,d]	4.0	±0.5 [f,g]	1.9	±0.7 [c,d]
	75	1.5	±1.0 [c,d]	2.7	±0.4 [e]	1.3	±1.0 [c]
	50	1.0	±1.0 [b]	3.4	±0.3 [e,f]	1.0	±1.0 [b,c]
	25	0.5	±0.6 [a,b]	1.1	±0.5 [b,c]	0.4	±0.5 [a,b]
	0	0.0	±0.0 [a]	0.0	±0.0 [a]	0.0	±0.0 [a]
AD 600	100	0.0	±0.0 [a]	7. 8	±2.6 [k]	3.3	±1.8 [e,f]
	75	0.0	±0.0 [a]	7.7	±2.3 [k]	2.8	±1.3 [e,f]
	50	0.0	±0.0 [a]	3.5	±0.3 [e,f]	0.8	±0.9 [b]
	25	0.0	±0.0 [a]	1.2	±1.0 [b,c]	0.0	±0.0 [a]
	0	0.0	±0.0 [a]	0.0	±0.0 [a]	0.0	±0.0 [a]
KNA 1/08	100	0.0	±0.0 [a]	10.4	±4.2 [m]	3.5	±1.2 [e,f]
	75	0.0	±0.0 [a]	8.6	±4.8 [l]	3.3	±0.7 [e,f]
	50	0.0	±0.0 [a]	5.5	±0.1 [i]	2.8	±0.7 [e,f]
	25	0.0	±0.0 [a]	4.8	±0.5 [g,h]	2.2	±1.1 [d,e]
	0	0.0	±0.0 [a]	0.0	±0.0 [a]	0.0	±0.0 [a]
Bb-12	100	0.0	±0.0 [a]	14.9	±2.1 [n]	4.6	±2.2 [g,h]
	75	0.0	±0.0 [a]	14.5	±3.0 [n]	5.6	±0.3 [i]
	50	0.0	±0.0 [a]	0.0	±0.0 [a]	5.1	±1.3 [h,i]
	25	0.0	±0.0 [a]	10.7	±2.9 [m]	4.9	±0.7 [h]
	0	0.0	±0.0 [a]	0.0	±0.0 [a]	0.0	±0.0 [a]
89	100	1.8	±1.1 [c,d]	5.8	±0.0 [j]	2.9	±1.0 [e,f]
	75	1.3	±1.0 [c]	3.6	±0.3 [e,f]	3.2	±1.1 [e,f]
	50	1.3	±1.0 [c]	2.0	±0.0 [d]	2.3	±1.0 [e]
	25	0.5	±0.6 [a,b]	4.5	±1.3 [g,h]	1.4	±1.0 [c]
	0	0.0	±0.0 [a]	0.6	±0.3 [a,b]	0.0	±0.0 [a]

Means with different lowercase letters (a–n) are significantly different ($p < 0.05$) using a three-way ANOVA.

Fisher and Philips [41] showed that the essential oils obtained from sweet oranges and lemons exhibited effective antibacterial activities compared to bergamot oil, citral, and linalool. The lemon essential oil contained the largest amount of limonene, but the lowest amount of citral compared to the essential oils obtained from sweet oranges and bergamot. Limonene did not show antibacterial action. The authors also stated that the Gram-positive bacteria were more sensitive to the antibacterial action of essential oils compared to the Gram-negative bacteria. Among the studied Gram-positive bacteria, *S. aureus* was found to be the most resistant. The results of previous research [41] confirmed that of our work, which showed that the lemon oil was characterized by weak antibacterial action. Prabuseenivasan et al. [42] tested the effect of 21 different essential oils on four species of Gram-negative bacteria (*E. coli*, *Klebsiella pneumoniae*, *Pseudomonas aeruginosa*, and *Proteus vulgaris*) and two species of Gram-positive bacteria (*Bacillus subtilis* and *S. aureus*) and showed that lemon oil moderately inhibited the growth of the tested bacteria.

Bevilacqua et al. [43] referred to the research of other authors on the influence of essential oils on the growth of microorganisms. Based on the studies that analyzed the antibacterial effects of essential oils and their active substances, these authors argued that the bioactivity of essential oils was generally associated with phenolic compounds. Phenolic compounds dissolve in a lipid diaphragm layer and change their membrane liquidity, as well as causing the leakage of intracellular components and the scattering of transmembrane gradient H^+. Thus, the inhibitory effect of the citrus extract on the growth of bacteria can be linked with the substances it contains. Citrus extract was found to be a natural source of antioxidants, flavonoids (rutin, naringin, quercetin, and naringenin), and other phenolic compounds, such as limonene, linalool, and citral. Therefore, it has been proposed that the bioactivity of the citrus extract may be the effect of the synergistic activity of these components. The results reported by Prabuseenivasana et al. [42] on of the antibacterial effects of lemon essential oils differed from that obtained in our work. Prabuseenivasan et al. [42] showed that lemon oil exhibited a stronger antibacterial effect than that observed in the present work. This could be due to the differences in the methods used for analyses, as well as the bacterial species tested. In their study on the influence of lemon essential oils on anaerobic bacteria that affect the teeth and oral cavity, which were grown from materials collected from patients, Kędzia et al. [44] found that the essential oil decreased the number of the analyzed bacteria. These results confirmed that Gram-positive bacteria were more sensitive to the antibacterial effects of lemon essential oils compared to Gram-negative bacteria. Among the Gram-positive bacteria group analyzed by Kędzia et al. [44], there were two *B. breve* strains, and these bacteria were inhibited by the citric essential oil. The minimum inhibitory concentration (MIC) values determined for both strains were the same, and amounted to 15 mg/mL. In summary, lemon oil inhibited the growth of certain types of bacteria, with a stronger effect on Gram-positive bacteria. However, its activity was weaker compared to the essential oils obtained from other plants such as cloves, cinnamon, and bergamot.

In our study, the clove bud essential oil was particularly distinguished from other tested essential oils due to the size of the bacterial growth zone (Table 2). This essential oil had a statistically significantly higher inhibitory effect on the growth of all tested *Bifidobacterium* strains. Compared to the other two essential oils, clove oil appeared to have significantly higher amounts of carnation oil.

Bevilacqua et al. [43] showed that eugenol did not inhibit the growth of *Lactobacillus* bacteria (*L. plantarum* and *L. brevis*). Badei et al. [45] compared the antimicrobial activity of three spices, cardamom, cinnamon, and clove, and essential oils obtained from them, using 13 strains of Gram-positive and Gram-negative bacteria, seven fungal strains, and two yeast strains. They showed that the clove essential oil had the strongest effect. Clove oil also showed stronger antimicrobial activity than phenol. Furthermore, this essential oil was one of the three tested essential oils that showed antibacterial effects on pathogenic *S. aureus* bacteria. These results confirm the strong antibacterial effect of clove bud essential oils observed in our work. It can be assumed that the antibacterial effect of clove bud essential oils may be related to the content of the active substances present in it. The most active and abundant component in the clove essential oil is eugenol, which has proven antibacterial, as well as antifungal, effects. Makuch et al. [46] also confirmed the antimicrobial action of the clove essential oil on the yeast-like fungi *Candida albicans*, the Gram-negative bacteria *E. coli*, and the Gram-positive bacteria *Staphylococcus epidermidis*, as well as the high content of eugenol in this oil. The average size of the growth inhibition zones observed in their study on *C. albicans*, *E. coli*, and *S. epidermidis* was 11.76 ± 0.7, 5.16 ± 0.2, and 5.10 ± 0.3 mm, respectively. The results presented by Makuch et al. [46] are in line with that of our work on the activity of the clove bud essential oil relative to Gram-positive bacteria. Liu et al. [47] performed many studies confirming the strong antimicrobial activity of clove essential oils. Ayoola et al. [48] and Ali et al. [49] also studied the composition and antimicrobial activity of clove essential oils. Similar to our study, the results of Ayoola et al. [48] showed that eugenol is the main component of the clove essential oil, and caryophyllene, eugenol acetate, and α-humulene are also abundant. The cited researchers also showed the inhibitory effects of oil on Gram-positive and Gram-negative bacteria, as well as yeast-like fungi, with the smallest growth inhibition zone observed for *Enterobacter cloacae* and *E. coli* bacteria (10 mm) and the largest for the yeast-like fungi *C. albicans* (35 mm). Gram-positive bacteria (*S. aureus*) were shown to be more sensitive to clove essential oils (the largest growth inhibition zone was 23 mm), where the largest inhibition zones were recorded for samples treated with the largest concentration of the clove essential oil.

Based on the strength of the inhibitory effects on the growth of *Bifidobacterium* strains studied in in vitro conditions, juniper berry essential oil ranked second among the three tested essential oils. The highest size of the bifidobacterial growth inhibition zone observed with this essential oil was 5.6 mm (Table 2). Statistical analyses revealed that the size of growth inhibition zones depends on the *Bifidobacterium* bacterial strain tested. The most sensitive strain was *B. animalis* subsp. *lactis* Bb-12. With the addition of 25% juniper berry essential oil, the size of the growth inhibition zone of the strain observed was 4.9 mm and it increased with the increase in the concentration of the essential oil. At a concentration of 50% and 75%, the average growth inhibition zone was 5.1 and 5.6 mm, respectively. A slight reduction in the growth inhibition zone was noted at a concentration of 100% and the size of the zone was 4.6 mm. However, HN 019 strains did not show sensitivity to the antibacterial effects of the juniper berry essential oil, because the average growth inhibition zones observed with the use of each concentration was 0 mm.

Şerban et al. [50] used two species of bacteria (*E. coli* and *S. aureus*) and one species of yeast (*C. albicans*) to determine the properties of the juniper berry essential oil and several other essential oils. The authors recorded growth inhibition zones of 14, 12, and 17 mm for *E. coli*, *S. aureus*, and *C. albicans*, respectively, using juniper berry essential oils. They observed larger growth inhibition zones of the analyzed microorganisms with the use of the juniper berry essential oil, compared to our work. This may be due to differences in the methods used for assessing growth inhibition, the sensitivity of microorganisms, and the composition of individual essential oils. Rezvani et al. [51] analyzed the composition of the juniper berry essential oil and the antibacterial activity of this oil by the well diffusion method using *S. aureus*, *P. Aeruginosa*, and *E. coli*. They identified 30 compounds but the GC–MS analysis showed that the main component of the juniper berry essential oil was α-piren (45.63%), which is in line with the research of other authors [52]. The results reported by Pepeljnjak et al. [10] also confirmed the results obtained in our work. In the study by Rezvani et al. [51], the juniper berry essential oil showed an antibacterial effect on all three tested bacteria, with *P. aeruginosa* found to be the most resistant. A growth inhibition zone of 4.5 mm was observed only with the use of a 100% concentration of essential oil. The most sensitive among the tested bacteria was *E. coli* and its growth inhibition zones that were observed for concentrations of essential oils of 0%, 10%, 20%, 50%, and 100% were measured at 0, 6.1, 7.2, 8.3, and 2.2 mm, respectively. The inhibition zones of the Gram-positive *S. aureus*, similar to the *Bifidobacterium* strains tested in this work, that were observed with 0%, 10%, 20%, 50%, and 100% concentrations of essential oils were measured at 0, 2.1, 4.8, 5.0, and 3.5 mm, respectively. These results on *S. aureus* are very similar to those found in our work for *B. animalis* subsp. *lactis* Bb-12, which turned out to be the most sensitive.

Some researchers have shown that essential oils obtained from juniper berries did not show any antibacterial action and only exhibited an antifungal effect [11], or weak antibacterial and antifungal activities [53]. Glišić et al. [54] examined the antibacterial activity of the juniper berry essential oil, with different fractions, in their study. These authors reported that the native juniper berry essential oil did not show a strong antibacterial effect compared to fractions containing a high amount of α-pinene and a mixture of sabinene and α-pinene, which showed the highest antimicrobial activity on almost all tested microorganisms (*B. cereus*, *E. coli*, *Listeria monocytogenes*, *Corynebacterium* sp., *P. aeruginosa*, *S. aureus*, *C. albicans*, *Alternaria* sp., *Aspergillus nidulans*, and *Aspergillus niger*). The MIC values estimated by Prabuseenivasana et al. [42] for 21 different essential oils and four species of Gram-negative bacteria (*E. coli*, *K. pneumoniae*, *P. aeruginosa*, and *P. vulgaris*) and two species of Gram-positive bacteria (*B. subtilis* and *S. aureus*) indicated that these results are similar to those obtained in our work. For Gram-positive bacteria, the MIC value of the lemon essential oil (12.8 and 12.8 mg/mL, respectively) was higher than that of the clove essential oil (6.4 and 3.2 mg/mL, respectively) which confirms that the clove essential oil has a stronger antibacterial effect than the lemon essential oil. The results obtained in our work, to some extent, coincide with those of Bevilacqua et al. [43], who studied the antimicrobial activity of a eugenol extract (the main component of the clove essential oil, limonene),the main component of the lemon essential oil, and a citrus extract (a mixture of different components such as citric acid, sodium acid, citrusic polyphenols, bioflavoproteins, citrus sugars, and citric acid) on three bacterial species, including two lactic acid bacteria (LAB), and three yeast species. The MIC values determined by Bevilacqua et al. [43] indicated that limonene showed less antimicrobial activity than eugenol on the tested bacteria (>1000 and >600 ppm, respectively). It should be noted that several compounds that were identified as strong antioxidants belong to the group of flavonoids found in the citrus extract, as well as the groups of other phenolic compounds. Bevilacqua et al. [43] showed that these substances acted synergistically. It can be argued that limonene, as a single compound, did not exhibit as strong a antimicrobial effect as the citrus extract that contained many additional compounds that synergistically increased its antimicrobial effect. This suggests that the action of a particular essential oil on a given bacterial species depends on the composition of the essential oil, the interaction between its components, and the concentration of the essential oil used. The effect of an essential oil is also influenced by the conditions under which its activity is tested and by the bacterial species and strains used for the analysis, as each strain is different and exhibits different sensitivities. Hammer et al. [55] emphasized that it is difficult to compare the results obtained by various authors regarding the antibacterial action of different essential oils. This is due to the fact that the composition of oils obtained from the same plants may differ due to changes that occur in plants under the influence of climatic and environmental factors. Essential oils with the same name can often be produced from different plant species. Moreover, the type of method used to measure the sensitivity of microorganisms to the essential oil is of great importance. Depending on the methods used, the tested microorganisms may experience different growth conditions and may be exposed to the essential oil to different degrees. The solubility of the emulsifier and essential oil, as well as its components in a given environment, may also vary. To the best of our knowledge, studies such as those performed in our work, i.e., using a well-diffusion method and exploring the influence of various concentrations of clove bud, lemon peel, and juniper berry essential oils on the survival of *Bifidobacterium* bacteria have not been conducted earlier (the well-diffusion method is applied in experiments concerning plant extracts rather than essential oils).

3.3. The Effect of Essential Oil Concentrations on the Viability of B. animalis Subsp. lactis Bb-12 in Cream during Fermentation and after 21 Days of Storage

The strain *B. animalis* subsp. *lactis* Bb-12 that was selected for the experiments was identified as the most sensitive to the tested essential oils. If there was no influence of the tested essential oils and their concentrations on the viability of the most sensitive strain (*B. animalis* subsp. *Lactis* Bb-12) in the cream, we assumed, with a high degree of certainty, that the other tested bifidobacterial strains would also not show a sensitivity to the tested essential oils added to the fermented cream. The number of *B. animalis* subsp. *lactis* Bb-12 cells present in our samples of fresh cream (before fermentation) ranged from 8.1 to 8.3 log(CFU/mL), regardless of the type of oil used and the concentration of the oil tested (Table 3). After fermentation, a change in the population of bifidobacterial cells was recorded and the statistical analyses showed that the final population of *B. animalis* subsp. *Lactis* Bb-12 cells in the obtained sour cream was not influenced by the type of essential oils, or the concentration used. The sour cream samples were stored for 21 days under refrigeration conditions to investigate the deferred or cumulative influence of the tested essential oils on the viability of *B. animalis* subsp. *lactis* Bb-12. It was noted that refrigerated storage caused a statistically significant reduction in the population of *B. animalis* subsp. *lactis* in the sour cream samples; however, the dynamics of changes did not depend on the type of essential oils, or the concentration of the oils used. In all stored sour cream samples, the number of live *B. animalis* subsp. *lactis* Bb-12 cells ranged from 7.5 to 7.8 log(CFU/mL).

Table 3. Effects of selected essential oils on *B. animalis* subsp. *lactis* Bb-12 population (expressed as log(CFU/mL)) in cream and sour cream samples (n = 4).

The Concentration of the Essential Oil Added to Cream	Samples	Lemon Peel [Log(CFU/mL)]		Clove Bud [Log(CFU/mL)]		Juniper Berry [Log(CFU/mL)]	
		Mean	±SD	Mean	±SD	Mean	±SD
100%	Cream	8.3	±0.4 [a]	8.2	±0.4 [a]	8.3	±0.4 [a]
	Sour cream	9.3	±0.1 [b]	9.3	±0.1 [b]	9.1	±0.1 [b]
	Stored sour cream	7.8	±0.1 [c]	7.6	±0.1 [c]	7.5	±0.1 [c]
50%	Cream	8.3	±0.4 [a]	8.3	±0.4 [a]	8.1	±0.4 [a]
	Sour cream	9.3	±0.1 [b]	9.3	±0.1 [b]	9.2	±0.0 [b]
	Stored sour cream	7.7	±0.2 [c]	7.7	±0.1 [c]	7.6	±0.2 [c]
25%	Cream	8.2	±0.4 [a]	8.3	±0.4 [a]	8.2	±0.4 [a]
	Sour cream	9.5	±0.1 [b]	9.3	±0.1 [b]	9.4	±0.0 [b]
	Stored sour cream	7.6	±0.4 [c]	7.7	±0.2 [c]	7.6	±0.3 [c]
12.5%	Cream	8.3	±0.4 [a]	8.2	±0.4 [a]	8.3	±0.4 [a]
	Sour cream	9.4	±0.2 [b]	9.4	±0.0 [b]	9.3	±0.0 [b]
	Stored sour cream	7.7	±0.2 [c]	7.5	±0.3 [c]	7.7	±0.1 [c]
0%	Cream	8.1	±0.4 [a]	8.3	±0.4 [a]	8.2	±0.4 [a]
	Sour cream	9.3	±0.2 [b]	9.3	±0.2 [b]	9.3	±0.2 [b]
	Stored sour cream	7.8	±0.1 [c]	7.7	±0.1 [c]	7.8	±0.1 [c]

Means with different lowercase letters (a–c) are significantly different ($p < 0.05$) using a three-way ANOVA.

Our previous study examined the effect of commercial essential oils on the growth of selected LAB [56,57], as well as the effect of plant extracts on the populations of LAB, which showed that LAB cultures tolerated the extracts present in milk well, at an amount of up to 3.0% [58]. The effects of coriander essential oils (concentrations from 1% to 100%) on the growth inhibition of the studied *Lactobacillus* bacteria (7 *L. acidophilus* strains, 7 *L. casei* strains, 3 *L. delbrueckii* strains, 3 *L. plantarum* strains, 3 *L. rhamnosus* strains, *L. fermentum*, *L. helveticus*, and *L. paracasei*) were tested using the well diffusion method. The growth inhibition zones did not exceed 6.3 mm. The low sensitivity of LAB to coriander oil favored the production of fermented products by these bacteria [58]. Until now, the influence of essential oils or plant extracts on the viability of bifidobacteria has not been investigated. The available data in the literature mainly related to LAB and other essential oils, not those studied in our work, and only a few studies included bifidobacteria in this subject of research.

Shahdadi et al. [59] determined the survival of *L. acidophilus* LA-5 and *B. animalis* subsp. *lactis* Bb-12 in stored probiotic yogurt drinks enriched with essential oils from green mint (*Mentha spicata*), eucalyptus (*Eucalyptus camaldulensis*), long-leaf mint (*Mentha longifolia*), and *Ziziphora tenuior* L. The authors found that the number of bacterial cells decreased during the storage of yogurts, and yogurts containing the eucalyptus essential oil showed the highest inhibitory effect. It is worth mentioning that the largest number of *L. acidophilus* LA-14 and *B. animalis* subsp. *lactis* Bb-12 cells were observed in yogurts containing the green mint essential oil. Interestingly, *B. animalis* subsp. *lactis* showed better survival rates in all yogurt samples compared to *L. acidophilus*. The results presented by Shahdadi et al. [59] confirmed those obtained in our work despite the differences, among others, in the types of essential oils and dairy products tested. Voosogh et al. [60] also found a decrease in the number of *B. lactis* Bb-12 and *L. acidophilus* LA-5 during storage in an Iranian milk drink (doogh) containing essential oils from mint. The essential oil was used at two concentrations (1% and 2%) and its effect on the number of bacterial cells was found to depend on its concentration. Similar to Shahdadi et al. [59], Voosogh et al. [60] noticed that *L. acidophilus* were more sensitive than *B. lactis*. After 8 weeks of storage, the number of bifidobacterial cells in the milk drink decreased by 2 log(CFU/mL), while the number of lactobacilli cells was below the detection limit. This confirms the higher viability of *B. animalis* subsp. *lactis* Bb-12, as observed in our work. Panesar and Shinde [61] also studied the viability of probiotic *L. acidophilus* and *B. bifidum* bacteria, along with cultures of yogurt bacteria in the yogurt-containing *Aloe vera* extract, and found that the *Bifidobacterium* population was higher than the lactobacilli population. Bayoumi [62] analyzed the influence of four essential oils, clove, cardamom, cinnamon, and peppermint, on the growth of various LAB in yogurts. They found that the tested bacteria were quite sensitive to all the studied essential oils and that the sensitivity of the tested microorganisms depended on the concentration of oil and the bacteria used.

Smith-Palmer et al. [63] proposed several theories explaining the influence of essential oils on microorganisms in various food products. One of the theories was based on fat content. Fat has the ability to form a protective layer around the bacterial cells in food and protects them against the action of essential oils. According to another theory, fat can absorb the essential oil used and can reduce its content in the aqueous fraction and, thus, its antimicrobial effect. This can explain the difference between our observations and the results of Khodaparast et al. [64] and Behard et al. [65] who used yogurt in studies, while our work was based on cream. The fat content in yogurt varies from 0.0% to 4.5%, while the cream used in the present work contained 36% fat. The number of probiotic bacteria cells found by Behard et al. [65] is smaller than the number of bacterial cells observed in sour cream samples in our work. Behard et al. [65] added cinnamon extract and liquorice to yogurt containing *L. acidophilus* LA-5 and found that the extract caused a reduction in the number of probiotic bacterial cells.

3.4. Fatty Acid Composition

GC–MS was used as a reliable method for the analysis of the fatty acid composition and the content of polyunsaturated, monounsaturated, and saturated fatty acids in the fat extracted from the cream samples fermented with *B. animalis* subsp. *lactis* Bb-12 and enriched with essential oils from the lemon peel, clove bud, and juniper berry. The fatty acid profile of the studied fat samples was determined after 21 days of storage at 6 °C and is presented in Table S1. The GC–MS analysis revealed the presence of 20 fatty acids. Saturated fatty acids were the dominant group in all the tested fat samples. In our work, their content ranged from 67.36% (fat extracted from cream not subjected to fermentation) to 68.32% (stored sour cream without essential oils). The most abundant saturated fatty acids were myristic acid (C14:0), palmitic acid (C16:0), and stearic acid (C18:0). The content of myristic acid (C14:0) was the greatest in fermented cream, containing 50% clove bud essential oils amounting to 11.84%, while in the remaining fat samples, it ranged between 11.58% and 11.78%. No statistically significant differences were noted in the content of this acid between the sour cream and stored sour cream samples. Moreover, the content of palmitic acid was slightly higher in samples fermented with *Bifidobacterium* compared to the nonfermented cream samples, except in the case of those samples enriched with lemon peel, clove bud, or juniper berry essential oils at their highest concentrations (100%). Similar observations were made regarding the content of stearic acid (C18:0) in the analyzed fat samples. Short-chain fatty acids (C4:0–C10:0) accounted for 6% of saturated fatty acids. Along with butyric (C4:0), caproic (C6:0), and caprylic acid (C8:0), capric acid (C10:0) was also found and its content was comparable in the analyzed fat samples, with the greatest amount noted in the fat extracted from sour cream samples enriched with 100% lemon peel and juniper berry essential oils and with 50% and 100% clove bud essential oils. The presence of C6:0, C8:0, and C10:0 fatty acids in ruminant milk may cause a specific aroma in milk and may exert beneficial effects on human health by inhibiting bacterial and viral growth, as well as by dissolving cholesterol deposits [66].

Compared to saturated fatty acids, the content of monounsaturated fatty acids in the analyzed samples ranged from 28.24% to 29.07%. Among these acids, oleic acid (C18:1 *n*-9) was the most abundant and its share in the total composition of fatty acids was, on average, 21.85%. In the fat samples extracted from sour cream enriched with three essential oils at the level of 100%, compared to the fat samples from sour cream not containing essential oils, a slightly lower oleic acid content was found in the cream sample not treated with *B. animalis* subsp. *lactis* Bb-12 and in the sample enriched with essential oils, both after storage for 21 days and immediately after fermentation. The highest content of oleic acid was noted in the sour cream samples containing juniper berry and clove bud essential oils that were added to the fermented cream at concentrations of 12.5% and 25%, and the oleic acid content in these samples exceeded 22%. The other monounsaturated acids found were myristoleic (C14:1), palmitoleic (C16:1), gadoleic (C20:1), C18:1 11c, and C18:1 12c acids, in minor amounts.

Similarly, the content of polyunsaturated fatty acids in the fat separated from sour cream samples accounted for 3.22–3.58% of the total fatty acid composition. The highest total content of polyunsaturated fatty acids was found in the fat separated from nonfermented cream (3.58%) and the lowest was found in the sour cream samples enriched with juniper berry essential oils that were stored for 21 days. Among the polyunsaturated fatty acids, linoleic acid (C18:2 *n*-6) was dominant in all tested fat samples, with an average amount of 1.59%. The lowest content of this fatty acid was found in the sour cream samples enriched with the juniper berry essential oil, except for the samples enriched with the highest concentration of this essential oil (100%). In all the studied fat samples, conjugated linoleic acid C18:2 9c11t (CLA) was also found. Due to its unique structure, it inhibits the enzymes involved in the deposition of adipose tissue. It also reduces the synthesis of adipose tissue, intensifies lipolysis, and has been proven to exhibit anticancer, antidiabetic, anti-inflammatory, and anti-atherosclerotic properties [66,67]. In our study, the content of CLA in the extracted fat samples varied within narrow limits (0.85–0.93%), regardless of the type of fat studied. Both CLA and vaccenic acid (C18:1 11t), which is a precursor of CLA in human organisms, are the main trans fatty acids present in ruminant milk. Their content slightly exceeded 3% in the sour cream samples. The study conducted by Izsó et al. [68] showed that trans fatty acids accounted for slightly more than 2% of the fatty acid content of the sour cream samples.

Comparing the fatty acid profile of the cream fermented by the probiotic *Bifidobacterium* bacteria to that of nonfermented cream, one can notice the variations in the content of four monounsaturated fatty acids, as well as three fatty acids that differ in the degree of unsaturation. The cream fermented with *Bifidobacterium* was characterized by a higher content of myristic, palmitic, and stearic acid compared to nonfermented cream. However, it contained a lower amount of lauric, oleic, linoleic, and α-linolenic acid. The content of these acids also did not change significantly during their 21-day storage. In turn, Yilmaz-Ersan et al. [69] observed that during storage, cream fermented with *B. lactis* contained a higher amount of linoleic and α-linolenic acid compared to the control sample and the cream fermented with *B. lactis* had increased amounts of mono- and polyunsaturated fatty acids compared to the cream fermented with *L. acidophilus*. In turn, Laučienė et al. [70] observed that the profile of individual fatty acids did not change during the processing of sour cream samples, or their storage at 5 °C.

3.5. The Volatile Compound Profile of Cream and Sour Cream Samples

Some fermented foods are a rich source of bioactive components which may have beneficial effects on health. LAB used as starter cultures may allow the transformation of lactic acid and citrate, lactate, protein, and fats into volatile compounds which, along with amino acids and other products, may play a critical role in the flavor of the resulting products [71]. In our work, volatile compounds in cream samples enriched with essential oils from the lemon peel, clove bud, and juniper berry that was fermented with *B. animalis* subsp. *lactis* Bb-12 were identified by SPME-GC–TOF-MS and the data are presented in Table 4. The results are presented as the average relative peak area of three replicates (the peak area of the compound divided by the peak area of internal standard naphthalene d8). These did not correspond with the absolute amount of a compound identified in the studied samples, but were calculated and used only for evaluating the differences between the analyzed samples. The use of the SPME-GC–TOF-MS method allowed us to identify 22 volatile compounds belonging to the classes of aldehydes, ketones, acids, sulfur compounds, terpenes, and phenolic constituents. Two sulfur compounds, five ketones, and one aldehyde were found in the cream samples not treated with LAB. The ketones were represented by the following compounds: 2-butanone, 2-pentanone, 2-hexanone, 2-heptanone, and 2-nonanone. They were identified as the most important aroma components of cheeses with mold growth [72]. Among them, 2-pentanone, 2-heptanone, and 2-nonanone are characterized by a fruity floral fragrance, while 2-butanone and 2-hexanone have a sharp sweet smell. This class of carbonyl compounds is formed by the enzymatic oxidation of free fatty acids to β-keto acids and their decarboxylation, although 2-butanone is formed from diacetyl (2,3-butanedione), which is a product of lactose fermentation and citrate transformation. Only the content of 2-heptanone and 2-nonanone was found to be changed in the cream samples fermented with *B. animalis* subsp. *lactis* Bb-12, while the level of the remaining ketones was comparable to the nonfermented cream samples. In general, except for the stored sour cream samples containing the highest concentration of clove bud essential oils, all stored sour cream samples showed no increase in the identified ketones compared to the samples at the initial day. Moreover, 2,3-butanedione (diacetyl), a diketone responsible for the buttery flavor of the dairy product, was present in a lower amount in stored sour cream samples compared to unstored sour cream samples. However, the sour cream samples enriched with 12.5% essential oils contained a similar amount of 2,3-butanedione as that of unstored sour cream. It should be noted that diacetyl is produced by some species of the LAB family, including *Streptococcus, Leuconostoc, Lactobacillus*, and *Pediococcus*. Lew et al. [73] found higher diacetyl production in lactobacilli compared to bifidobacteria. The strain *L. casei* BT 1268 contained higher amounts of diacetyl in both intracellular and extracellular extracts, while *B. animalis* subsp. *lactis* BB 12 and *B. longum* BL 8643 showed a lower concentration in both fractions. In addition, the content of diacetyl in a product may increase or decrease during fermentation. Such changes can be explained by the fact that diacetyl is highly volatile and can evaporate at low temperatures, as well as from dry products. Among the volatile compounds identified in all tested samples, there were two sulfur compounds, namely, hexanal (aldehyde) and dimethyl sulfide. Dimethyl sulfide determines the flavor of cheese and may be present in yogurts [74] in varying amounts depending on the technique used for yoghurt-making and whether skim milk or full-fat milk is used for yogurt production. Sulfur compounds can be formed from the degradation of methionine by enzymes released by LAB. In our study, the stored sour cream samples containing the essential oils of clove buds and juniper berries had a higher amount of hexanal than the stored sour cream samples without additives. On the other hand, essential oil-enriched sour cream samples that were fermented and stored for 21 days had an increased content of dimethyl sulfide.

Table 4. Volatile compounds identified in cream samples using SPME-GC–TOF-MS system.

Compound	IK[1]	Ion[2]	Cream	Sour Cream	Storage Sour Cream	Lemon Peel				Clove Bud				Juniper Berry			
						12.5%	25%	50%	100%	12.5%	25%	50%	100%	12.5%	25%	50%	100%
Acetaldehyde	438	43	-[3]	-	-	32 ± 1 [a,b]	32 ± 1 [a,b]	33 ± 1 [a,b]	29 ± 0 [a,b]	247 ± 23 [c]	34 ± 2 [a,b]	33 ± 2 [a,b]	28 ± 0 [a]	24 ± 2 [a]	35 ± 1 [b]	29 ± 1 [a,b]	27 ± 1 [a]
Methanethiol	446	47	31 ± 2 [c]	90 ± 7 [g]	52 ± 4 [f]	35 ± 2 [c]	34 ± 4 [c]	35 ± 2 [c]	28 ± 0 [b]	37 ± 1 [c,d]	42 ± 2 [e]	35 ± 2 [c]	25 ± 1 [b]	16 ± 1 [a]	38 ± 2 [d,e]	16 ± 3 [a]	16 ± 2 [a]
Dimethyl sulfide	525	47	130 ± 13 [e]	113 ± 9 [d,e]	68 ± 4 [a]	123 ± 2 [e]	117 ± 2 [d,e]	96 ± 3 [c]	78 ± 4 [b]	104 ± 2 [c,d]	124 ± 13 [e]	121 ± 7 [e]	80 ± 2 [b]	111 ± 4 [d]	112 ± 1 [d]	84 ± 8 [b]	56 ± 4 [a]
2-Butanone	579	72	376 ± 39 [f]	372 ± 49 [f]	251 ± 19 [e]	165 ± 4 [c,d]	183 ± 7 [d]	181 ± 4 [d]	134 ± 6 [b]	273 ± 1 [e]	223 ± 10 [e]	144 ± 16 [b,c]	189 ± 14 [d]	138 ± 9 [b]	179 ± 4 [d]	142 ± 3 [b]	94 ± 1 [a]
2,3-Butanedione	601	86	-	496 ± 35	387 ± 14	461 ± 5 [d]	263 ± 18 [b]	337 ± 13 [c]	170 ± 10 [a]	487 ± 23 [d]	422 ± 26 [d]	341 ± 31 [c]	165 ± 11 [a]	474 ± 41 [d]	330 ± 9 [c]	258 ± 10 [b]	144 ± 5 [a]
Acetic acid	638	60	-	931 ± 14 [a]	1542 ± 19 [c]	16,808 ± 1633 [c]	9939 ± 1001 [a,b]	7110 ± 5272 [a]	8360 ± 1635 [a,b]	13,098 ± 878 [b,c]	6162 ± 691 [a]	6159 ± 432 [a]	5761 ± 380 [a]	16717 ± 511	10,091 ± 7724 [a,b]	16,394 ± 540 [c]	13,825 ± 281 [b,c]
2-Pentanone	689	58	446 ± 23 [c]	414 ± 39 [b,c]	339 ± 10 [c]	318 ± 80 [a]	369 ± 3 [b]	302 ± 6 [a]	284 ± 7 [a]	345 ± 10 [a]	607 ± 17 [e]	499 ± 134 [c]	582 ± 73 [d]	341 ± 17 [a,b]	348 ± 4 [a,b]	280 ± 5 [a]	279 ± 3 [a]
Dimethyl disulfide	718	94	-	93 ± 8 [d]	23 ± 1 [c]	23 ± 2 [c]	10 ± 0 [a]	12 ± 0 [a,b]	11 ± 0 [a]	23 ± 1 [c]	25 ± 7 [c]	25 ± 2 [c]	14 ± 1 [b]	22 ± 1 [c]	11 ± 0 [a]	8 ± 0 [a]	12 ± 0 [a,b]
Hexanal	780	56	166 ± 24 [d]	414 ± 35 [f]	132 ± 7 [b]	142 ± 8 [b,c]	129 ± 7 [a,b]	122 ± 7 [a]	116 ± 4 [a]	188 ± 7 [e]	183 ± 8 [e]	184 ± 1 [e]	178 ± 4 [d,e]	158 ± 7 [c,d]	166 ± 4 [d]	153 ± 10 [c]	179 ± 6 [e]
2-Hexanone	785	58	336 ± 45 [d]	364 ± 16 [d]	300 ± 15 [d]	242 ± 8 [c]	173 ± 8 [b]	165 ± 1 [b]	139 ± 6 [a]	306 ± 8 [d]	304 ± 14 [d]	294 ± 10 [c,d]	295 ± 23 [c,d]	253 ± 15 [c]	218 ± 11 [c]	162 ± 1 [b]	136 ± 8 [a]
Butanoic acid	821	60	-	1187 ± 161 [a]	1954 ± 23 [a,b]	2899 ± 163 [c]	2074 ± 212 [b]	2522 ± 226 [b]	2133 ± 1670 [b]	6087 ± 354 [e]	3176 ± 180 [c,d]	3281 ± 120 [c]	3128 ± 226 [c]	3526 ± 493 [d]	2541 ± 137 [b]	3562 ± 118 [d]	2746 ± 47 [b,c]
2-Heptanone	896	71	485 ± 42 [c]	850 ± 90 [f]	702 ± 85 [e]	364 ± 5 [b]	466 ± 73 [b,c]	57 ± 10 [d]	559 ± 10 [d]	813 ± 196 [e]	762 ± 50 [e,f]	736 ± 43 [e]	584 ± 61 [c,d]	345 ± 27 [b]	584 ± 37 [c,d]	732 ± 26 [e]	212 ± 19 [a]
Sabinene	976	136	-	-	-	-	-	-	-	-	-	-	-	199 ± 22 [a]	272 ± 31 [b]	316 ± 11 [b,c]	374 ± 8 [c]
β-pinene	981	136	-	-	-	40 ± 1 [a]	71 ± 3 [b]	78 ± 2 [b]	86 ± 7 [b]	-	-	-	-	194 ± 7 [c]	24 ± 4 [c]	248 ± 22 [c,d]	290 ± 16
Myrcene	986	136	-	-	-	-	-	-	-	-	-	-	-	196 ± 16 [a]	252 ± 14 [a,b]	318 ± 14 [b]	365 ± 13 [b]
Limonene	1030	136	-	-	-	680 ± 45 [a]	1207 ± 31 [b]	1277 ± 12 [b]	1294 ± 24 [b]	-	-	-	-	-	-	-	-
2-Nonanone	1052	142	36 ± 3 [d]	53 ± 4 [e]	37 ± 5 [d]	30 ± 5 [c]	29 ± 2 [b,c]	21 ± 2 [a]	28 ± 5 [b]	44 ± 1 [e]	29 ± 4 [b]	23 ± 3 [a,b]	33 ± [c,d]	32 ± 8 [c]	29 ± 3 [b,c]	31 ± 2 [c]	42 ± 7 [e]
γ-Terpinene	1057	136	-	-	-	100 ± 11 [a]	207 ± 12 [b]	243 ± 5 [b]	23 ± 11 [b]	-	-	-	-	-	-	-	-
Eugenol	1351	164	-	-	-	-	-	-	-	318 ± 2 [a]	1512 ± 66 [b]	2319 ± 70 [b,c]	1800 ± 16 [b,c]	-	-	-	-
Caryophyllene	1428	189	-	-	-	-	-	-	-	1 ± 0 [a]	3 ± 0 [b]	5 ± 0 [c]	7 ± 1 [c]	-	-	-	-
Humulene	1465	93	-	-	-	-	-	-	-	70 ± 4 [a]	174 ± 6 [b]	252 ± 16 [b,c]	246 ± 11 [b,c]	-	-	-	-
Eugenol acetate	1552	164	-	-	-	-	-	-	-	15 ± 1 [a]	37 ± 2 [a]	85 ± 2 [b]	81 ± 2 [b]	-	-	-	-

[1] Kovats' Retention Index on SLB-5 + Supelcowax10 columns; [2] ions used for semi-quantification; [3] not detected. Data are presented as an average relative peak area of triplicates (peak area of the compound divided by peak area of internal standard) with standard deviation value. Data are displayed as mean ± SD. Mean values marked by the different lower-case superscript letters (a–g) within a row denote statistically significant differences ($p < 0.05$) using a two-way ANOVA.

Short-chain fatty acids, such as acetic and butyric (butanoic) acid, were present in all samples fermented with *B. animalis* subsp. *lactis* Bb-12. The levels of both acids did not decrease during the storage of samples for 21 days. These acids can be produced by the lipolysis of milk fat, or they can be specifically formed by the action of LAB or from the deamination of amino acids as products of lactose metabolism or even lipid oxidation. The representative *Bifidobacterium* species used in this study is mainly responsible for producing acetate through the fermentation pathway. In contrast, lactobacilli can produce end products, such as pyruvate by carbohydrate fermentation, during the glycolytic metabolic pathway. The nonfermented samples did not have these acids, which confirms the significant influence of the fermentation process on their formation. The highest increase in the content of acetic and butanoic acids was found in sour cream samples enriched with essential oils at the lowest concentration (12.5%). Volatile organic acids, including acetic and butanoic acids, are important components determining the overall palatability of fermented milk beverages. A significant increase in the content of butyric acid during the storage of products may result in undesirable changes in the smell and taste, characteristic of rancidity. The addition of essential oils to cream subjected to fermentation resulted in the formation of the identified volatile components, which were part of the chemical composition of the used essential oils. Sour cream samples containing lemon peel essential oils contained β-pinene, limonene, and γ-terpinene. On the other hand, samples enriched with clove bud essential oils contained eugenol, eugenol acetate, humulene, and caryophyllene, and those supplemented with juniper berry essential oils contained three terpenes, namely sabinene, β-pinene, and myrcene. The content of these compounds in sour cream samples stored for 21 days increased in proportion to the concentration of essential oils added.

4. Conclusions

The essential oils obtained from lemon peels, clove buds, and juniper berries by the steam distillation method showed antimicrobial activity against the tested strains of *Bifidobacterium* in in vitro conditions. The size of the growth inhibition zones of these bacteria was influenced by both the type of essential oil used and the strain tested. All the analyzed strains of *Bifidobacterium* bacteria were sensitive to clove oil. On the other hand, the largest growth inhibition zones were recorded for the *B. animalis* subsp. *lactis* Bb-12 strain with the use of clove and juniper berry essential oils. However, neither the type of essential oil nor its concentration had any influence on the number of *B. animalis* subsp. *lactis* Bb-12 cells in the samples of cream directly fermented with this strain and those stored for 21 days at 6 °C. The main components of the lemon peel oil were identified to be limonene and γ-terpinene, while the juniper berry essential oil was rich in sabinene and β-myrcene and the clove bud oil was rich in eugenol and eugenol acetate. These compounds were also found among the volatile aroma compounds detected in sour cream samples. The results of this study suggest that the tested essential oils could serve as a natural additive for dairy products, conferring them with additional health-promoting and sensory properties. However, this should be confirmed by further research.

Supplementary Materials: The following supporting information can be downloaded at: https://www.mdpi.com/article/10.3390/app12031067/s1. Table S1 and Table S1 (continued) entitled "Fatty acid composition (%) of fat extracted from cream, sour cream, and stored sour cream samples enriched with essential oil from lemon peel, clove bud, and juniper berry."

Author Contributions: Conceptualization, M.K. and M.Z.; methodology, M.K., M.Z., M.R. and M.M.; investigation, M.K., M.Z., M.R., M.M., J.M. and K.M.; data curation, M.K. and M.Z.; writing—original draft preparation, review and editing, M.K. and M.Z.; project administration, M.K. and M.Z. All authors have read and agreed to the published version of the manuscript.

Funding: The study was financially supported by sources of the Ministry of Education and Science within funds of the Institute of Food Sciences of Warsaw University of Life Sciences (WULS), for scientific research.

Institutional Review Board Statement: Not applicable.

Informed Consent Statement: Not applicable.

Data Availability Statement: Not applicable.

Conflicts of Interest: The authors declare no conflict of interest.

References

1. Guyonnet, D.; Chassany, O.; Ducrotte, P.; Picard, C.; Mouret, M.; Mercier, C.-H.; Matuchansky, C. Effect of a fermented milk containing *Bifidobacterium animalis* DN–173 010 on the health—related quality of life and symptoms in irritable bowel syndrome in adults in primary care: A multicentre, randomized, double—blind, controlled trial. *Alimet. Pharmacol. Ther.* **2007**, *26*, 475–486. [CrossRef]
2. Ridwan, B.U.; Koning, C.J.M.; Besselink, M.G.H.; Timmerman, H.M.; Brouwer, E.C.; Verhoefl, J.; Gooszen, H.G.; Akkermans, L.M.A. Antimicrobial activity of a multispecies probiotic (Ecologic 641) against pathogens isolated from infected pancreatic necrosis. *Lett. Appl. Microbiol.* **2008**, *46*, 61–67. [CrossRef]
3. Begley, M.; Hill, C.; Gahan, C.G.M. Bile salt hydrolase activity in probiotics. *Appl. Environ. Microbiol.* **2006**, *72*, 1729–1738. [CrossRef] [PubMed]
4. Arunachalam, K.D. Role of bifidobacteria in nutrition, medicine and technology. *Nutr. Res.* **1999**, *19*, 1559–1597. [CrossRef]
5. Arboleya, S.; Ruas–Madiedo, P.; Margolles, A.; Solís, G.; Salminen, S.; de los Reyes–Gavilán, C.G.; Gueimonde, M. Characterization and in vitro properties of potentially probiotic *Bifidobacterium* strains isolated from breast-milk. *Int. J. Food Microbiol.* **2011**, *149*, 28–36. [CrossRef]
6. Cosentino, S.; Tuberoso, C.I.; Pisano, B.; Satta, M.; Mascia, V.; Arzedi, E.; Palmas, F. *In-vitro* antimicrobial activity and chemical composition of *Sardinian Thymus* essential oils. *Lett. Appl. Microbiol.* **1999**, *29*, 130–135. [CrossRef]
7. Hussain, A.I.; Anwar, F.; Hussain Sherazi, S.T.; Przybylski, R. Chemical composition, antioxidant and antimicrobial activities of basil (*Ocimum basilicum*) essential oils depends on seasonal variations. *Food Chem.* **2008**, *108*, 986–995. [CrossRef]
8. Kozłowska, M.; Żbikowska, A.; Szpicer, A.; Półtorak, A. Oxidative stability of lipid fractions of sponge-fat cakes after green tea extracts application. *J. Food Sci. Technol.* **2019**, *56*, 2628–2638. [CrossRef] [PubMed]
9. Kozłowska, M.; Żbikowska, A.; Marciniak-Łukasiak, K.; Kowalska, M. Herbal extracts incorporated into shortbread cookies: Impact on color and fat quality of the cookies. *Biomolecules* **2019**, *9*, 858. [CrossRef]
10. Pepeljnjak, S.; Kosalec, I.; Kalodera, Z.; Blazević, N. Antimicrobial activity of juniper berry essential oil (*Juniperus communis* L., Cupressaceae). *Acta Pharm.* **2005**, *55*, 417–422.
11. Schelz, Z.; Molnar, J.; Hohmann, J. Antimicrobial and antiplasmid activities of essential oils. *Fitoterapia* **2006**, *77*, 279–285. [CrossRef]
12. Omidbeygi, M.; Barzegar, M.; Hamidi, Z.; Naghdibadi, H. Antifungal activity of thyme, summer savory and clove essential oils against *Aspergillus flavus* in liquid medium and tomato paste. *Food Control* **2007**, *18*, 1518–1523. [CrossRef]
13. Baik, J.S.; Kim, S. –S.; Lee, J.–A.; Oh, T.–H.; Kim, J.–Y.; Lee, N.H.; Hyun, C.-G. Chemical composition and biological activities of essential oils extracted from Korean endemic citrus species. *J. Microbiol. Biotechnol.* **2008**, *18*, 74–79.
14. Viuda-Martos, M.; Ruiz-Navajas, Y.; Fernández-López, J.; Pérez-Álvarez, J. Antifungal activity of lemon (*Citrus lemon* L.), mandarin (*Citrus reticulata* L.), grapefruit (*Citrus paradisi* L.) and orange (*Citrus sinensis* L.) essential oils. *Food Control* **2008**, *19*, 1130–1138. [CrossRef]
15. Pinto, E.; Vale-Silva, L.; Cavaleiro, C.; Salgueiro, L. Antifungal activity of the clove essential oil from *Syzygium aromaticum* on *Candida*, *Aspergillus* and dermatophyte species. *J. Med. Microbiol.* **2009**, *58*, 1454–1462. [CrossRef] [PubMed]
16. Gülçin, I.; Elmastaş, M.; Aboul-Enein, H.Y. Antioxidant activity of clove oil—A powerful antioxidant source. *Arab. J. Chem.* **2012**, *5*, 489–499. [CrossRef]
17. Dimić, G.R.; Kocić–Tanackov, S.D.; Jovanov, O.O.; Cvetković, D.D.; Markov, S.L.; Velićanski, A.S. Antibacterial activity of lemon, caraway and basil extracts on *Listeria* ssp. *Acta Period. Technol.* **2012**, *43*, 239–246. [CrossRef]
18. Guerra, F.Q.S.; Mendes, J.M.; Oliveira, W.A.; Souza, F.S.; Trajano, V.; Douglas, H.; Coutinho, M.; Lima, E. Antibacterial activity of the essential oil of *Citrus limon* against multidrug resistant *Acinetobacter* strains. *Rev. Bras. Farm.* **2013**, *94*, 142–147.
19. Kozłowska, M.; Ścibisz, I.; Przybył, J.; Ziarno, M.; Żbikowska, A.; Majewska, E. Phenolic contents and antioxidant activity of extracts of selected fresh and dried herbal materials. *Pol. J. Food Nutr. Sci.* **2021**, *71*, 269–278. [CrossRef]
20. Fu, Y.; Zu, Y.; Chen, L.; Shi, X.; Wang, Z.; Sun, S.; Efferth, T. Antimicrobial activity of clove and rosemary essential oils alone and in combination. *Phytother. Res.* **2007**, *21*, 94–99. [CrossRef]
21. Kozłowska, M.; Ziarno, M.; Rudzińska, M.; Tarnowska, K.; Majewska, E.; Kowalska, D. Chemical composition of coriander essential oil and its effect on growth of selected lactic acid bacteria. *Zywn. Nauka Technol. Jakosc/Food Sci. Technol. Qual.* **2018**, *25*, 97–111. [CrossRef]
22. American Oil Chemist's Society (AOCS) Official Method Ce 1k-07. In *Direct Methylation of Lipids for the Determination of Total Fat, Saturated, cis-Monounsaturated, cis-Polyunsaturated, and Trans Fatty Acids by Gas Chromatography*; American Oil Chemist's Society (AOCS): Urbana, IL, USA, 2007.
23. American Oil Chemist's Society (AOCS) Official Method Ce 1h-05. In *Determination of cis-, trans-, Saturated, Monounsaturated and Polyunsaturated Fatty Acids in Vegetable or Non-ruminant Animal Oils and Fats by Capillary GLC*; American Oil Chemist's Society (AOCS): Urbana, IL, USA, 2005.

24. Falasca, A.; Caprari, C.; De Velice, V.; Fortini, P.; Saviano, G.; Zollo, F.; Iorizzi, M. GC-MS analysis of the essential oils of *Juniperus communis* L. berries growing wild in the Molise region: Seasonal variability and in vitro antifungal activity. *Biochem. Syst. Ecol.* **2016**, *69*, 166–175. [CrossRef]

25. Sela, F.; Karapandzova, M.; Stefkov, G.; Kulevanova, S. Chemical composition of berry essential oils from *Juniperus communis* L. (Cupressaceae) growing wild in Republic of Macedonia and assessment of the chemical composition in accordance to European Pharmocopoeia. *Maced. Pharm. Bull.* **2011**, *1-2*, 43–51. [CrossRef]

26. Höferl, M.; Stoilova, I.; Schmidt, E.; Wanner, J.; Jirovetz, L.; Trifonova, D.; Krastev, L.; Krastanov, A. Chemical composition and antioxidant properties of juniper berries (*Juniperus communis* L.) essential oils. Action of the essential oil on the antioxidant protection of *Saccharomyces cerevisiae* model organism. *Antioxidants* **2014**, *3*, 81–98. [CrossRef]

27. Bourgou, S.; Rahali, F.Z.; Ourghemmi, I.; Tounsi, M.S. Changes of peel essential oil composition of four Tunisian citrus during fruit maturation. *Sci. World J.* **2012**, *2012*, 528593. [CrossRef]

28. Hsouna, A.B.; Halima, N.B.; Smaoui, S.; Hamdi, N. *Citrus lemon* essential oil: Chemical composition, antioxidant and antimicrobial activities with its preservative effect against *Listeria monocytogenes* inoculated in minced beef meat. *Lipids Health Dis.* **2017**, *16*, 146. [CrossRef]

29. Bozkurt, T.; Gülnaz, O.; Kaçar, Y.A. Chemical composition of the essential oils from some citrus species and evaluation of the antimicrobial activity. *IOSR J. Environ. Sci. Toxicol. Food Technol.* **2017**, *11*, 29–33. [CrossRef]

30. Aguilar-Hernández, M.G.; Sánchez-Bravo, P.; Hernández, F.; Carbonell-Barrachina, A.A.; Pastor-Pérez, J.J.; Legua, P. Determination of the volatile profile of lemon peel oils as affected by rootstock. *Foods* **2020**, *9*, 241. [CrossRef]

31. Chatterjee, D.; Bhattacharjee, P. Use of eugenol-lean clove extract as a flavoring agent and natural antioxidant in mayonnaise: Product characterization and storage study. *J. Food Sci. Technol.* **2015**, *52*, 4945–4954. [CrossRef]

32. Carpena, M.; Nuñez-Estevez, B.; Soria-Lopez, A.; Garcia-Oliveira, P.; Prieto, M.A. Essential oils and their application on active packaging systems: A Review. *Resources* **2021**, *10*, 7. [CrossRef]

33. Chaieb, K.; Hajlaoui, H.; Zmantar, T.; Kahla-Nakbi, A.B.; Rouabhia, M.; Mahdouani, K.; Bakhrouf, A. The chemical composition and biological activity of clove essential oil, *Eugenia caryophyllata* (*Syzigium aromaticum* L. Myrtaceae): A short review. *Phytother. Res.* **2007**, *21*, 501–506. [CrossRef]

34. Alma, M.H.; Ertaş, M.; Nitz, S.; Kollmannsberger, H. Chemical composition and content of essential oil from the bud of cultivated Turkish clove (*Syzygium aromaticum* L.). *BioRes* **2007**, *2*, 265–269. [CrossRef]

35. Sulistyoningrum, A.S.; Saepudin, E.; Cahyana, A.H.; Rahayu, D.U.C.; Amelia, B.; Haib, J. Chemical profiling of clove bud oil (*Syzygium aromaticum*) from Toli-Toli and Bali by GC-MS analysis. In *International Symposium on Current Progress in Mathematics and Sciences 2016*; AIP Conference Proceeding 1862, 030089-1–030089-6; AIP Publishing: College Park, MD, USA, 2016. [CrossRef]

36. Bhuiyan, M.N.I.; Begum, J.; Nandi, N.C.; Akte, F. Constituents of the essential oil from leaves and buds of clove (*Syzigium caryophyllatum* (L.) Alston). *Afr. J. Plant Sci.* **2010**, *4*, 451–454.

37. Alfikri, F.N.; Pujiarti, R.; Wibisono, M.G.; Hardiyanto, E.B. Yield, quality, and antioxidant activity of clove (*Syzygium aromaticum* L.) bud oil at the different phenological stages in young and mature trees. *Scientifica* **2020**, *2*, 9701701. [CrossRef]

38. Sohilait, H.J. Chemical composition of the essential oils in *Eugenia caryophyllata*, Thunb from Amboina Island. *Sci. J. Chem.* **2015**, *3*, 95–99. [CrossRef]

39. Amelia, B.; Saepudin, E.; Cahyana, A.H.; Rahayu, D.U.; Sulistyoningrum, A.S.; Haib, J. GC-MS analysis of clove (*Syzygium aromaticum*) bud essential oil from Java and Manado. In *International Symposium on Current Progress in Mathematics and Sciences 2016*; AIP Conf. Proc. 1862, 030082-1–030082-9; AIP Publishing: College Park, MD, USA, 2016. [CrossRef]

40. Kamal, G.M.; Anwar, F.; Hussain, A.I.; Sarri, N.; Ashraf, M.Y. Yield and chemical composition of *Citrus* essential oils as affected by drying pretreatment of peels. *Int. Food Res. J.* **2011**, *18*, 1275–1282.

41. Fisher, K.; Phillips, C.A. The effect of lemon, orange and bergamot essentials oils and their components on the survival of the *Campylobacter jejuni*, *Escherichia coli* O157, *Listeria monocytogenes*, *Bacillus cereus* and *Staphylococcus aureus* in vitro and in food systems. *J. Appl. Microbiol.* **2006**, *101*, 1232–1240. [CrossRef]

42. Prabuseenivasan, S.; Jayakumar, M.; Ignacimuthu, S. In vitro antibacterial activity of some plant essential oils. *BMC Complement. Altern. Med.* **2006**, *6*, 39. [CrossRef]

43. Bevilacqua, A.; Corbo, M.R.; Sinigaglia, M. In vitro evaluation of the antimicrobial activity of eugenol, limonene, and citrus extract against bacteria and yeasts, representative of the spoiling microflora of fruit juices. *J. Food Prot.* **2010**, *73*, 888–894. [CrossRef]

44. Kędzia, A.; Ziółkowska–Klinkosz, M.; Włodarkiewicz, A.; Kusiak, A.; Kędzia, A.W.; Kochańska, B. Sensitivity anaerobic bacteria to lemon oil (*Oleum Citri*). *Post. Fit.* **2013**, *2*, 71–75.

45. Badei, A.Z.M.; El–Akel, A.T.M.; Faheid, S.M.M.; Mahmoud, B.S.M. Application of some spices in flavoring and preservation of cookies: 1—antioxidant properties of cardamom, cinnamon and clove. *Dtsch. Lebensmitt. Rundsch.* **2002**, *98*, 176–183.

46. Makuch, E.; Wróblewska, A.; Markowska-Szczupak, A.; Kucharski, Ł.; Klimowicz, A. Investigation of the antimicrobial properties of the cream with the addition of clove oil using the disc diffusion method. In *Innovations in Polish Science: An Overview of the Current Research Topics in the Chemical Industry*; Kujawski, J., Doskocz, J., Eds.; Wyd. Nauk. Sophia: Katowice, Poland, 2016; pp. 77–85, ISBN 9788365357281. (In English Abstract)

47. Liu, Q.; Meng, X.; Li, Y.; Zhao, C.; Zhao, C.-N.; Tang, G.-Y.; Li, H.-B. Antibacterial and antifungal activities of spices. *Int. J. Mol. Sci.* **2017**, *18*, 1283. [CrossRef] [PubMed]

48. Ayoola, G.A.; Lawore, F.M.; Adelowotan, T.; Aibinu, I.E.; Adenipekun, E.; Coker, H.A.B.; Odugbemi, T.O. Chemical analysis and antimicrobial activity of the essential oil of *Syzigium aromaticum* (clove). *Afr. J. Microbiol. Res.* **2008**, *2*, 162–166.

49. Ali, H.S.; Kamal, M.; Mohamed, S.B. In vitro clove oil activity against periodontopathic bacteria. *J. Sci. Technol.* **2009**, *10*, 1–7.

50. Şerban, E.S.; Ionescu, M.; Matinca, D.; Maier, C.S.; Bojiţă, M.T. Screening of the antibacterial and antifungal activity of egiht volatile essentials oils. *Farmacia* **2011**, *59*, 440–446.

51. Rezvani, S.; Rezai, M.A.; Mahmoodi, N.O. Analysis and antimicrobial activity of the plant *Juniperus communis*. *Rasayan J. Chem.* **2009**, *2*, 257–260.

52. Ložienė, K.; Venskutonis, P.R. Juniper (*Juniperus communis* L.) Oils. In *Essential Oils in Food Preservation, Flavor and Safety*; Preedy, V.R., Ed.; Elsevier: Amsterdam, The Netherlands, 2016; pp. 495–500. [CrossRef]

53. Cosetino, S.; Barra, A.; Pisano, B.; Cabizza, M.; Pirisi, F.M.; Palmas, F. Composition and antimicrobial properties of Sardinian *Juniperus* esstential oils against foodborne pathogens and spoilage microorganisms. *J. Food Prot.* **2003**, *66*, 1288–1291. [CrossRef]

54. Glišić, S.B.; Milojević, S.Ž.; Dimitrijević-Brankovic, S.I.; Orlović, A.M.; Skala, D.U. Antimicrobial activity of the essential oil and different fractions of *Juniperus communis* L. and a comparison with some commercial antibiotics. *J. Serb. Chem. Soc.* **2007**, *72*, 311–320. [CrossRef]

55. Hammer, K.A.; Carson, C.F.; Riley, T.V. Antimicrobial activity of essential oils and other plant extracts. *J. Appl. Microbiol.* **1999**, *86*, 985–990. [CrossRef]

56. Kozłowska, M.; Ścibisz, I.; Zaręba, D.; Ziarno, M. Antioxidant properties and effect on lactic acid bacterial growth of spice extracts. *CyTA-J. Food* **2015**, *13*, 573–577. [CrossRef]

57. Kozłowska, M.; Ziarno, M.; Gruczyńska, E.; Kowalska, D.; Tarnowska, K. Effect of coriander essential oil on lactic acid bacteria growth. *Brom. Chem. Toksykol.* **2016**, *3*, 341–345.

58. Ziarno, M.; Kozłowska, M.; Ścibisz, I.; Kowalczyk, M.; Pawelec, S.; Stochmal, A.; Szleszyński, B. The effect of selected herbal extracts on lactic acid bacteria activity. *Appl. Sci.* **2021**, *11*, 3898. [CrossRef]

59. Shahdadi, F.; Mirzaie, H.; Kashaninejad, M.; Khomeiri, M.; Ziaiifar, A.M.; Akbarian, A. Effect of various essentials oils on chemical and sensory characteristic and activity of probiotic bacteria in drinking yoghurt. *Agric. Commmunic.* **2015**, *3*, 16–21.

60. Vousough, A.S.; Khomeyri, M.; Kashaninezhad, M.; Jafari, S.M. Effects of mint extract on the viability of probiotic bacteria in a native Iranian dairy drink (Doogh). *J. Agric. Sci. Nat. Resour.* **2009**, *16*, 156–164.

61. Panesar, P.S.; Shinde, C. Effect of storage on syneresis, pH, *Lactobacillus acidophilus* count, *Bifidobacterium bifidum* count of *Aloe vera* fortified probiotic yoghurt. *Cur. Res. Dairy Sci.* **2011**, *4*, 17–23. [CrossRef]

62. Bayoumi, S. Bacteriostatic effect of some spices and their utilization in the manufacture of yoghurt. *Chem. Mikrobiol. Technol. Lebensmit.* **1992**, *14*, 21–26.

63. Smith-Palmer, A.; Stewart, J.; Fyfe, L. The potential application of plant essential oils as natural food preservatives in soft cheese. *Food Microbiol.* **2001**, *18*, 463–470. [CrossRef]

64. Khodaparast, H.; Hosein, M.; Sangatash, M.; Masoumeh Habii Najafi, K.R.; Bagher, M.; Shahram, B.T. Effects of essential oil and extract of *Ziziphora clinopodioides* on yoghurt starter culture activity. *Iran. J. Food Sci. Technol.* **2007**, *3*, 47–55.

65. Behrad, S.; Yusof, M.Y.; Goh, K.L.; Baba, A.S. Manipulation of probiotics fermentation of yogurt by cinnamon and licorice: Effects on yogurt formation and inhibition of *Helicobacter pylori* growth in vitro. *World Acad. Sci. Eng. Technol.* **2009**, *60*, 590–594.

66. Markiewicz-Kęszycka, M.; Czyżak-Runowska, G.; Lipińska, P.; Wójtowski, J. Fatty acid profile of milk—A review. *Bull. Vet. Inst. Pulawy.* **2013**, 135–139. [CrossRef]

67. Dachev, M.; Bryndová, J.; Jakubek, M.; Moučka, Z.; Urban, M. The effects of conjugated linoleic acids on cancer. *Processes* **2021**, *9*, 454. [CrossRef]

68. Izsóa, T.; Kasza, G.Y.; Somogyı, L. Differences between fat-related characteristics of sour cream and sour cream analogues. *Acta Aliment.* **2020**, *49*, 390–397. [CrossRef]

69. Yilmaz-Ersan, L. Fatty acid composition of cream fermented by probiotic bacteria. *Mljekarstvo* **2013**, *63*, 132–139.

70. Laučienė, L.; Andrulevičiūtė, V.; Sinkevičienė, I.; Kašauskas, A.; Urbšienė, L.; Šernienė, L. Impact of technology and storage on fatty acids profile in dairy products. *Mljekarstvo* **2019**, *69*, 229–238. [CrossRef]

71. Sévérin, D.B.D.; Renu, A.; François, Z.N.; Fonteh, A.F. Effect of lactic acid bacteria concentrations on the composition of bioactive compounds in a fermented food formulation. *Saudi J. Med. Pharm. Sci.* **2018**, *4*, 1096–1102. [CrossRef]

72. McSweeney, P.H.L.; Sousa, M.J. Biochemical pathways for the production of flavour compounds in cheeses during ripening: A review. *Lait* **2000**, *80*, 293–324. [CrossRef]

73. Lew, L.-C.; Gan, C.-Y.; Liong, M.-T. Dermal bioactives from lactobacilli and bifidobacteria. *Ann. Microbiol.* **2013**, *63*, 1047–1055. [CrossRef]

74. Aljewicz, M.; Majcher, M.; Nalepa, B. A comprehensive study of the impacts of oat-glucan and bacterial curdlan on the activity of commercial starter culture in yogurt. *Molecules* **2020**, *25*, 5411. [CrossRef]

Article

Aroma Profiles of Dry-Hopped Ciders Produced with Citra, Galaxy, and Mosaic Hops

Matthew T. Bingman [1], Josephine L. Hinkley [2], Colin P. Bradley II [3] and Callie A. Cole [4,*]

[1] Department of Chemistry & Biochemistry, University of Oregon, 1585 E 13th Ave., Eugene, OR 97403, USA; mbingman@uoregon.edu

[2] Rocky Vista College of Osteopathic Medicine, 255 E Center St., Ivins, UT 84738, USA; josephine.hinkley@ut.rvu.edu

[3] Department of Chemistry, Columbia Basin College, 2600 N 20th Ave., Pasco, WA 99301, USA; cbradley@columbiabasin.edu

[4] Department of Chemistry & Biochemistry, Fort Lewis College, 1000 Rim Drive, Durango, CO 81301, USA

* Correspondence: ccole@fortlewis.edu

Abstract: Cider quality and consumer acceptance are greatly influenced by its aroma. With the continued expansion of the craft cider industry, cider producers are employing techniques such as dry hopping to develop unique flavor profiles. Few studies, however, have explored the VOCs of dry-hopped cider. Herein, we monitor the development of VOCs from pressed apple juice, through fermentation and dry hopping by HS–SPME–GC–MS, to elucidate when and how aroma compounds arise in cider production. In all, 89 VOCs were detected, spanning eight classes of organic compounds. Racking events decreased ester concentrations by $10 \pm 1\%$, but resting on the lees allowed these pleasant, fruity aromas to be reestablished. Dry hopping was conducted with three types of hops (Citra, Galaxy, and Mosaic). The varied development of terpenes and esters between hop varieties supports the use of this technique to diversify the aroma profiles of ciders. Herein, we report that both the variety of hops and the timing of key processing steps including racking and hop addition significantly alter the identity and concentration of aroma-important VOCs in dry-hopped cider.

Keywords: cider; dry hopping; gas chromatography; mass spectrometry; solid phase microextraction; volatiles

Citation: Bingman, M.T.; Hinkley, J.L.; Bradley, C.P., II; Cole, C.A. Aroma Profiles of Dry-Hopped Ciders Produced with Citra, Galaxy, and Mosaic Hops. *Appl. Sci.* **2022**, *12*, 310. https://doi.org/10.3390/app12010310

Academic Editor: Agata Górska

Received: 24 November 2021
Accepted: 20 December 2021
Published: 29 December 2021

Publisher's Note: MDPI stays neutral with regard to jurisdictional claims in published maps and institutional affiliations.

1. Introduction

Cider consumption over the past two decades in the United States has seen substantial growth. From 2010 to 2018, consumption of specialty beverages, including cider, grew 487%, and in 2018, represented an industry worth USD 2.2 billion in the USA [1]. Although most commercially-produced cider is still fermented from imported apple juice concentrates or processing fruits (fruit that did not make it for consumption sale) [1], the demand for beverages crafted naturally and locally is high [2]. Before Prohibition, southwest Colorado, USA, had a diverse population of apple varietals used for fresh consumption and cider production. The Montezuma Orchard Restoration Project (MORP) has documented 436 apple varietals planted in Colorado prior to 1930 [3]. MORP is working with local orchards to restore Colorado's rich history in cider production by genetically profiling apple trees across the state. With an increased understanding of which apple varietals are currently growing in the region and what organoleptic cider qualities they produce, cider makers can make informed decisions on selections of fruit to harvest and process to a final product based on measurable chemical characteristics. This work focuses on the Gravenstein apple which is common to southwest Colorado and is considered a quality cider apple [4,5]. Results reported herein have been added to the MORP cider chemistry database to inform cider makers about the properties of southwest Colorado cider apples.

With increased cider consumption and production, cider makers are looking for ways to diversify their products. One such approach that is gaining traction is the use of hops within the cider-making process. Dry hopping is a technique that is commonly used in the beer brewing industry [6]. This process involves the addition of whole cone hops or dried hop pellets to the beverage after fermentation has been completed. Dry hopping is desirable in the brewing industry, because the hops impart appealing aroma compounds without the bitterness or astringency typically associated with hops. Although dry hopping has been well studied within the beer brewing industry, little work has been carried out to understand how dry hopping impacts the aroma profile of cider. Hops have been shown to contain numerous aroma compounds which include esters, ketones, short-chain fatty acids, and terpenes [7]. Previous studies have shown that the three most prominent "woody" terpenes [8] in hops, comprising up to 80% of the total volatile organic compounds (VOCs), are β-myrcene, caryophyllene, and humulene [7]. The story is not quite this simple, however. The VOCs that dominate the headspace of hops are often quite different from those in the headspace of dry-hopped beer, due to biotransformation and other variables involved in the fermentation process [9,10]. For this reason, it is important to monitor the aroma development after dry hopping, to differentiate which aroma active compounds were efficiently extracted from the hops into the beverage, and how they may have changed. In this work, we identify when and to what extent these and other aroma-important species, outlined in Figure 1, are incorporated into the cider headspace during cider making and over the dry hopping process. These previous studies suggest that not only the variety of hops, but the timing of key processing steps including racking and hop addition, significantly alter the identity and concentration of aroma-important VOCs in fermented beverages.

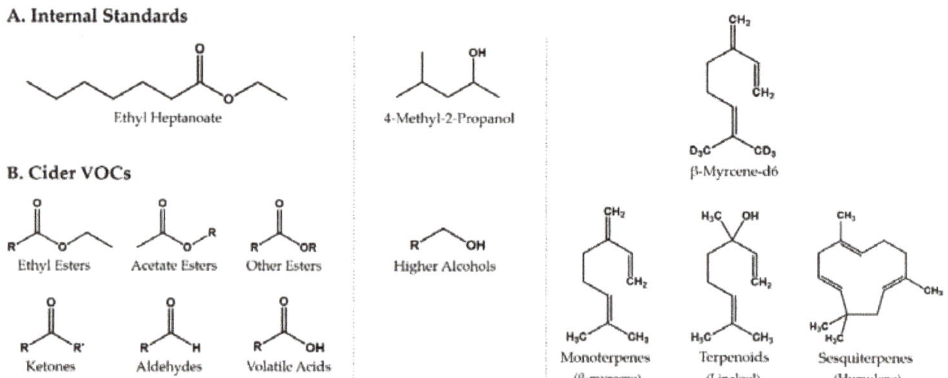

Figure 1. The structures of internal standards (**A**) and key cider volatile organic compounds of importance to this study (**B**).

Volatile compounds in cider have been successfully analyzed by sampling the headspace above the beverage [11]. Headspace–solid phase micro extraction–gas chromatography–mass spectrometry (HS–SPME–GC–MS) offers a simple and robust approach to analyze cider VOCs that has been used extensively in recent publications [11–15]. Headspace sampling requires a SPME fiber that will efficiently preconcentrate volatiles of interest. Due to the large variety of compounds in a cider matrix, a divinyl-benzene/carboxen/polydimethylsiloxane (DVB/CAR/PDMS) fiber was chosen for this study to select for compounds with varying degrees of polarity. Importantly, this study does not focus on thiols, which can be an important contributor to hopped beverage aroma [16]. These species may be sampled more efficiently by a different SPME fiber or sorptive extraction method than the method that we report herein.

Chemical analysis of dry-hopped cider is severely lacking in the published literature. We seek to fill this knowledge gap by monitoring the evolution of VOCs in the headspace of a novel, single varietal (Gravenstein) apple juice inoculated with a popular strain of *saccharomyces cervisiae* (Lalvin QA23), to perform three replicate and parallel fermentations. By measuring the VOC development throughout the 135-day fermentation and maturation, we elucidate how key processing steps such as racking influence the development of cider aroma over time. After fermentation completes, three different varieties of hops (Citra, Mosaic, and Galaxy) are added to the three ciders (one hop varietal per fermentation), to examine how the dry hopping process creates aroma differentiation among otherwise identical cider samples over the course of 8 days. Herein, we test the hypothesis that both hop variety and the timing of cider processing events (racking and dry hopping) are central contributors to final cider aroma development.

2. Materials and Methods

2.1. Apple Fruit Collection

Gravenstein apples were handpicked from a small, privately-owned orchard in Hermosa, Colorado. Fruit was picked from six different trees that have been genetically profiled as the Gravenstein varietal by MORP (Tree ID: ca2c010) [17] and thus used in this study for apple varietal consistency. Fruit was only collected if free of damage and visible disease. Flesh softness and skin color (ground color) observations were used to determine apple ripeness, and both orchard owners and local cider makers (EsoTerra Cider, Dolores, CO, USA) were consulted to verify consistency across apple types. The fruit was stored in a temperature-controlled (1.7 °C) environment immediately after collection for two and one half weeks to await further processing.

2.2. Apple Fruit Processing and Juice Analysis

Whole Gravenstein apples were rinsed and ground using an electric fruit mill, and then pressed into a juice with a hand press. Fruit was washed with dechlorinated tap water and ground and pressed at room temperature. One week prior to total sample processing, a small preliminary sample of Gravenstein apples was juiced and analyzed for tannin concentration (0.0467% ± 0.4), titratable acidity (0.770% ± 0.008 malic acid equivalents), specific gravity (1.049 ± 0.005), and pH (3.28 ± 0.02), as these are pre-fermentation parameters important to the cider-making process.

Specific gravity and pH were also measured before sampling the apple juice for HS–SPME–GC–MS analysis. All measurements were conducted in triplicate. The limit of detection (LOD) and limit of quantitation (LOQ) values were calculated based on the ASBC method for low-level detection [18]. LOD is defined as $LOD = \bar{x}_B + 3\sigma$ and LOQ is defined as $LOQ = \bar{x}_B + 10\sigma$ ($\bar{x}_B$ = mean signal of replicate reagent blanks, N = 10; and σ = standard deviation of the blank). The reported precision of our method is one standard deviation of the three trials.

Tannin concentration and titratable acidity were determined by the well-established Lowenthal permanganate titration and the malic acid titration, respectively [4]. The Lowenthal permanganate titration was carried out by adding 1 g of indigo carmine (Millipore Sigma, Saint Louis, MO, USA, 85%) and 50 mL concentrated sulfuric acid (Fisher Scientific, Waltham, MA, USA, 98.0%) to 1000 mL of deionized water. A total of 5 mL of the indigo carmine solution and 1 mL of Gravenstein apple juice were then added to 150 mL of deionized water. An indicator blank was prepared by combining 5 mL indigo carmine solution and 150 mL of deionized water. These solutions were titrated with 0.005 M potassium permanganate (Millipore Sigma, ACS Reagent, ≥99.0%) until a golden yellow end point was reached. The malic acid titration was performed by diluting 25 mL of Gravenstein apple juice to 100 mL with deionized water and titrating with 0.2 M NaOH (Fisher Scientific, ≥97.0%) until a visible end point was reached. Phenol red indicator (Millipore Sigma, ACS reagent) was added to observe the color change. The concentration of malic acid equivalents was calculated according to the published expression by Miles et al. [5].

A Vernier PH-BTA pH probe was used throughout this study to measure the pH of juice and cider. Before each measurement, it was calibrated with buffers of known pH. Specific gravity was determined using an EasyDens portable density meter from Anton

Paar. In addition to the juice analysis, both measurements were taken with every sample throughout fermentation.

2.3. Fermentation

The bulk batch of Gravenstein apples were milled and pressed at room temperature. Juice was collected in three 18.9 L glass carboys for simultaneous triplicate fermentations. Each carboy was inoculated with 5.00 ± 0.03 g of dry Lalvin QA23 *Saccharomyces cerevisiae* yeast that was reconstituted with filtered water (35–40 °C) and then immediately capped with airlocks. Carboys were then transferred to a dark, 14.4 ± 1.7 °C temperature-controlled chamber for fermentation. Approximately 20 mL aliquots from each fermentation vessel were used for HS-SPME-GC-MS analysis, pH, and specific gravity readings every 48 h for 22 days. After 22 days, samples were monitored once a week until all specific gravities plateaued (47 days). On day 25 of alcoholic fermentation, the cider was cold-crashed (24 h at 2 °C) and racked.

2.4. Dry Hopping

After 135 days of fermentation and maturation, the three ciders were racked into six 11.4 L carboys (two per fermentation: one control sample and one dry-hopped sample). Totals of 7 g of Galaxy, Citra, and Mosaic hop pellets (Artisan Hops) were added to 3 of the ciders (one hop varietal per cider). This ratio of hops to cider volume was recommended on the Artisan Hops packaging. Ciders were kept in contact with the hops for eight days at 14.4 ± 1.7 °C, a common temperature range for fermentation and maturation [19]. HS-SPME-GC-MS samples were taken on day 5 and day 8 of dry hopping and analyzed for VOC development. Control samples that were not dry hopped were sampled on day eight for comparison.

2.5. Headspace–Solid Phase Microextraction (HS–SPME)

Cider sample aliquots (10 mL), 2.6 g of NaCl (Sigma Aldrich, St. Louis, MO, USA), and internal standards were transferred to 20 mL, 22.5×75 mm glass headspace vials (Sigma Aldrich) for solid phase microextraction (SPME). A volume of 30 µL of 4-methyl-2-propanol (0.80 g mL^{-1}) and 1 µL ethyl heptanoate (0.87 g mL^{-1}) were added as internal standards (Millipore Sigma, 99% and 98%, respectively). Terpenes were semi-quantified by spiking 0.1 µL of 0.01 g mL^{-1} β-myrcene-d6 in ethyl acetate (Axios Research, Washington, DC, USA) [20] into the cider samples after dry hopping. The structures of the internal standards are displayed in Figure 1A and the 8 classes of compounds that were semi-quantified by these internal standards are shown in Figure 1B. Ethyl hepatanoate was used for semi-quantification of ethyl esters, acetate esters, other esters, ketones, aldehydes, and volatile acids. 4-Methyl-2-propanol was used for higher alcohol semi-quantification. B-Myrcene-d6 was used to semi-quantify three classes of terpenes that developed after the addition of hops (monoterpenes, terpenoids, and sesquiterpenes). Vials were then capped with 20 mm aluminum release seal PTFE/silicone liner cap (Sigma Aldrich) with a manual vial crimper. Vials were held in a 40 °C water bath for ten minutes for headspace equilibration. A conditioned DVB/CAR/PDMS 50/30 µm SPME fiber (Supelco, Bellefonte, PA, USA) was then exposed to the headspace for 30 min for volatile preconcentration. Samples were analyzed by GC-MS within 24 h of exposure and stored at 4 °C in the interim. SPME fibers were conditioned after every sample according to manufacturer recommendations.

2.6. Gas Chromatography—Mass Spectrometry (GC-MS) Analysis

An Agilent Technologies 7820A Gas Chromatograph (GC) and a 5977E Mass Spectrometer Detector (MSD) with Mass Hunter GC-MS acquisition Software (B.07.00.1203 Agilent, Santa Clara, CA, USA) were used for all analyses. SPME fibers were introduced to the GC inlet at 260 °C for 1 min in splitless mode to thermally desorb analytes and introduce them onto the column. Helium gas was used as the carrier gas at a flow rate of 1 mL min^{-1} through an Agilent DB-5, 30 m $\times$ 0.25 mm $\times$ 0.25 µm column. The oven was held at 50 °C for 5 min, increased to 190 °C at a rate of 3 °C min^{-1}, increased again to 230 °C at a rate of 70 °C min^{-1}, and then held at 230 °C for 5 min. The transfer line to the MS was held at 230 °C.

Eluates were ionized and fragmented in a 70 eV electron ionization (EI) source and separated with a quadrupole mass spectrometer set to a mass range of *m/z* 50–550. The quadrupole interface was held at 150 °C. As Figure 2 illustrates, the deuterated internal standard (β-myrcene-d6) was distinguished from the non-deuterated β-myrcene in the dry-hopped cider samples by selecting for *m/z* 75 and *m/z* 69, respectively. These ions likely form via C_5H_7 loss from their corresponding molecular ions, and their structures are predicted in Figure 2. Identification of analytes was accomplished by referencing the National Institute of Standards and Technology (NIST) library (Version 2.2. Agilent Technologies, Santa Clara, CA, USA) of EI spectra. Column and fiber blanks were run regularly, and siloxanes originating from the column and SPME fiber were identified and removed from the analysis [21].

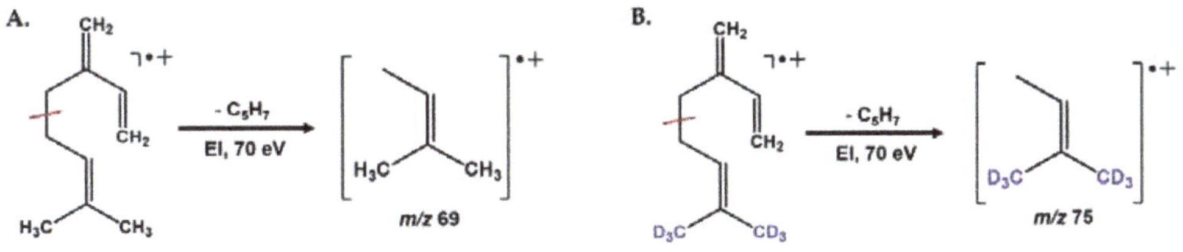

Figure 2. The electron ionization fragmentation observed for the molecular ions of β-myrcene (**A**) and β-myrcene-d6 (**B**).

3. Results and Discussion

3.1. Fermentation Monitoring

Cider makers monitor the progression of fermentation to inform their decisions regarding the timing of racking events. For the three ciders monitored in this study, specific gravity (SG) was measured with an Anton Parr EasyDens digital gravity and alcohol tester. SG and pH were measured every 48 h throughout fermentation and again at the end of maturation. Between the end of alcoholic fermentation and maturation, the three ciders were not sampled to reduce the potential for contamination. As Figure 3 depicts, an overall increase in pH (3.28–3.72) and a decrease in SG from 1.049 (juice) to 0.996 (cider) was observed, as expected. After the cider was racked (day 25), specific gravity continued to drop, and the pH of the ciders rose slightly to 3.51. The end of primary fermentation was on day 47 when the averaged specific gravity was ≤1.003 and the average pH was 3.56. The three ciders were not racked through maturation and as a result rested on the remaining lees for 88 days to increase the concentration of fruity, volatile esters. During maturation, the specific gravity decreased from 1.003 to 0.996 on day 135. The pH increased from 3.51 to 3.70 during this same time frame. This increase in pH is due to the production of ethanol throughout fermentation, which has been shown to increase the dissociation of weak acids present in the cider [22]. Similarly, the SG results were consistent with a successful fermentation. As the yeast consumes the sugars present, the density of the solution decreases, and SG values below 1.000 indicate that the fermentable sugars have been converted to ethanol and other fermentation products.

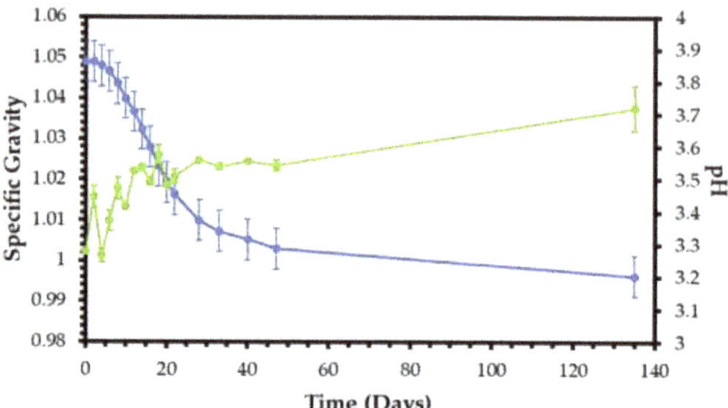

Figure 3. Specific gravity (**blue**) and pH (**green**) are shown for the entire fermentation and maturation process. On the left y-axis, the average specific gravity of three Gravenstein apple cider fermentations is plotted against time. The right y-axis shows pH measurements for the fermentations. Day 0 corresponds to measurements taken of pre-inoculated apple juice in triplicate and Day 135 represents the end of cider maturation (prior to dry hopping).

3.2. Higher Alcohol and Ester Analysis

Three concurrent trials of Gravenstein apple juice were fermented to a finished cider product and monitored for changes in VOC concentrations over the course of 135 days. The chemical composition of the headspace of these ciders changed dramatically over this time frame as a result of the changing byproducts of yeast metabolism. In total, 89 VOCs were identified by HS–SPME–GC–MS throughout fermentation and maturation: 11 higher alcohols, 7 acetate esters, 10 ethyl esters, 32 other esters, 5 aldehydes, 8 ketones, 5 volatile acids, and 11 terpenes. The details of these detections are included in Supplementary Table S1. Five esters and four higher alcohols that were consistently measured above the LOQ throughout fermentation have published odor perception thresholds [14,23] that were used to display the long running trends of ester and higher alcohol development throughout fermentation and maturation in both Table 1 and Figure 4.

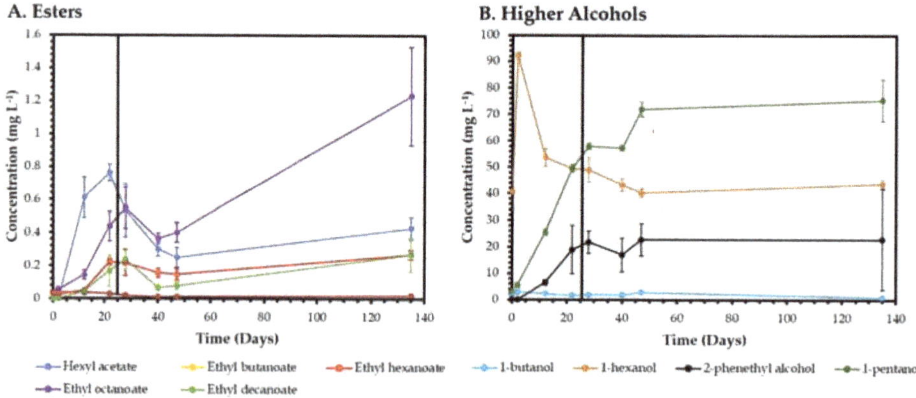

Figure 4. HS-SPME-GCMS measurements of headspace ester (**A**) and higher alcohol (**B**) concentrations that represent the overall trends observed from juice (day 0) to mature cider (day 135). Day 25 (vertical black line) is the date of cold crashing and racking the cider samples. Error bars and are reported as $\bar{x} \pm 1$ s of triplicate samples.

A comparison of these esters and alcohols is provided in Table 1a,b in order of their chromatographic retention times (RT), along with their published odor descriptors and perception thresholds (PT). Concentrations of four alcohols (1-butanol, 1-propanol, 1-hexanol, and 2-phenethylethanol) and five esters (ethyl butanoate, ethyl hexanoate, hexyl acetate, ethyl octanoate, and ethyl decanoate) are shown for the 22-day period before racking (Table 1a) and the period after racking until the end of maturation (Table 1b). Prior to day 12, higher alcohols, primarily originating from the juice, dominated the volatile composition of the ciders. By day 12 of fermentation, esters had risen significantly in concentration. In fact, from juice to cider (135 days post inoculation), total ester concentration increased by a factor of 71, whereas the total higher alcohol concentration increased by only a factor of 3. This agrees with the literature, as the esterification of higher alcohols efficiently occurs in fermentations with *saccharomyces cerevisiae* [24]. A practical insight for cider makers comes from a direct comparison from day 22 (pre-racking) to day 28 (post-racking). The total concentration of higher alcohols increased by $14 \pm 1\%$, and the total concentration of esters decreased by $10 \pm 1\%$ across this 6-day period. Previous research indicates that esters (including ethyl hexanoate and ethyl octanoate) are influenced by the time of contact with lees [25], but this is the first time to our knowledge that a quantifiable decrease in ester concentration has been linked directly to a racking event in cider making. Similarly, the simultaneous increase in higher alcohol concentration after racking highlights the importance of this process in tuning the cider aroma. As Table 1 shows, many of the major higher alcohols contribute floral green notes, whereas the esters are primarily perceived as fruity. These data suggest that in the days immediately following a racking event, the cider aroma may lose fruity notes. If allowed to rest on the lees for longer time periods without racking, the fruity esters have a chance to grow in.

Table 1. (a) Representative higher alcohol and ester concentrations (mg L^{-1}) from pre-inoculation through racking. (b) Representative higher alcohol and ester concentrations (mg L^{-1}) from racking to the end of maturation.

(a)

Higher Alcohols	RT (min)	Juice Concentration [a]	Day 2 Concentration	Day 12 Concentration	Day 22 Concentration	Odor Descriptor	PT (mg L^{-1})
1-butanol	2.3	1.920 ± 0.286	3.008 ± 0.358	2.300 ± 0.295	1.767 ± 0.021	Medicinal [b]	150 [b]
1-pentanol [b]	3.0	3.070 ± 0.723	5.452 ± 0.934	25.477 ± 1.275	49.65 ± 1.367	Alcohol, fruity [c]	0.030–45 [c]
1-hexanol	6.6	40.533 ± 6.665	92.153 ± 1.417	53.733 ± 3.387	49.433 ± 1.037	Flower, green [e]	8 [d]
2-phenylethanol	17.7	-	-	6.500 ± 1.019	18.927 ± 9.112	Floral, rose [e]	14 [d]
Esters							
Ethyl butanoate	4.3	0.028 ± 0.017	0.039 ± 0.008	0.038 ± 0.006	0.031 ± 0.002	Fruity, sweet [e]	0.4 [b]
Ethyl hexanoate	12.3	0.031 ± 0.018	0.027 ± 0.007	0.050 ± 0.001	0.224 ± 0.021	Fruity, apple [e]	0.014 [d]
Hexyl acetate	12.9	-	-	0.612 ± 0.121	0.763 ± 0.050	Fruity, herb [e]	0.67 [b]
Ethyl octanoate	22.0	-	0.057 ± 0.005	0.146 ± 0.029	0.440 ± 0.091	Fruity, fatty [e]	0.005 [d]
Ethyl decanoate	30.8	-	-	0.044 ± 0.013	0.170 ± 0.094	Fruity, grape [b,e]	0.2 [b]

(b)

Higher Alcohols	RT (min)	Day 28 Concentration [a]	Day 40 Concentration	Day 47 Concentration	Day 135 Concentration	Odor Descriptor	PT (mg L^{-1})
1-butanol	2.3	2.030 ± 0.135	1.953 ± 0.132	2.870 ± 0.305	0.930 ± 0.268	Medicinal [b]	150 [b]
1-pentanol [b]	3.0	57.923 ± 1.207	57.350 ± 1.084	71.967 ± 2.764	75.452 ± 7.844	Alcohol, fruity [c]	0.030–45 [c]
1-hexanol	6.6	49.100 ± 1.037	43.167 ± 4.691	40.267 ± 2.301	43.633 ± 1.584	Flower, green [e]	8 [d]
2-phenylethanol	17.7	21.670 ± 4.229	17.000 ± 6.318	22.700 ± 6.239	22.857 ± 19.067	Floral, rose [e]	14 [d]
Esters							
Ethyl butanoate	4.3	0.021 ± 0.011	0.012 ± 0.002	0.013 ± 0.003	0.020 ± 0.003	Fruity, sweet [e]	0.4 [b]
Ethyl hexanoate	12.3	0.218 ± 0.078	0.157 ± 0.025	0.149 ± 0.034	0.270 ± 0.028	Fruity, apple [e]	0.014 [d]
Hexyl acetate	12.9	0.534 ± 0.161	0.302 ± 0.043	0.251 ± 0.060	0.430 ± 0.068	Fruity, herb [e]	0.67 [b]
Ethyl octanoate	22.0	0.550 ± 0.124	0.365 ± 0.032	0.403 ± 0.061	1.230 ± 0.229	Fruity, fatty [e]	0.005 [d]
Ethyl decanoate	30.8	0.239 ± 0.018	0.069 ± 0.001	0.079 ± 0.023	0.270 ± 0.102	Fruity, grape [b,e]	0.2 [b]

[a] All semi-quantified analyte concentrations are reported as $\bar{x} \pm 1$ s. [b] [26] [c] 1-pentanol (balsamic, fruity, PT = 0.47 mg L^{-1}) coeluted with 3-methyl-1-butanol (alcohol, PT = 45 mg L^{-1}) and 2-methyl-1-butanol (banana, green, PT = 0.030 mg L^{-1}) [24,27,28]. A sum of these three unresolved species is reported here. [d] [29] [e] [14].

Figure 4A depicts the ester development from juice to the end of maturation for the five esters described previously. Ethyl butanoate (fruity [24]) and ethyl hexanoate (green apple, brandy, fruity [24]) were present in the apple juice, but their concentrations increased by 26% and 38%, respectively, by day 12 of fermentation. Similarly, hexyl acetate (fruity, sweet [30]), ethyl octanoate (fruity, candy, pineapple [11]), and ethyl decanoate (brandy, fruity [24]), which were not detected in the juice, were established as major species within the headspace by fermentation day 12. Maximum concentrations of ethyl hexanoate (0.224 ± 0.021 mg L^{-1}) and hexyl acetate (0.763 ± 0.05 mg L^{-1}) during the fermentation phase were observed on day 22. Before the ciders were cold crashed and racked on day 25 (marked in Figure 4 with a vertical black line), ethyl octanoate and ethyl decanoate reached fermentation concentration maxima of 0.55 ± 0.12 mg L^{-1} and 0.55 ± 0.124 mg L^{-1}, which are above or within error of the published perception thresholds for these species [11]. All five esters experienced a sizeable decrease to minimum concentrations following the racking event on day 25. On day 40, ethyl butanoate, ethyl octanoate, and ethyl decanoate had diminished by 69%, 33%, and 71%, respectively. Ethyl hexanoate and hexyl acetate decreased by 33% and 67% from their maximum concentrations by day 47. HS–SPME–GC–MS analysis was conducted again at the end of maturation on day 135. At the end of maturation, ester concentrations increased, most notably ethyl octanoate and ethyl decanoate. This experimental evidence supports previous observations of ester growth when cider is held over yeast lees for extended periods of time. All five of these esters have a predominately fruity aroma and were all detected above perception thresholds and likely contribute pleasant fruity notes to the ciders [14].

The development and changes in four higher alcohols are shown in Figure 4B. 1-Pentanol (balsamic, fruity [24]) coelutes with two other species (3-methyl-1-butanol and 2-methyl-1-butanol) that were not baseline resolved with our GC method. The integrated peak area of the three chromatographic peaks were summed and reported together as 1-pentanol. These species together were quantified to be, perhaps unsurprisingly, at the highest concentration of any other analyte observed in this study. 1-pentanol and 2-methyl-1-butanol were both detected in the juice, whereas 3-methyl-1-butanol was first detected within the first 12 days of fermentation indicating the possibility of either isomerization or the development of a new fermentation byproduct. Better separation of the species could be achieved by beginning the separation at a lower temperature (<50 °C) or slowing the temperature ramping.

Trends for the four higher alcohols monitored over the 135-day process were much more varied than the esters. 1-Hexanol (green, herbaceous [24]) was the predominant alcohol in the juice and two days after fermentation had increased in concentration by 56%. Interestingly, by day 12, 1-hexanol decreased by 42%. From day 12 to the end of maturation (day 135), 1-hexanol concentration decreased another 19%, but maintained its position as the second most abundant higher alcohol. 1-Butanol remained relatively constant throughout the 135-day process, at a concentration considerably below the perception threshold. 2-Phenylethyl alcohol (floral, honey-like [11,14]) was not present in the juice but was the third most abundant higher alcohol by the end of maturation. Although a rise was observed for many higher alcohols after racking (day 25), the alcoholic and floral aroma imparted by these species remained steady once the specific gravity reached levels close to 1.001 (day 47).

3.3. Dry-Hopped Cider Analysis

Each Gravenstein apple cider was dry hopped with a different variety of pressed pellet hop: Citra, Galaxy, and Mosaic. HS–SPME–GC–MS analysis was conducted after samples were in contact with the various hops for five days and eight days. This timeline was chosen because it falls within the published industry standard of 4–12 days for dry-hopping beer. Figure 5 shows the experimental chromatograms labelled with selected VOCs that were consistently observed across all samples. As illustrated by these experimental data, after only 5 days, a wide array of new VOCs is already present in the cider headspace. From day 5 to day 8 (3-day period) of dry hopping, total ester concentrations in Citra and Galaxy doubled. Over the same time span, total ester concentrations increased by a factor of 18 for Mosaic hops. As expected, the Mosaic hops sample also had the highest overall concentration of esters of all samples studied after 8 days of dry hopping. Esterification of hop-derived monocarboxylic acids with ethanol is likely the synthetic route to many of these volatile esters during dry hopping [6]. Terpenes, however, were added to the cider headspace most efficiently by Citra hops. Total terpene concentration doubled in the Citra sample from day 5 to day 8, whereas a negligible increase was observed for both Galaxy and Mosaic hopped samples. Not only did Citra hops yield the highest concentration of terpenes after 8 days, they also generated the largest variety of unique terpenes at concentrations greater than the LOQ. More investigation into the timeline for dry hopping cider is needed; however, these preliminary results suggest that dry hopping time could be shorter than many current beer industry practices, and added VOCs are heavily dependent on hop type.

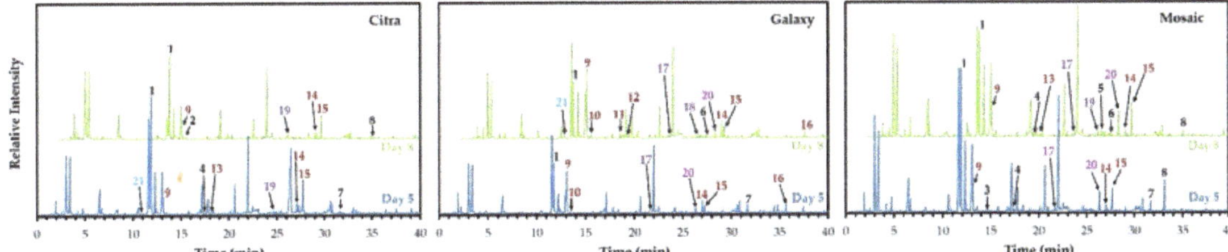

Figure 5. HS–SPME–GC–MS chromatograms are shown after 5 (**blue**) and 8 (**green**) days of dry hopping with three hop varieties (Citra, Galaxy, and Mosaic). Select VOCs that originated from the hops and were consistently detected across samples are numbered here. Terpenes are in black 1–8 (1: β-myrcene; 2: D-limonene; 3: ocimene; 4: linalool; 5: geraniol; 6: cyclotene; 7: caryophyllene; 8: humulene), esters are denoted in red 9–16 (9: 3-methylbutyl-2-methylpropanoate, 10: methyl heptanoate, 11: methyl nonanoate, 12: 2-ethylhexyl pentanoate, 13: 2-methylbutyl-2-methylbutanoate, 14: methyl 4-decanoate, 15: methyl geranoate, 16: 2-methyl propanoate), ketones are purple 17–20 (17: 2-dodecanone, 18: 2-tridecanone, 19: 2-decanone, 20: 2-undecanone) and propanoic acid is light blue (21).

Although no sensory work was conducted for this study, many relevant perception threshold analyses have been previously published for the VOCs observed in these experiments. We have summarized those VOCs and their previously published aromas here. Supplementary Table S2 includes every hop-derived VOC with published odor descriptors and perception thresholds. For a complete list of all detected species during the dry hopping process, please refer to Supplementary Table S1. Five VOCs were semi-quantified above published perception thresholds in these samples, as shown in bold in Supplementary Table S2: β-myrcene, linalool, methyl heptanoate, 2-nonanone, and propanoic acid. β-Myrcene (citrus, spicy [23,31]) was the only terpene found in all 6 hopped samples, and was also present at the highest concentration of all the terpenes. As its name suggests, Citra hops produced the highest concentration of this citrusy terpene. β-Myrcene has also been identified as the predominant aroma compound in other hop varieties [8]. Linalool (citrus, floral [23,31]) was also semi-quantified above the perception threshold in both Citra- and Mosaic-hopped samples, but was not generated from Galaxy hops in detectable quantities. Previously established as one of the most important hop-derived odorants in dry hopped beer [6], linalool also contributes greatly to dry-hopped cider aroma. Overall, citrus aroma-contributors were much more concentrated in Citra- and Mosaic-hopped samples when compared with those hopped with Galaxy. Rather, Galaxy hops created more fruity VOCs, as evidenced by methyl heptanoate existing in quantifiable abundance only in this hop variety (fruity [32]). One floral ketone, 2-nonanone, was semi-quantified above the perception threshold in Mosaic hopped samples only. Lastly, propanoic acid may add a spicy, soy fragrance in ciders hopped with Citra and Galaxy. Each hop variety generated a unique set of aroma-contributing VOCs when added to identical cider samples: Citra-hopped cider was citrus and spicy, Galaxy-hopped cider was fruity, and Mosaic-hopped cider had both citrus and floral VOCs based on these perception threshold analyses.

4. Conclusions

Quantitative analysis of triplicate Gravenstein apple cider fermentations has revealed a large array of VOCs. The 89 total VOCs detected by HS–SPME–GC–MS throughout fermentation and dry hopping included higher alcohols (11), acetate esters (7), ethyl esters (10), other esters (32), aldehydes (5), ketones (8), volatile acids (5), and terpenes (11). Specifically, species above the published perception thresholds were monitored carefully, revealing three unique ciders with appealing citrus (Citra hops), fruity (Galaxy hops), and floral (Mosaic hops) aromas. During the fermentation process, esters develop quickly, but drop in concentration (−10%) immediately following racking events. This is of practical interest to cider makers when planning the timeline of racking and bottling. By contrast, a single cold crash and racking event increased higher alcohol concentrations (+14%) for 3 days afterwards. Although few studies have focused on dry hopping in a cider context, we hypothesized based on previous beer publications that not only the variety of hops, but the timing of hop addition significantly alter the identity and concentration of aroma-important VOCs in cider. This hypothesis was confirmed. Citra hops added the highest concentration of terpenes in the cider headspace, and Mosaic hops added the highest concentration of esters. From day 5 to day 8, a significant increase in ester concentration was observed for all hop varieties. Although further research is needed to fully characterize the dry hopping process for cider making through detailed sensory analysis, this work provides novel analytical insight into how cider makers might make informed decisions on the application of hops to diversify the aroma profile of their beverages. Similarly, we have provided useful information on racking events and insight into how a single varietal apple cider's aroma profile changes throughout fermentation, maturation, and dry hopping. Relevant data discussed in this work have been added to the MORP cider chemistry database to inform cider makers about the properties of southwest Colorado cider apples.

There are myriad variables involved in cider making, all of which are worthy of further exploration. Important future work should include optimizing the dry hopping process to limit the quantity of hops used while maximizing desirable aroma contributors, thereby decreasing cost for cider makers and decreasing the environmental impact of excessive hop use.

Supplementary Materials: The following are available online at https://www.mdpi.com/article/10.3390/app12010310/s1, Table S1: Semi-quantification of volatile compounds (mg/L) in juice, cider, and dry-hopped cider and Table S2: Semi-quantified VOCs a (mg L^{-1}) added to cider headspace as a result of dry hopping for 5 and 8 days with Citra, Galaxy, and Mosaic hops.

Author Contributions: C.A.C., undergraduate students M.T.B. and J.L.H., and collaborator C.P.B.II implemented the study and interpreted the results together. Undergraduate students M.T.B. and J.L.H. collected and analyzed all reported experimental data under the mentorship of C.A.C. and C.P.B.II, M.T.B. prepared the manuscript draft and all manuscript editing was performed by C.A.C. C.A.C. acquired all funding that supported this work and led project administration and supervision. All authors have read and agreed to the published version of the manuscript.

Funding: This work was made possible thanks to the 2019–2020 Undergraduate Analytical Research Program (UARP) grant of the Society for the Analytical Chemists of Pittsburgh (SACP) and the 2019–2020 American Society for Mass Spectrometry Research at PUIs Award funded by Agilent Technologies. We thank these organizations for their financial support of undergraduate research.

Data Availability Statement: Results of this work have been reported in the publicly available Montezuma Orchard Restoration Project (MORP) database linked below and can also be provided by the corresponding author upon request. https://airtable.com/shrixCDelEJbp0SrV/tblUWhAm7 Mtbh6QUJ/viwEQfYwXwQJISdEa (accessed on 1 June 2021).

Acknowledgments: We thank Jared Scott, Elizabeth Philbrick, and Avery Scott of EsoTerra Cider, for their collaborative insights, cider samples, and assistance picking and pressing the apples used in this study. We also thank them greatly for the uses of their facilities for sample preparation. Special thanks go to the orchard owners in Hermosa, CO for their Gravenstein apples. We acknowledge and thank Sam Bingman, Marty Emmes, Kelly Bleck, Zach Myers, and Michael Grubb for their discussions, edits, advice, and assistance throughout the project. Lastly, our laboratory would not function without the continual support of several good dogs, cats, and rabbits: thank you especially to O'Malley, Liessel, Tilli, Ivy, and Stuart for your continual support and cuteness.

Conflicts of Interest: The authors declare no conflict of interest.

References

1. Miles, C.A.; Alexander, T.R.; Peck, G.; Galinato, S.P.; Gottschalk, C.; Van Nocker, S. Growing Apples for Hard Cider Production in the United States—Trends and Research Opportunities. *HortTechnology* **2020**, *30*, 148–155. [CrossRef]
2. Watson, B. 2020 Points and 2021 Predictions. Available online: https://www.brewersassociation.org/insights/2020-points-and-2021-predictions/ (accessed on 25 June 2021).
3. Schuenemeyer, A.; Schuenemeyer, J. Montezuma Orchard Resoration Project. Available online: https://montezumaorchard.org/ (accessed on 24 May 2021).
4. Lafontaine, S.R.; Shellhammer, T.H. Impact of static dry-hopping rate on the sensory and analytical profiles of beer. *J. Inst. Brew.* **2018**, *124*, 434–442. [CrossRef]
5. Miles, C.A.; King, J.; Alexander, T.; Scheenstra, E. Evaluation of Flower, Fruit, and Juice Characteristics of a Multinational Collection of Cider Apple Cultivars Grown in the U.S. Pacific Northwest. *HortTechnology* **2017**, *27*, 431–439. [CrossRef]
6. Brendel, S.; Hofmann, T.; Granvogl, M. Dry-Hopping to Modify the Aroma of Alcohol-Free Beer on a Molecular Level—Loss and Transfer of Odor-Active Compounds. *J. Agric. Food Chem.* **2020**, *68*, 8602–8612. [CrossRef] [PubMed]
7. Duarte, L.M.; Amorim, T.L.; Grazul, R.M.; de Oliveira, M.A.L. Differentiation of aromatic, bittering and dual-purpose commercial hops from their terpenic profiles: An approach involving batch extraction, GC–MS and multivariate analysis. *Food Res. Int.* **2020**, *138*, 109768. [CrossRef] [PubMed]
8. Van Opstaele, F.; De Causmaecker, B.; Aerts, G.; De Cooman, L. Characterization of Novel Varietal Floral Hop Aromas by Headspace Solid Phase Microextraction and Gas Chromatography–Mass Spectrometry/Olfactometry. *J. Agric. Food Chem.* **2012**, *60*, 12270–12281. [CrossRef]
9. Rettberg, N.; Biendl, M.; Garbe, L.-A. Hop Aroma and Hoppy Beer Flavor: Chemical Backgrounds and Analytical Tools—A Review. *J. Am. Soc. Brew. Chem.* **2018**, *76*, 1–20. [CrossRef]
10. Werrie, P.-Y.; Deckers, S.; Fauconnier, M.-L. Brief Insight into the Underestimated Role of Hop Amylases on Beer Aroma Profiles. *J. Am. Soc. Brew. Chem.* **2021**, 1–9. [CrossRef]
11. Bingman, M.T.; Stellick, C.E.; Pelkey, J.P.; Scott, J.M.; Cole, C.A. Monitoring Cider Aroma Development throughout the Fermentation Process by Headspace Solid Phase Microextraction (HS-SPME) Gas Chromatography–Mass Spectrometry (GC-MS) Analysis. *Beverages* **2020**, *6*, 40. [CrossRef]
12. Perestrelo, R.; Silva, C.L.; da Silva, P.M.C.; Medina, S.; Pereira, R.; Câmara, J.S. Untargeted fingerprinting of cider volatiles from different geographical regions by HS-SPME/GC-MS. *Microchem. J.* **2019**, *148*, 643–651. [CrossRef]
13. Nešpor, J.; Karabín, M.; Štulíková, K.; Dostálek, P. An HS-SPME-GC-MS Method for Profiling Volatile Compounds as Related to Technology Used in Cider Production. *Molecules* **2019**, *24*, 2117. [CrossRef] [PubMed]
14. Englezos, V.; Torchio, F.; Cravero, F.; Marengo, F.; Giacosa, S.; Gerbi, V.; Rantsiou, K.; Rolle, L.; Cocolin, L. Aroma profile and composition of Barbera wines obtained by mixed fermentations of Starmerella bacillaris (synonym Candida zemplinina) and Saccharomyces cerevisiae. *LWT* **2016**, *73*, 567–575. [CrossRef]
15. Wei, J.; Zhang, Y.; Qiu, Y.; Guo, H.; Ju, H.; Wang, Y.; Yuan, Y.; Yue, T. Chemical composition, sensorial properties, and aroma-active compounds of ciders fermented with Hanseniaspora osmophila and Torulaspora quercuum in co- and sequential fermentations. *Food Chem.* **2020**, *306*, 125623. [CrossRef]
16. Ochiai, N.; Sasamoto, K.; Kishimoto, T. Development of a Method for the Quantitation of Three Thiols in Beer, Hop, and Wort Samples by Stir Bar Sorptive Extraction with in Situ Derivatization and Thermal Desorption–Gas Chromatography–Tandem Mass Spectrometry. *J. Agric. Food Chem.* **2015**, *63*, 6698–6706. [CrossRef] [PubMed]
17. Schuenemeyer, A.; Schuenemeyer, J. Montezuma Orchard Restoration Project Map. Available online: https://morp-acp.maps.arcgis.com/apps/webappviewer/index.html?id=91f531fb0d9843fab5d7455c15addefd&extent=-12208863.2258%2C4278369.2786%2C-11132629.8676%2C5105112.1766%2C102100 (accessed on 24 May 2021).
18. ASBC. *ASBC Methods of Analysis*, 14th ed.; ASBC: St. Paul, MN, USA, 2011.

19. Herrero, M.; García, L.A.; Díaz, M. Volatile Compounds in Cider: Inoculation Time and Fermentation Temperature Effects. *J. Inst. Brew.* **2006**, *112*, 210–214. [CrossRef]
20. Dennenlöhr, J.; Thörner, S.; Manowski, A.; Rettberg, N. Analysis of Selected Hop Aroma Compounds in Commercial Lager and Craft Beers Using HS-SPME-GC-MS/MS. *J. Am. Soc. Brew. Chem.* **2019**, *78*, 16–31. [CrossRef]
21. Curran, A.M.; Rabin, S.I.; Prada, P.A.; Furton, K.G. Comparison of the Volatile Organic Compounds Present in Human Odor Using Spme-GC/MS. *J. Chem. Ecol.* **2005**, *31*, 1607–1619. [CrossRef]
22. Akin, H.; Brandam, C.; Meyer, X.-M.; Strehaiano, P. A model for pH determination during alcoholic fermentation of a grape must by Saccharomyces cerevisiae. *Chem. Eng. Process. Process. Intensif.* **2008**, *47*, 1986–1993. [CrossRef]
23. Miyazawa, M.; Nagata, T.; Nakahashi, H.; Takahashi, T. Characteristic odor components of essential oil from Caesalpinia decapetala. *J. Essent. Oil Res.* **2012**, *24*, 441–446. [CrossRef]
24. He, W.; Liu, S.; Heponiemi, P.; Heinonen, M.; Marsol-Vall, A.; Ma, X.; Yang, B.; Laaksonen, O. Effect of Saccharomyces cerevisiae and Schizosaccharomyces pombe strains on chemical composition and sensory quality of ciders made from Finnish apple cultivars. *Food Chem.* **2021**, *345*, 128833. [CrossRef]
25. Antón-Díaz, M.J.; Valles, B.S.; Hevia, A.G.; Lobo, A.P. Aromatic Profile of Ciders by Chemical Quantitative, Gas Chromatography-Olfactometry, and Sensory Analysis. *J. Food Sci.* **2014**, *79*, S92–S99. [CrossRef]
26. Arcari, S.; Caliari, V.; Sganzerla, M.; Godoy, H.T. Volatile composition of Merlot red wine and its contribution to the aroma: Optimization and validation of analytical method. *Talanta* **2017**, *174*, 752–766. [CrossRef]
27. Yu, H.; Xie, T.; Xie, J.; Chen, C.; Ai, L.; Tian, H. Aroma perceptual interactions of benzaldehyde, furfural, and vanillin and their effects on the descriptor intensities of Huangjiu. *Food Res. Int.* **2020**, *129*, 108808. [CrossRef]
28. Tura, D.; Failla, O.; Bassi, D.; Pedò, S.; Serraiocco, A. Cultivar influence on virgin olive (*Olea europea* L.) oil flavor based on aromatic compounds and sensorial profile. *Sci. Hortic.* **2008**, *118*, 139–148. [CrossRef]
29. Escudero, A.; Gogorza, B.; Melús, M.A.; Ortín, N.; Cacho, A.J.; Ferreira, V. Characterization of the Aroma of a Wine from Maccabeo. Key Role Played by Compounds with Low Odor Activity Values. *J. Agric. Food Chem.* **2004**, *52*, 3516–3524. [CrossRef] [PubMed]
30. Aparicio, R.; Morales, M.T. Characterization of Olive Ripeness by Green Aroma Compounds of Virgin Olive Oil. *J. Agric. Food Chem.* **1998**, *46*, 1116–1122. [CrossRef]
31. Wijaya, C.H.; Hadiprodjo, I.T.; Apriyantono, A. Identification of Volatile Compounds and Key Aroma Compounds of Andaliman Fruit (Zanthoxylum Acanthopodium DC.). *Food Sci. Biotechnol.* **2002**, *2*, 680–683.
32. Pino, J.A.; Trujillo, R. Characterization of odour-active compounds of sour guava (Psidium acidum [DC.]Landrum) fruit by gas chromatography-olfactometry and odour activity value. *Flavour Fragr. J.* **2021**, *36*, 207–212. [CrossRef]

applied sciences

MDPI

Article

Phosphorus and Nitrogen Limitation as a Part of the Strategy to Stimulate Microbial Lipid Biosynthesis

Katarzyna Wierzchowska [1,*], Bartłomiej Zieniuk [1], Dorota Nowak [2] and Agata Fabiszewska [1]

[1] Department of Chemistry, Institute of Food Sciences, Warsaw University of Life Sciences-SGGW, 159c Nowoursynowska Street, 02-776 Warsaw, Poland; bartlomiej_zieniuk@sggw.edu.pl (B.Z.); agata_fabiszewska@sggw.edu.pl (A.F.)

[2] Department of Food Engineering and Process Management, Warsaw University of Life Sciences-SGGW, 159c Nowoursynowska Street, 02-776 Warsaw, Poland; dorota_nowak@sggw.edu.pl

[*] Correspondence: katarzyna_wierzchowska1@sggw.edu.pl

Abstract: Microbial lipids called a sustainable alternative to traditional vegetable oils invariably capture the attention of researchers. In this study, the effect of limiting inorganic phosphorus (KH_2PO_4) and nitrogen (($NH_4)_2SO_4$) sources in lipid-rich culture medium on the efficiency of cellular lipid biosynthesis by *Y. lipolytica* yeast has been investigated. In batch cultures, the carbon source was rapeseed waste post-frying oil (50 g/dm^3). A significant relationship between the concentration of KH_2PO_4 and the amount of lipids accumulated has been revealed. In the shake-flask cultures, storage lipid yield was correlated with lower doses of phosphorus source in the medium. In bioreactor culture in mineral medium with (g/dm^3) 3.0 KH_2PO_4 and 3.0 ($NH_4)_2SO_4$, the cellular lipid yield was 47.5% (*w/w*). Simultaneous limitation of both phosphorus and nitrogen sources promoted lipid accumulation in cells, but at the same time created unfavorable conditions for biomass growth (0.78 g$_{d.m.}$/dm^3). Increased phosphorus availability with limited cellular access to nitrogen resulted in higher biomass yields (7.45 g$_{d.m.}$/dm^3) than phosphorus limitation in a nitrogen-rich medium (4.56 g$_{d.m.}$/dm^3), with comparable lipid yields (30% and 32%). Regardless of the medium composition, the yeast preferentially accumulated oleic and linoleic acids as well as linolenic acid up to 8.89%. Further, it is crucial to determine the correlation between N/P molar ratios, biomass growth and efficient lipid accumulation. In particular, considering the contribution of phosphorus as a component of coenzymes in many metabolic pathways, including lipid biosynthesis and respiration processes, its importance as a factor in the cultivation of the oleaginous microorganisms was highlighted.

Keywords: *Yarrowia lipolytica*; microbial lipids; phosphorus limitation; nitrogen limitation

Citation: Wierzchowska, K.; Zieniuk, B.; Nowak, D.; Fabiszewska, A. Phosphorus and Nitrogen Limitation as a Part of the Strategy to Stimulate Microbial Lipid Biosynthesis. *Appl. Sci.* **2021**, *11*, 11819. https://doi.org/10.3390/app112411819

Academic Editor: Agata Górska

Received: 26 November 2021
Accepted: 10 December 2021
Published: 13 December 2021

Publisher's Note: MDPI stays neutral with regard to jurisdictional claims in published maps and institutional affiliations.

1. Introduction

Among oleaginous microorganisms capable of accumulating lipids exceeding 20% of cell dry weight, the species *Yarrowia lipolytica* stands out as a model organism [1]. Moreover, the non-conventional yeast *Y. lipolytica* is a permissive species with wide-ranging biotechnological applications. Considering well-studied metabolism, fully sequenced genome and secretion capabilities that provide opportunities to obtain new microbial products previously obtained through other routes, they have become a model for oleaginous species in many basics and applications studies [2–6]. Single cell oil (SCO) extracted from cells of oleaginous microorganisms is promising for food technology and nutrition notably due to its content of polyunsaturated fatty acids. It is reported that SCOs in addition to polysaccharides or single cell proteins (SCPs) could be a valuable food additive. Microbial lipids may increase the nutritional value of the final product enriched with these components, being at the same time a potential substitute for vegetable lipids, e.g., palm oil, cocoa butter and other fatty acids of industrial importance [7,8].

Nowadays, PUFA-rich SCOs are mainly extracted from microalgae due to the innate ability to synthesize valuable fatty acids and lipid production efficiency, such as *Schizochytrium* sp. 50–77% of dry weight, *Nitzschia* sp. 45–47%, *Nannochloropsis* sp. 31–68% or *Neochloris oleoabundans* 35–54% oil content [9]. Oleaginous yeasts, compared to microalgae, have a rather lower ability to synthesize SCO. Remarkably, wild yeast strains of *Y. lipolytica* metabolizing glucose are able to store lipids up to 36% of cell dry weight, and even up to 50–60% in the case of feeding biomass with hydrophobic substrates [6,10]. In SCO production, lipid yield per unit dry weight is a critical factor; thus, solutions are still being sought to improve it [11]. Fields such as synthetic and system biology, along with metabolic engineering techniques, are used to improve lipid storage yield and obtain lipids rich in polyunsaturated fatty acids (PUFAs). By using various molecular biology methods, it has been possible to obtain *Y. lipolytica* yeast mutants capable of accumulating up to 25% eicosapentenoic acid in the total content of the fatty acids [12].

Oleaginous yeasts, depending on the type of carbon substrate in the culture medium, accumulate lipids through two different biochemical pathways: de novo for hydrophilic and ex novo for hydrophobic substrates. In recent years, research has been conducted to define and analyze genes associated with the activity of enzymes specific to both biosynthetic pathways [2,13]. There are reports that sometimes cellular lipid accumulation may occur through two pathways simultaneously [2].

The yeast *Y. lipolytica* is well known for its ability to metabolize complex lipid substrates, including industrial waste products [14]. Studies on SCO synthesis using including pork lard, stearin-an industrial derivative of animal fat, waste glycerol, molasses and other lignocellulosic wastes have been reported in the literature [11,15–17]. Nowadays, processed vegetable oils (waste cooking oils—WCO) have been considered a waste product with high potential for biotechnological application [18]. *Y. lipolytica* strains are able to produce many metabolites in media with waste cooking oils: citric acid [19], erythritol [20], lipases [21,22], and microbial lipids [23–25]. In this manner, environmentally burdensome wastes have found application in the cell culture of *Y. lipolytica* species. The enormous potential of *Y. lipolytica* in the utilization and management of hydrophobic industrial wastes was presented in a review by Wierzchowska et al. [26].

The oleaginous yeast accumulates significant amounts of SCO under stress conditions caused by limited access to nutrients. Biosynthesis of microbial oil is enhanced in media depleted of nutrients other than carbon. The key issue for efficient lipid accumulation is limiting access to nitrogen [27]. Unlike nitrogen, access of microorganism cells to carbon should be unrestricted, at a constantly high level [28]. There is a notion that a high carbon/nitrogen ratio and a high carbon/phosphorus ratio in the medium composition promotes lipid accumulation in oleaginous yeast cells [29]. Many studies have been carried out on the potential of oleaginous microorganisms to accumulate high lipid contents in medium with a limiting element (nitrogen, magnesium, phosphorus, iron, zinc, etc.) [29–35]. However, most of works was concerned the analysis of the effects of nutrient limitation on the synthesis of storage lipids via the de novo pathway. Therefore, there is still some need to explore how its limitation involves the ex novo biosynthesis route.

The dynamics of microbial processes is inextricably linked to the response of microorganisms to changes in the environmental conditions in which they grow. To ensure optimal production efficiency on an industrial scale, tolerance of microorganisms to environmental stress is highly desirable. In particular, this is due to the growing interest in microbial biotechnological processes. Understanding how physiological systems respond to changing environmental conditions resulting in a specific level of growth or production of desired metabolites is critical to finding new and optimizing already developed applications [36].

This paper attempts to evaluate the effect of limiting inorganic phosphorus and nitrogen sources as well as duration of culture on the efficiency of microbial oil production, fatty acid composition and growth of *Y. lipolytica* yeast with simultaneous waste valorization. The work assumed the use of oleaginous yeasts to manage waste from the food industry, specifically rapeseed post-frying oil, as an essential carbon source for cell growth.

2. Materials and Methods

2.1. Yeast strain and Culture Conditions

In the current work, the yeast strain *Y. lipolytica* KKP 379 from the Collection of Industrial Microorganisms Cultures belonging to Professor Wacław Dąbrowski Institute of Agricultural and Food Biotechnology—State Research Institute in Warsaw (Poland) was used. The yeast culture was stored at $-20\,^{\circ}\text{C}$ using cryovials containing ceramic beads and cryoprotective agent (Protect Select, Technical Service Consultants Ltd., Heywood, UK).

Inoculation culture was conducted in YPG medium with the following composition: $10\,\text{g}/\text{dm}^3$ yeast extract (Y), $20\,\text{g}/\text{dm}^3$ peptone (P) and $20\,\text{g}/\text{dm}^3$ glucose (G). The flasks were incubated at $28\,^{\circ}\text{C}$ for 24 h with a rotation amplitude of 140 rpm.

The yeast cells were incubated in $500\,\text{cm}^3$ flasks at $28\,^{\circ}\text{C}$ with 140 rpm rotary shaker speed. Each flask contained $200\,\text{cm}^3$ of sterile medium. Batch cultures were also conducted in a BIOFLO 3000 laboratory bioreactor from New Brunswick Scientific (USA) with a working volume of $4\,\text{dm}^3$, fermentation temperature $28\,^{\circ}\text{C}$ and 0.025% (v/v) inoculum. Oxygenation control was applied using compressed air to maintain the relative degree of oxygenation in the culture medium at a level not lower than 30% of the initial oxygen concentration (variable agitation speed 300–600 rpm). The dissolved oxygen level was measured using an oxygen electrode. Changes in pH values were also monitored using a selective electrode. Parameters of cellular lipid biosynthesis in a batch culture of *Y. lipolytica* strain were calculated according to Fabiszewska et al. [2].

The yeast grew in a mineral medium containing $1.5\,\text{g}/\text{dm}^3$ $MgSO_4$, $0.16\,\text{g}/\text{dm}^3$ $FeSO_4 \cdot H_2O$, $0.15\,\text{g}/\text{dm}^3$ $CaCl_2$, $0.08\,\text{g}/\text{dm}^3$ $MnCl_2 \cdot 4H_2O$, $0.02\,\text{g}/\text{dm}^3$ $ZnSO_4$ and KH_2PO_4, Na_2HPO_4, and $(NH_4)_2SO_4$ at different levels. All inorganic chemicals were purchased from Avantor Performance Materials Poland S.A (Gliwice, Poland). In all batch cultures, the carbon source was waste rapeseed post-frying oil at $50\,\text{g}/\text{dm}^3$. Waste oil, in which cod (*Gadus morhua*) fillets were fried at $170\,^{\circ}\text{C}$ in full immersion, came from a fish processing company in Podlaskie Voivodeship (Poland). The preliminary batch shaking flask cultures in duplicate have been conducted with statistical planning of experiments using a Latin square plan (4 × 4) (Table 1).

Table 1. Latin Square plan 4 × 4-flask experiment design.

$KH_2PO_4\,(\text{g/dm}^3)$	Duration of the Cultivations (Days)				
	3	4	5	6	
3	4.0	6.0	8.0	10.0	$(NH_4)_2SO_4\,(\text{g/dm}^3)$
7	6.0	8.0	10	4.0	
9	8.0	10.0	4.0	6.0	
11	10.0	4.0	6.0	8.0	

Bioreactor batch cultures have been conducted in media containing variable concentrations of KH_2PO_4, Na_2HPO_4 and $(NH_4)_2SO_4$ (Table 2)

Table 2. Composition of culture media used in yeast cultivation in a laboratory bioreactor.

(g/dm^3)	M1	M2	M3	M4
KH_2PO_4	3.0	3.0	3.5	7.0
Na_2HPO_4	1.1	1.1	2.5	2.5
$(NH_4)_2SO_4$	3.0	8.0	10.0	4.0

2.2. General Analytical Techniques

Yeast biomass yield was determined by cell dry weight measured by thermogravimetric method. Cells were harvested by centrifugation at 8000 rpm, 4 °C for 10 min, washed using distilled water and dried at 105 °C to constant weight.

Determination of nitrogen content in the culture medium after biomass removal was carried out by the modified Kjeldahl method [37] with the sample mineralization step omitted. From 90 cm^3 of the culture medium sample, ammonia was distilled (50 cm^3 40% NaOH, distillation time—4 min) into 25 cm^3 4% boric acid. The solution was titrated with 0.1 M HCl in the presence of Tashiro indicator. The nitrogen content of the sample was converted to the level of ammonium sulfate remaining in the medium.

Extraction of cellular lipids from dry material was performed in Soxhlet extractor using *n*-hexane as a solvent. The authors used modified Folch et al. [38] method extraction for batch shaking cultures by treating the dried and washed biomass four times with portions of a chloroform and methanol (2:1) mixture (1 cm^3/1 g$_{d.m}$). The solvents were separated by distillation under reduced pressure of 360 mbar. Distillation was carried out in a Buchi Rotavapor R-200 evaporator (Flawil, Switzerland). The lipid substrate consumption in the culture was evaluated using a simple extraction method with 10 cm^3 portions of hexane. Magnesium sulfate was then added to the oil phase to remove water. After 10 min, the whole was filtered to remove the drying agent. The solvent was evaporated from the organic phase and the remaining oil was weighed.

2.3. Gas Chromatography

Microbial lipid samples extracted from yeast cells were derivatized with 14% solution of BF$_3$ (boron trifluoride) in methanol added in equal volumes and heated for 2 h at 60 °C. Fatty acid composition of microbial lipids was determined by gas chromatography using a flame ionization detector (GC-FID) in an Agilent Technology 7820 (Santa Clara, CA, USA) with Zebron ZB-FFAP Capillary GC Column (30 m × 0.25 mm × 0.25 μm). Nitrogen has been used as a carrier gas at a flow rate of 35 mL/min. The temperature program was as follows: 80 °C (2 min) to 200 °C (10 min) (5 °C/min). Injector temperature: 250 °C; detector temperature: 290 °C; injection volume: 1 μL.

2.4. Statistical Analyses

Statistical elaboration of the results was performed using Statistica 13.0 set plus software (Statsoft, Cracow, Poland). A design of experiment was applied as a tool in investigating the effect of selected culture conditions on *Y. lipolytica* growth and lipid biosynthesis yield. A 4 × 4 latin square design method was used in the experiment provided in shaken flasks and *p*-value was then estimated at 0.01 (Table 1).

3. Results

3.1. Flask Scale Experiments

The preliminary experiment has been designed to evaluate the effect of culture time as well as limiting components such as phosphorus and nitrogen source on the growth of *Y. lipolytica* yeast biomass, cellular lipid yield, and the utilization of hydrophobic substrate as a carbon source. The cultures have been conducted in 16 medium variants, in which the concentrations of KH$_2$PO$_4$ (3.0, 7.0, 9.0, 11.0 g/dm^3) and (NH$_4$)$_2$SO$_4$ (4.0, 6.0, 8.0, 10.0 g/dm^3) have been modified according to the experimental scheme (Table 1). The incubations were terminated after 3, 4, 5 and 6 days. As can be seen in Figure 1A, the value of biomass yield was closely related only to the time of culture (*p*-value 0.01). The effect of the duration of yeast cultivations was also significant for intracellular lipid content. The yeast cells accumulated higher amounts of lipids on the 5th and 6th days of culture. The level of KH$_2$PO$_4$ supplementation appeared to be a significant factor influencing lipid accumulation efficiency (*p*-value 0.01) (Figure 1B). Higher lipid biosynthesis yield was correlated with lower doses of inorganic phosphorus source in the medium. In the case of the level of the nitrogen source additive in the form of (NH$_4$)$_2$SO$_4$, no significant relationship was found in the studied concentration range. It is noteworthy that none of the three analyzed factors was significantly associated with changes in the content of waste post-frying oil (carbon source) in the medium. Nevertheless, it decreased with time of culture (Figure 1C).

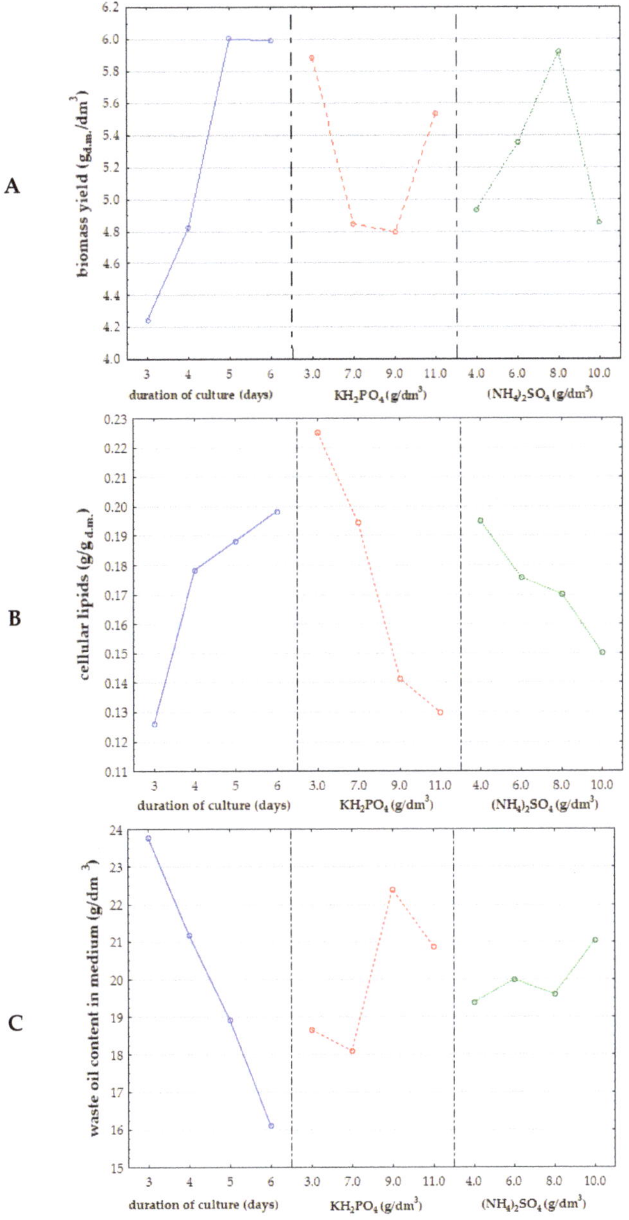

Figure 1. Statistical analysis of correlation between (**A**) biomass yield ($g_{d.m.}/dm^3$), (**B**) cellular lipid content ($g/g_{d.m.}$), (**C**) waste post-frying oil content in medium (g/dm^3) and concentration of KH_2PO_4 and $(NH_4)_2SO_4$, and culture time of *Y. lipolytica* KKP 379 yeast in medium with 5% addition of post-frying rapeseed oil.

3.2. Bioreactor Scale Experiments

Four batch cultures of *Y. lipolytica* KKP 379 strain in a laboratory bioreactor have been conducted to evaluate the effect of the previously selected factors. In each culture, changes in the pH of the culture medium and the rate of oxygen consumption have been monitored (Figure 2). Generally, in all variants of the culture medium, the yeast completed the logarithmic growth phase after nearly 16 h, thus entering the stationary phase associated with the production of metabolites such as acids, which was reflected in a significant decrease in pH, but most importantly the cells began to accumulate storage lipids.

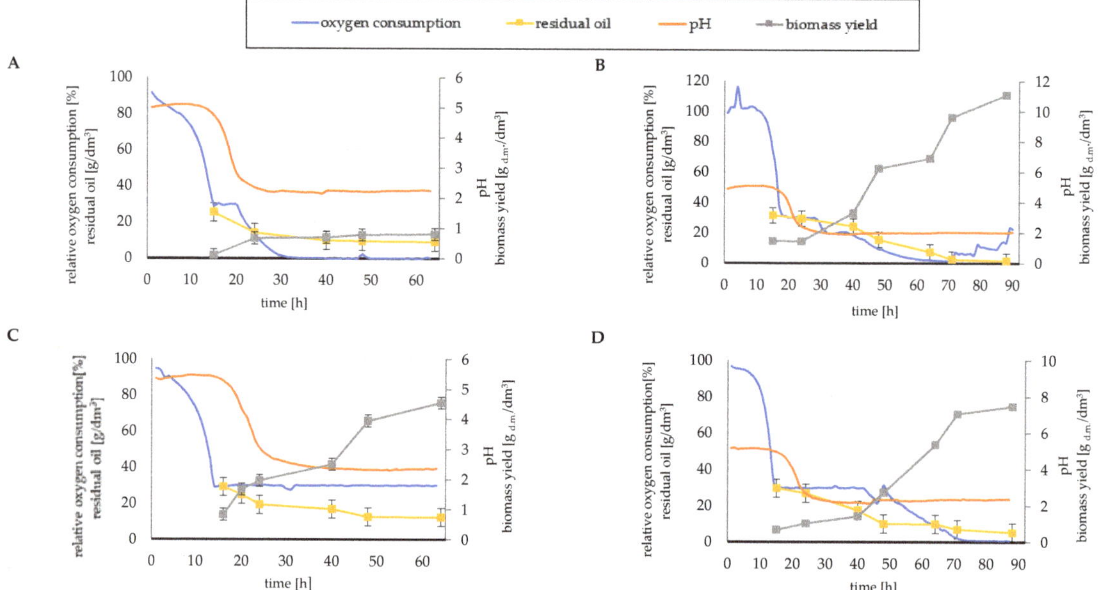

Figure 2. Changes in relative oxygen consumption and pH of the strain *Y. lipolytica* KKP 379 grown in (**A**) M1, (**B**) M2, (**C**) M3 and (**D**) M4 medium with 5% rapeseed waste post-frying oil as a carbon source.

For M1 medium, in addition to the modification related to KH_2PO_4 (3.0 g/dm^3) and $(NH_4)_2SO_4$ (3.0 g/dm^3) doses, the level of Na_2HPO_4 supplementation (1.1 g/dm^3) was also adjusted due to the buffering properties of the medium. The molar ratios C/N and C/P were 71.8:1 and 114.9:1, respectively. A significant increase in the oxygen demand of the cells has been observed from the 20th hour of culture onwards, resulting in a decrease in the dissolved oxygen content of the medium (Figure 2A). During the entire cultivation process, the biomass yield value was marginal. Therefore, the high oxygen demand could be related to the increased lipid synthesis process due to the beginning of the stationary growth phase rather than to the oxygen demand caused by the intensive biomass growth, which did not occur. After 63 h of cultivation in mineral medium M1 with phosphorus and nitrogen source limitation, the efficiency of intracellular lipid accumulation by wild-type *Y. lipolytica* cells was 47.44%, which is considered an astonishing result. The yeast cells also efficiently utilized the waste carbon source, leaving only 9.11 g/dm^3 unused. Despite the lipid yield result, considering the increase in biomass, the amount of SCO obtained per unit of substrate was very low (Table 3).

Table 3. Parameters of *Y. lipolytica* KKP 379 yeasts strain culture in media with 5% rapeseed waste post-frying oil as a carbon source.

Parameter	Unit	M1	M2	M3	M4
Molar ratio in the medium [C/N/P]	mol	114.9/1.6/1	114.9/4.3/1	81.6/3.8/1	53.1/1/1.07
Initial concentration of carbon source [S]	g/dm^3	50	50	50	50
Time [t]	h	63	88	63	88
Biomass yield [X]	$g_{d.m.}/dm^3$	0.78	11.10	4.56	7.45
Maximum concentration of lipids produced [L_{max}]	g/dm^3	0.37	2.32	1.49	2.24
Conversion yield of biomass per carbon substrate [$Y_{X/S}$]	$g_{d.m.}/g$	0.0156	0.2220	0.0912	0.1490
Conversion yield of storage lipids per biomass formed [$Y_{L/X}$]	$g/g_{d.m.}$	0.4744	0.2090	0.3267	0.3007
Conversion yield of storage lipids per carbon substrate [$Y_{L/S}$]	g/g	0.0074	0.0464	0.0298	0.0448
Volumetric rate of storage lipids production [q_{Lv}]	$g/dm^3/h$	0.0059	0.0256	0.0236	0.0254

The residual oil content of the substrate was a reflection of the biomass yield in the M2 substrate (C/N = 26.7:1, C/P = 114.9:1). A significant increase in biomass yield after 2 days of culture combined with a decrease in the residual carbon source in the medium was noticeable (Figure 2B). After 88 h of culture, the cells had used almost all of the waste post-frying oil, reducing its initial content in the culture medium from 50 to 1.56 g/dm^3, achieving at the same time the highest biomass yield (11.10 $g_{d.m.}/dm^3$) among the other yeast culture variants (Table 3). In addition, there was a gradual decrease in oxygen demand at the end of the culture (71 h). Cells no longer multiplied intensively; moreover, lipids were not efficiently accumulated, gaining a final yield of 20.9%.

Oxygen turned out to be a factor worth observing also in the case of the next batch culture in M3 medium. The addition of $(NH_4)_2SO_4$ was further increased to 10 g/dm^3 (C/N = 21.5:1). Similarly, the level of the addition of phosphorus sources was slightly increased, KH_2PO_4 3.5 g/dm^3 and Na_2HPO_4 2.5 g/dm^3 (C/P = 81.6:1). In culture medium, the dissolved oxygen content remained constant from the 15th hour until the end of the cultivation process. The amplitude range of the agitation speed appeared to be sufficient to maintain the oxygenation level of the culture medium at a level close to 30%. The constantly low level of oxygen consumption by the cells indicated low oxygen demand, which could be associated with moderate growth of yeast biomass (Figure 2C). Although, the biomass yield was only 4.56 $g_{d.m.}/dm^3$. The final waste post-frying oil content in the medium was relatively low at 12.44 g/dm^3. Despite the low biomass yield, yeast consumption of the oily substrate was at a relatively high level, similar to the M1 medium. This is another example of the diversion of the available carbon source to the needs of storage lipid accumulation, not to the energy needs associated with growth. After 63 h, the efficiency of lipid storage was lower in this medium, but still at a satisfactory level of 32.67%.

When *Y. lipolytica* yeast was cultured in M4 medium, significant changes were observed after 2 days of yeast growth. After about 40 h, dissolved oxygen consumption increased due to the increase in biomass yield, which reached 7.45 $g_{d.m.}/dm^3$ (Figure 2D). Thus, this resulted in a decrease in the lipid carbon source content of the culture medium; the final result after 88 h of culturing was a reduction to 5.33 g/dm^3. Molar ratios of C/P = 49.6:1 and C/N = 53.1:1 yielded 2.24 g of microbial oil per dm^3 of culture medium after 88 h (30.07% cell dry weight). As could be expected, the C/N ratio changed with the duration of the culture. After 24 h, when the most intensive period of cell growth was already over, the ratio decreased to C/N = 30.6:1, after the following hours it was already half (15.5:1), and at the end of the culture the molar C/N ratio was equal to 8.4:1.

The marked observation to emerge from the data comparison (Figure 3) was the dominance of oleic acid (C18:1) in the total fatty acids content of the cellular oils from each medium variant. When analyzing the composition of extracted oils in the context of their potential use as nutritionally valuable products, the content of linoleic (C18:2) and linolenic acids (C18:3) should be considered. The oleic acid content of the oils ranged from 53.89% (M2) to 60.44% (M3), while the linoleic acid content ranged from 17.39% (M4) to

22.58% (M2). For cellular lipids derived from culture in M2 medium, at the same time with the lowest proportion of oleic acid, the highest contents of linoleic acid and linolenic acid (8.89%) were noted. Importantly, considering all variants of the medium, the linolenic acid was at least, 5.82%. Saturated fatty acids were also determined in each of the cellular lipid samples. The highest percentage of stearic acid was found in samples from medium M1-8.40% and M4-10.93%. Palmitic acid (C16:0) from 1.40% for medium M4 to 4.76% for M2, palmitoleic acid (C16:1): 1.06% (M1)—2.62% (M2), arachidic acid (C20:0): 0.98% (M2)—1.55% (M4) and behenic acid (C22:0): 0.90% (M4)—2.42% (M3) were also identified. The content of myristic acid (C14:0) in the total pool of all fatty acids did not exceed 0.91%.

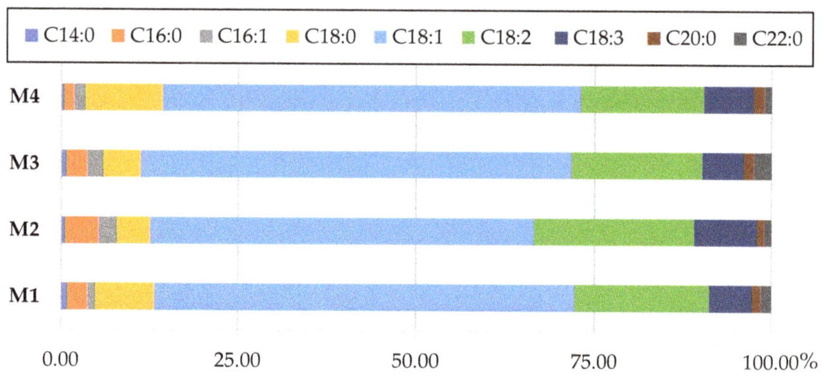

Figure 3. Fatty acid composition [% w/w] of cellular lipids in *Y. lipolytica* KKP 379 grown in **M1**, **M2**, **M3** and **M4** medium with 5% rapeseed waste post-frying oil as a carbon source.

4. Discussion

4.1. Lipid Production in Phosphorus-Limited Media

Phosphorus is one of the basic elements for the cultivation of microorganisms, including oleaginous ones. Due to its incorporation into phospholipids, nucleic acids or coenzymes, the analyzed component has been considered crucial for the structure of yeast cells as well as their functioning. Phosphorus may also be stored as polymetaphosphate in biomass. Interestingly, the non-conventional yeast *Y. lipolytica* contains nearly half as much phosphorus compared to the common yeast species *Sacharomyces cerevisiae* [39].

According to Wu et al. [31], phosphorus limitation has the potential to be used as a regulator of lipid production by microorganisms, also in the presence of nitrogen-rich media. In other words, this approach provides new opportunities for microbial lipid biosynthesis in a more economical manner, due to the use of complex waste materials as a carbon source. Before the rapeseed oil after fish frying was used as substrate, in the present experiment, it was mechanically purified from the residue of the fried product suspended in it. In view of the above, the waste was not a source of nitrogen, but only a lipid carbon source. Taskin et al. [29] were the first to attempt to investigate the potential of cheese whey as a substrate for microbial oil production by *Y. lipolytica* B9. The analyzed by-product consisting of nearly 93% water and about 7% dry matter is particularly rich in phosphorus. One of the factors examined was the effect of supplementation of KH_2PO_4 as an additional phosphorus source on biomass growth and lipid accumulation levels. Similar to the current study, the authors found that phosphorus limitation increased the lipid content in yeast cells. The additional phosphorus source significantly decreased the efficiency of lipid substance accumulation in yeast cells compared to the medium without supplementation, when the efficiency was 44% on a dry weight basis. The addition of 1 and 2 g/dm^3 KH_2PO_4 contributed to a cellular lipid content of 37 and 26%, respectively.

The effect of limiting access to mineral salts including potassium phosphate on SCO production by *Y. lipolytica* yeast was also studied by Hoarau et al. [35]. In the study, waste produced by ethanol distilleries named Distillery Spent Wash (DSW) was used. After all, the lipid accumulation efficiency was only 9.15% using a medium with 7.0 g/dm^3 KH$_2$PO$_4$ and DSW. Moreover, phospholipids accounted for 34% of all lipids. However, this accumulation efficiency result was higher compared to the culture when raw DSW was used without additional phosphate supplementation (7.18%). The addition of 7 g/dm^3 KH$_2$PO$_4$ and 1.5 g/dm^3 MgSO$_4$ to the culture medium resulted in the highest biomass yield of 7.34 g$_{d.m.}$/dm^3. It is worth noting that the best growth of *Y. lipolytica* yeast was in DSW with the addition of KH$_2$PO$_4$ in the range of 7.5–8.0 g/dm^3. The results of the current study appear to be consistent. Using M4 medium with 7.0 g/dm^3 KH$_2$PO$_4$, 1.5 g/dm^3 MgSO$_4$ and 4 g/dm^3 (NH$_4$)$_2$SO$_4$, the biomass yield was 7.45 g$_{d.m.}$/dm^3, but it was not the highest score. The yeast showed the best growth in medium with limiting additions of phosphorus source, 3.0 g/dm^3 KH$_2$PO$_4$ and 1.1 g/dm^3 Na$_2$HPO$_4$, with a relatively higher nitrogen source addition.

A valuable experiment was conducted by Wu et al. [31], whose results, similar to the current work, showed that better microbial lipid production efficiency correlated with low phosphorus source doses and higher C/P molar ratio. When yeast *R. toruloides* Y4 was cultivated in the medium with 3.6 g/L of KH$_2$PO$_4$ and 70 g/dm^3 initial glucose content, which corresponded to a C/P molar ratio of 72:1, after 96 h of batch shaking the flask biomass yield was 18.6 g$_{d.m.}$/dm^3 with a lipid content of 21.2%. Moreover, enhancing the limitation of phosphorus source to C/P= 9552:1 resulted in improved SCO production to 62.1% and better biomass growth of 19.4 g$_{d.m.}$/dm^3. Surprisingly, when the nitrogen source was additionally restricted (C/N = 22.3:1), the accumulation efficiency was 63.3% with a biomass yield of 19.9 g$_{d.m.}$/dm^3. Under phosphorus-limiting conditions, lipid efficiency was consistently close to 60%, even when C/N = 6.1:1 [31].

In the present experiments, the results of the flask-shaken cultures show that the level of KH$_2$PO$_4$ supplementation appeared to be a significant factor affecting the lipid accumulation efficiency, in contrast to the addition of nitrogen source in the form of (NH$_4$)$_2$SO$_4$, for which no significant relationship was found over the concentration range studied. In a batch culture in a bioreactor, the highest efficiency of cellular lipid production by *Y. lipolytica* yeast has been observed for M1 medium, in which C/P was 114.9:1 and C/N = 71.8:1. However, in bioreactor culture mode, a comparable relationship was not found for the impact of the molar ratio C/P. At a C/P of 114.9, both the highest and lowest microbial lipid production among all substrates was obtained. Under the established conditions, this nitrogen addition was significant. In the experiment of Fabiszewska et al. [2], *Y. lipolytica* yeast cultured in YPG medium with high nitrogen content and therefore low C/N ratio did not produce SCO. This is in contrast to mineral media with glucose (MG7) and olive oil (MO7), where the conversion yield of storage lipids per biomass was 0.116 and 0.207 g/g$_{d.m.}$. These could be further examples that confirm the role of nitrogen limitation in culture medium.

4.2. Lipid Production in Nitrogen-Limited Media

Nitrogen is considered to be a special component, the limited amount of which in the medium has the greatest effect on the induction of lipogenesis in the cells of oil-producing species. In other words, when the culture medium is deficient in nitrogenous compounds or when its pool is depleted, the synthesis of nucleic acids and proteins is inhibited and the rate of biomass growth decreases [40]. According to Bellou et al. [32], in addition to magnesium limitation, the restriction of cellular access to nitrogen is the factor for efficient cellular lipid biosynthesis. Continuous culture of the wild yeast strain *Y. lipolytica* ACA-DC50109 resulted in an accumulation efficiency of 47.5% with a biomass yield of 12.2 g/L. Furthermore, lipid accumulation at high levels was provided by low nitrogen doses, but at levels that allowed *Y. lipolytica* yeast cells to grow. This correlation was related to the maintenance of the pentose-phosphate pathway properly function, which provides the coenzyme NADPH for the mechanism of lipogenesis. For this reason, understanding the physiology of oleaginous microorganisms is of great importance in efficient cellular lipids production.

Inorganic $(NH_4)_2SO_4$ was the only source of nitrogen in the present study. The current experiment assumed the use of a medium containing only mineral components, excluding the carbon source. There are reports that the presence of organic nitrogen sources in the medium may stimulate microbial synthesis of lipid components [41,42]. Neither the preliminary study stage nor the bioreactor experiments proved a clear relationship between efficient cellular lipid accumulation via ex novo pathway and the level of nitrogen supplementation. In M3 medium (10 g/dm^3 $(NH_4)_2SO_4$), the lipid accumulation efficiency was 32.67%, while in M2 (8 g/dm^3 $(NH_4)_2SO_4$) it was only 20.90%. It is worth noting that in the medium with a lower dose of ammonium sulfate the biomass yield reached more than twice the value (Table 3). When the low addition of a nitrogen source in the M4 culture medium was compensated by a higher level of a phosphorus source, the accumulation efficiency was close to that for the phosphorus-limited M3 medium. This result further strengthened the hypothesis that the ratio of the two analyzed elements nitrogen and phosphorus can be considered an important factor for both yeast cell growth and lipid accumulation capacity.

The oil-producing species of the yeast *Trichosporon oleaginosus*, in the presence of xylose and limited access to nitrogen compounds, accumulated over 50% of lipids in the dry mass of cells. The type of limitation used (-N, -C and -CN) had a significant effect on the fatty acid profile of the oil extracted from yeast cells. The content of linoleic acid (C18:2 in the -cis configuration) showed the greatest variation depending on the conditions applied. When the researchers limited the access of the cells to nitrogen and carbon, the C18:2 content was 7.31% and 7.45% (w/w), respectively. When carbon source limitation was applied alone, the content of linoleic acid (as the main component of the cell membrane) increased threefold, accounting for 21.58% of all fatty acids [43]. In the current study, the linoleic acid content of the extracted oils was a maximum of 22.58% of total fatty acids. Notably, nutritionally valuable linolenic acid was present in each of the oil samples, with the highest percentage from M2 cultivation.

In the experiment, the highest percentage of SCO production by *Y. lipolytica* yeast was obtained for culture medium with 3.0 g/dm^3 $(NH_4)_2SO_4$ and 3.0 g/dm^3 KH_2PO_4 and molar ratios of C/P = 114.9:1 and C/N = 71.8:1. A molar C/N ratio of 20:1 is considered to be the minimum for promoting lipid accumulation in oleaginous yeast cells [44], in the experiment, was not lower than 21.5. As found, simultaneous limited access to phosphorus and nitrogen promoted lipid accumulation, but inhibited biomass growth. When the nitrogen dose was increased, with the phosphorus dose unchanged (M2), yeast growth was improved, but the efficiency of the biosynthesis of lipids was significantly reduced. The decreased nitrogen-to-phosphorus ratio (N/P = 3,8:1) in M3 medium, compared to M2 medium (N/P = 4.3:1), resulted in half the biomass yield with 32% cellular lipid accumulation efficiency. Under different conditions, when the reduced nitrogen rate was accompanied by increased cell access to phosphorus sources in the culture medium (M4—N/P = 1:1), growth needs were met, resulting in one of the best biomass yields of 7.45 g$_{d.m.}$/dm^3 with satisfactory SCO production efficiency. It can be summarized that phosphorus is needed by cells for growth, but in this context nitrogen is crucial.

Nevertheless, a greater addition of phosphorus source to a nitrogen-deficient medium may lead to the accumulation of higher storage lipid yields without negative effects on growth caused by reduced access to nitrogen. Both bioreactor experiments and preliminary studies emphasized the role of phosphorus as a substrate component in metabolic processes related to the basic metabolism of the yeast cell and SCO biosynthesis. Still, the role of phosphorus in combination with the issue of media supplementation with a nitrogen source needs further studies.

Using a C/N ratio of 30:1 led to an increase in saturated fatty acid production and a decrease in polyunsaturated fatty acids in *Y. lipolytica* yeast cells. The application of additional phosphorus limitation (C/P = 1043:1) in the medium did not alter the fatty acid group content of total lipid acids observed previously. Similar to the current experiment, total lipid content increased under limiting conditions, but did not exceed 30%. The highest biomass yield (over 7 $g_{d.m.}/dm^3$) was achieved for culture in medium with C/N = 30:1, without limiting phosphorus source (C/P = 6:1) [33].

Microbial oil accumulated in lipid bodies consists mainly of TAG and steryl-esters in smaller amounts (neutral fractions), which are enclosed by polar fractions, e.g., phospholipids, glycolipids, and sphingolipids. Generally, the growth of oleaginous microorganisms may be divided into two phases. When cells grow in media rich in all the nutrients necessary for growth, biomass, also free of lipids, is produced. During this phase, lipids with a higher proportion of polar fractions, corresponding to cell membrane lipids, are accumulated. In a situation of limited access to sources such as nitrogen, phosphorus or sulfate, the lipid accumulation phase is stimulated, mainly of neutral fractions [40].

4.3. Cellular Oxygen Demand in Microbial Lipid Production

Based on the authors' knowledge, there is no clear statement on the effect of the oxygenation degree on the efficiency of lipid accumulation. The link between dissolved oxygen in the medium and accumulation capacity by *Y. lipolytica* yeast is inconclusive. However, as a result of thought-provoking observations, the issue of oxygen demand by *Y. lipolytica* yeast appeared to be worth exploring. The concentration of oxygen available in the culture medium is a crucial variable in both the context of growth and efficient microbial lipid production by strictly aerobic yeast *Y. lipolytica*. It is predicted that an increase in substrate oxygenation may result in the stimulation of cells for growth [45].

The observations highlight the importance of the cellular oxygen demand in relation to growth, but also to the process of cellular lipid accumulation. Despite the negligible increase in biomass yield, the yeast still efficiently used oxygen from the substrate, which could be related to the intensively occurring process of lipid biosynthesis. Interestingly, it may be reasonably assumed that the cells should be exposed to a sufficiently high concentration of oxygen in the medium. Nevertheless, too high a concentration of dissolved oxygen may cause inhibition of cellular metabolism resulting from oxidative stress. Given the need to maintain an appropriate degree of oxygenation of the culture medium, agitation rate is an important parameter. Under conditions of increased cellular oxygen demand, high agitation rate is used. As is widely known, this kind of approach may have negative consequences. High shear forces and mechanical stress may make culture development impossible [46]. Therefore, despite the increase in cellular oxygen demand in bioreactor cultures, the set amplitude of the agitator rotation was not able to meet the demand of the yeast.

According to Bellou et al. [47], high dissolved oxygen (1.5 mg/dm^3) in the culture media stimulates cellular lipid synthesis. Upregulation of ATP-citrate lyase (ATP-CL) and malic enzyme (ME), which are involved in microbial lipid synthesis, was observed under such conditions. The importance of malic enzyme for cellular metabolism of oleaginous microorganisms including *Y. lipolytica* yeast has been recognized as crucial. The product of reaction catalyzed by ME gives the product necessary for fatty acid synthesis, namely, coenzyme NADPH. Under conditions of limited access of cells to nitrogenous compounds, NADPH is produced only by enzymatic reactions involving ME. Conversely, increasing the dissolved oxygen concentration resulted in a decrease in NAD-dependent isocitrate dehydrogenase enzyme activity. When the Krebs cycle is inhibited due to the depletion of nitrogen sources by the cells and allosteric AMP (adenosinemonophosphate) activator concentrations being reduced, ICDH is deactivated. This phenomenon is considered to be characteristic of the de novo biosynthesis pathway [48]. When yeast cultures were conducted in glucose medium, it was observed that limited cellular access to phosphorus, as with nitrogen, results in a reduction in AMP levels to activate cellular phosphorus reserves.

There are two strategies to maximize cellular lipid accumulation. The first one, by deleting genes encoding for acyl-oxidase activity, assumes the prevention of the degradation of storage lipids [49]. The second one aims to support the production via the overexpression of genes encoding for ATP-CL and ME activity [50,51]. Wasylenko et al. [52] using 13C-Metabolic Flux Analysis singled the oxidative pentose-phosphate pathway out as the main source of lipogenic coenzyme NADPH rather than malic enzyme activity. In view of the above, the purpose for research on the role of phosphorus as a coenzyme component in processes related to cellular respiration and provision of energy necessary for SCO biosynthesis has been justified.

Many mechanisms and interactions accompanying the accumulation of lipid compounds via the de novo pathway have been recognized. Nevertheless, there is still a need to search for physiological and biochemical relationships that can explain, and thus support, efficient ex novo biosynthesis. The authors of this paper hypothesized a way in which the de novo and ex novo pathways could be related. The dissimilated fatty acids are degraded in the first step, which is mitochondrial β-oxidation. This process provides the energy necessary for cell growth and development by generating 1 mole of FADH and 1 mole of NADH per every 1 mole of acetyl-CoA produced before it is diverted to the Krebs cycle. Energy is also required for the production of intermediate metabolites [40]. Dissolved fatty acids after entering the cell are degraded in the first step, which is β-oxidation in peroxisomes (Figure 4). β-oxidation is a cyclic process that will be repeated until the lipid substrate is completely broken down. However, there is an anomaly to the cycle, leakage of certain fatty acids from the pathway could be observed, which may be stored in lipid bodies in the next step [48].

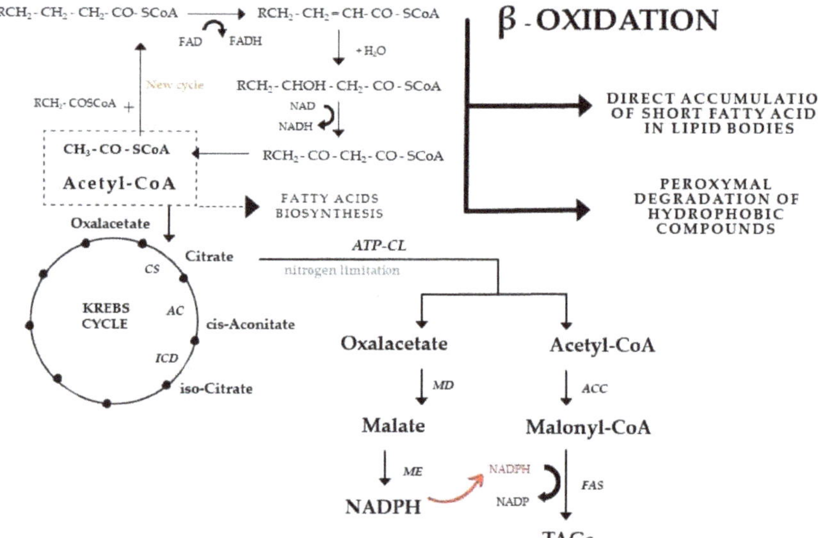

Figure 4. A hypothetical combination of de novo and ex novo pathways of cellular lipid biosynthesis in media with hydrophobic carbon sources. CS—citrate synthase, AC—aconitase, ICD—iso-citrate dehydrogenase, ATP-CL—ATP-citrate lyase, MD—malate dehydrogenase, ME—malic enzyme, ACC—acetyl-CoA carboxylase, FAS—fatty acid synthetase. From [40], adapted.

5. Conclusions

To conclude, the results of the current research suggest that nitrogen limitation in culture media is crucial, but additional phosphorus limitation may further improve the efficiency of microbial lipid biosynthesis. To the authors' current knowledge, phosphorus is a component of the culture medium, which still needs to be investigated in relation to the optimal level of its limitation in media stimulating SCO biosynthesis. In particular, evaluating the effect of the level of simultaneous substrate supplementation with phosphorus, nitrogen and the ratio of these two components seems to be one of the key issues in the context of efficient microbial lipid production combined with satisfactory biomass yield. Knowledge of the role of phosphorus as a component of coenzymes essential for a number of biochemical transformations occurring in yeast cells may prove crucial to improving microbial lipid biosynthesis and simultaneously supporting biomass growth, as well as considering respiratory processors. Further studies aimed at looking for strategies to stimulate microbial lipid biosynthesis via the ex novo pathway are worthy of consideration, as most of the available research work involves the de novo pathway. The approach is important for the basic research and from the practical points of view, especially if the possibility of oily waste applicability as a substrate in microbial culture is considered.

Author Contributions: Conceptualization, K.W.; Methodology, K.W., A.F., D.N. and B.Z.; Formal analysis: K.W. and A.F., Investigation, K.W., A.F., D.N. and B.Z.; Resources, A.F. and D.N.; Data Curation, K.W.; Writing—Original Draft Preparation, K.W.; Writing—Review & Editing, A.F. and B.Z.; Visualization, K.W.; Supervision, D.N. and A.F. All authors have read and agreed to the published version of the manuscript.

Funding: The study was financially supported by sources of the Ministry of Education and Science within funds of the Institute of Food Sciences of Warsaw University of Life Sciences (WULS), for scientific research.

Conflicts of Interest: The authors declare no conflict of interest.

References

1. Beopoulos, A.; Cescut, J.; Haddouche, R.; Uribelarrea, J.L.; Molina-Jouve, C.; Nicaud, J.M. *Yarrowia lipolytica* as a model for bio-oil production. *Prog. Lipid Res.* **2009**, *48*, 375–387. [CrossRef]
2. Fabiszewska, A.; Misiukiewicz-Stępień, P.; Paplińska-Goryca, M.; Zieniuk, B.; Białecka-Florjańczyk, E. An insight into storage lipid synthesis by *Yarrowia lipolytica* yeast relating to lipid and sugar substrates metabolism. *Biomolecules* **2019**, *9*, 685–697. [CrossRef] [PubMed]
3. Groenewald, M.; Boekhout, T.; Neuvéglise, C.; Gaillardin, C.; van Dijck, P.W.; Wyss, M. *Yarrowia lipolytica*: Safety assessment of an oleaginous yeast with a great industrial potential. *Crit. Rev. Microbiol.* **2014**, *40*, 187–206. [CrossRef] [PubMed]
4. Martínez, E.J.; Raghavan, V.; González-Andrés, F.; Gómez, X. New biofuel alternatives: Integrating waste management and single cell oil production. *Int. J. Mol. Sci.* **2015**, *16*, 9385–9405. [CrossRef]
5. Pawar, P.P.; Odaneth, A.A.; Vadgama, R.N.; Lali, A.M. Simultaneous lipid biosynthesis and recovery for oleaginous yeast *Yarrowia lipolytica*. *Biotechnol. Biofuels* **2019**, *12*, 237. [CrossRef] [PubMed]
6. Papanikolaou, S.; Chevalot, I.; Galiotou-Panayotou, M.; Komaitis, M.; Marc, I.; Aggelis, G. Industrial derivative of tallow: A promising renewable substrate for microbial lipid, single-cell protein and lipase production by *Yarrowia lipolytica*. *Electron. J. Biotechnol.* **2007**, *10*, 425–435. [CrossRef]
7. Cheirsilp, B.; Louhasakul, Y. Industrial wastes as a promising renewable source for production of microbial lipid and direct transesterification of the lipid into biodiesel. *Bioresour. Technol.* **2013**, *142*, 329–337. [CrossRef]
8. Bharathiraja, B.; Sridharan, S.; Sowmya, V.; Yuvaraj, D.; Praveenkumar, R. Microbial oil—A plausible alternate resource for food and fuel application. *Bioresour. Technol.* **2017**, *233*, 423–432. [CrossRef] [PubMed]
9. Subramaniam, R.; Dufreche, S.; Zappi, M.; Bajpai, R. Microbial lipids from renewable resources: Production and characterization. *J. Ind. Microbiol. Biotechnol.* **2010**, *37*, 1271–1287. [CrossRef]
10. Ratledge, C.; Wynn, J.P. The biochemistry and molecular biology of lipid accumulation in oleaginous microorganisms. *Adv. Appl. Microbiol.* **2002**, *51*, 1–52. [PubMed]
11. Lopes, M.; Gomes, A.S.; Silva, C.M.; Belo, I. Microbial lipids and added value metabolites production by *Yarrowia lipolytica* from pork lard. *J. Biotechnol.* **2018**, *265*, 76–85. [CrossRef]
12. Ji, X.J.; Ledesma-Amaro, R. Microbial lipid biotechnology to produce polyunsaturated fatty acids. *Trends Biotechnol.* **2020**, *38*, 832–834. [CrossRef]
13. Patel, A.; Pruthi, V.; Pruthi, P.A. Synchronized nutrient stress conditions trigger the diversion of CDP-DG pathway of phospholipids synthesis towards de novo TAG synthesis in oleaginous yeast escalating biodiesel production. *Energy* **2017**, *139*, 962–974. [CrossRef]
14. Fabiszewska, A.U.; Kotyrba, D.; Nowak, D. Assortment of carbon sources in medium for *Yarrowia lipolytica* lipase production: A statistical approach. *Ann. Microbiol.* **2015**, *65*, 1495–1503. [CrossRef] [PubMed]
15. Papanikolaou, S.; Muniglia, L.; Chevalot, I.; Aggelis, G.; Marc, I. Accumulation of a cocoa-butter-like lipid by *Yarrowia lipolytica* cultivated on agro-industrial residues. *Curr. Microbiol.* **2003**, *46*, 124–130. [CrossRef] [PubMed]

16. Gajdos, P.; Nicaud, J.M.; Rossignol, T.; Čertík, M. Single cell oil production on molasses by *Yarrowia lipolytica* strains overexpressing DGA2 in multicopy. *Appl. Microbiol. Biotechnol.* **2015**, *99*, 8065–8074. [CrossRef] [PubMed]
17. Qin, L.; Liu, L.; Zeng, A.P.; Wei, D. From low-cost substrates to single cell oils synthesized by oleaginous yeasts. *Bioresour. Technol.* **2017**, *245*, 1507–1519. [CrossRef]
18. Lopes, M.; Miranda, S.M.; Belo, I. Microbial valorization of waste cooking oils for valuable compounds production—A review. *Crit. Rev. Environ. Sci. Technol.* **2020**, *50*, 2583–2616. [CrossRef]
19. Liu, X.; Lv, J.; Xu, J.; Zhang, T.; Deng, Y.; He, J. Citric acid production in *Yarrowia lipolytica* SWJ-1b yeast when grown on waste cooking oil. *Appl. Biochem. Biotechnol.* **2015**, *175*, 2347–2356. [CrossRef]
20. Xiaoyan, L.; Yu, X.; Lv, J.; Xu, J.; Xia, J.; Wu, Z.; Zhang, T.; Deng, Y. A cost-effective process for the coproduction of erythritol and lipase with *Yarrowia lipolytica* M53 from waste cooking oil. *Food Bioprod. Process.* **2017**, *103*, 86–94. [CrossRef]
21. Nunes, P.M.B.; Martins, A.B.; Brigida, A.I.S.; Rocha Leao, M.H.M.; Amaral, P. Intracellular lipase production by *Yarrowia lipolytica* using different carbon sources. *Chem. Eng. Trans.* **2014**, *38*, 421–426.
22. Lopes, M.; Miranda, S.M.; Alves, J.M.; Pereira, A.S.; Belo, I. Waste cooking oils as feedstock for lipase and lipid-rich biomass production. *Eur. J. Lipid Sci. Technol.* **2019**, *121*, 1800188–1800196. [CrossRef]
23. El Bialy, H.; Gomaa, O.M.; Azab, K.S. Conversion of oil waste to valuable fatty acids using oleaginous yeast. *World J. Microbiol. Biotechnol.* **2011**, *27*, 2791–2798. [CrossRef]
24. Katre, G.; Joshi, C.; Khot, M.; Zinjarde, S.; RaviKumar, A. Evaluation of single cell oil (SCO) from a tropical marine yeast *Yarrowia lipolytica* NCIM 3589 as a potential feedstock for biodiesel. *AMB Express* **2012**, *2*, 36–49. [CrossRef] [PubMed]
25. Tzirita, M.; Papanikolaou, S.; Chatzifragkou, A.; Quilty, B. Waste fat biodegradation and biomodification by *Yarrowia lipolytica* and a bacterial consortium composed of *Bacillus* spp. and *Pseudomonas putida*. *Eng. Life Sci.* **2018**, *18*, 932–942. [CrossRef]
26. Wierzchowska, K.; Zieniuk, B.; Fabiszewska, A. Use of Non-Conventional Yeast *Yarrowia lipolytica* in Treatment or Upgradation of Hydrophobic Industry Wastes. *Waste Biomass Valorization* **2021**, 1–23. Available online: https://link.springer.com/article/10.1007/s12649-021-01516-9 (accessed on 9 December 2021).
27. Ratledge, C. Regulation of lipid accumulation in oleaginous micro-organisms. *Biochem. Soc. Trans.* **2002**, *30*, 1047–1050. [CrossRef]
28. Ratledge, C.; Cohen, Z. Microbial and algal oils: Do they have a future for biodiesel or as commodity oils? *Lipid Technol.* **2008**, *20*, 155–160. [CrossRef]
29. Taskin, M.; Saghafian, A.; Aydogan, M.N.; Arslan, N.P. Microbial lipid production by cold-adapted oleaginous yeast *Yarrowia lipolytica* B9 in non-sterile whey medium. *Biofuels Bioprod. Biorefin.* **2015**, *9*, 595–605. [CrossRef]
30. Gill, C.O.; Hall, M.J.; Ratledge, C. Lipid accumulation in an oleaginous yeast (*Candida* 107) growing on glucose in single-stage continuous culture. *Appl. Environ. Microbiol.* **1977**, *33*, 231–239. [CrossRef]
31. Wu, S.; Hu, C.; Jin, G.; Zhao, X.; Zhao, Z.K. Phosphate-limitation mediated lipid production by *Rhodosporidium toruloides*. *Bioresour. Technol.* **2010**, *101*, 6124–6129. [CrossRef]
32. Bellou, S.; Triantaphyllidou, I.E.; Mizerakis, P.; Aggelis, G. High lipid accumulation in *Yarrowia lipolytica* cultivated under double limitation of nitrogen and magnesium. *J. Biotechnol.* **2016**, *234*, 116–126. [CrossRef]
33. Kolouchová, I.; Maťátková, O.; Sigler, K.; Masák, J.; Řezanka, T. Lipid accumulation by oleaginous and non-oleaginous yeast strains in nitrogen and phosphate limitation. *Folia Microbiol.* **2016**, *61*, 431–438. [CrossRef]
34. Huang, X.; Luo, H.; Mu, T.; Shen, Y.; Yuan, M.; Liu, J. Enhancement of lipid accumulation by oleaginous yeast through phosphorus limitation under high content of ammonia. *Bioresour. Technol.* **2018**, *262*, 9–14. [CrossRef]
35. Hoarau, J.; Petit, T.; Grondin, I.; Marty, A.; Caro, Y. Phosphate as a limiting factor for the improvement of single cell oil production from *Yarrowia lipolytica* MUCL 30108 grown on pre-treated distillery spent wash. *J. Water Process Eng.* **2020**, *37*, 101392. [CrossRef]
36. Timoumi, A.; Guillouet, S.E.; Molina-Jouve, C.; Fillaudeau, L.; Gorret, N. Impacts of environmental conditions on product formation and morphology of *Yarrowia lipolytica*. *Appl. Microbiol. Biotechnol.* **2018**, *102*, 3831–3848. [CrossRef] [PubMed]
37. AOAC. *International Official Methods of Analysis*, 17th ed.; AOAC International: Gaithersburg, MD, USA, 2000; Volume 2.
38. Folch, J.; Lees, M.; Stanley, G.S. A simple method for the isolation and purification of total lipides from animal tissues. *J. Biol. Chem.* **1957**, *226*, 497–509. [CrossRef]
39. Michalik, B.; Biel, W.; Lubowicki, R.; Jacyno, E. Chemical composition and biological value of proteins of the yeast *Yarrowia lipolytica* growing on industrial glycerol. *Can. J. Anim. Sci.* **2014**, *94*, 99–104. [CrossRef]
40. Papanikolaou, S.; Aggelis, G. Lipids of oleaginous yeasts. Part I: Biochemistry of single cell oil production. *Eur. J. Lipid Sci. Technol.* **2011**, *113*, 1031–1051. [CrossRef]
41. Beopoulos, A.; Mrozova, Z.; Thevenieau, F.; Le Dall, M.T.; Hapala, I.; Papanikolaou, S.; Chardot, T.; Nicaud, J.M. Control of lipid accumulation in the yeast *Yarrowia lipolytica*. *Appl. Environ. Microbiol.* **2008**, *74*, 7779–7789. [CrossRef]
42. Tsigie, Y.A.; Wang, C.Y.; Truong, C.T.; Ju, Y.H. Lipid production from *Yarrowia lipolytica* Po1g grown in sugarcane bagasse hydrolysate. *Bioresour. Technol.* **2011**, *102*, 9216–9222. [CrossRef] [PubMed]
43. Santek, M.I.; Miskulin, E.; Petrovic, M.; Beluhan, S.; Santek, B. Effect of carbon and nitrogen source concentrations on the growth and lipid accumulation of yeast *Trichosporon oleaginosus* in continuous and batch culture. *J. Soc. Chem. Ind.* **2017**, *92*, 1620–1629.
44. Dobrowolski, A.; Mituła, P.; Rymowicz, W.; Mirończuk, A.M. Efficient conversion of crude glycerol from various industrial wastes into single cell oil by yeast *Yarrowia lipolytica*. *Bioresour. Technol.* **2016**, *207*, 237–243. [CrossRef] [PubMed]
45. Magdouli, S.; Brar, S.K.; Blais, J.F. Morphology and rheological behaviour of *Yarrowia lipolytica*: Impact of dissolved oxygen level on cell growth and lipid composition. *Process Biochem.* **2018**, *65*, 1–10. [CrossRef]

46. Alonso, F.O.M.; Oliveira, E.B.L.; Dellamora-Ortiz, G.M.; Pereira-Meirelles, F.V. Improvement of lipase production at different stirring speeds and oxygen levels. *Braz. J. Chem. Eng.* **2005**, *22*, 9–18. [CrossRef]
47. Bellou, S.; Makri, A.; Triantaphyllidou, I.E.; Papanikolaou, S.; Aggelis, G. Morphological and metabolic shifts of *Yarrowia lipolytica* induced by alteration of the dissolved oxygen concentration in the growth environment. *Microbiology* **2014**, *160*, 807–817. [CrossRef]
48. Beopoulos, A.; Nicaud, J.M. Yeast: A new oil producer? *Oléagineux Corps Gras Lipides* **2012**, *19*, 22–28. [CrossRef]
49. Dulermo, T.; Nicaud, J.M. Involvement of the G3P shuttle and oxidation pathway in the control of TAG synthesis and lipid accumulation in *Yarrowia lipolytica*. *Metab. Eng.* **2011**, *13*, 482–491. [CrossRef]
50. Blazeck, J.; Hill, A.; Liu, L.; Knight, R.; Miller, J.; Pan, A.; Otoupal, P.; Alper, H.S. Harnessing *Yarrowia lipolytica* lipogenesis to create a platform for lipid and biofuel production. *Nat. Commun.* **2014**, *5*, 3131. [CrossRef]
51. Zhang, H.; Zhang, L.; Chen, H.; Chen, Y.Q.; Chen, W.; Song, Y.; Ratledge, C. Enhanced lipid accumulation in the yeast *Yarrowia lipolytica* by over-expression of ATP: Citrate lyase from Mus musculus. *J. Biotechnol.* **2014**, *192*, 78–84. [CrossRef]
52. Wasylenko, T.M.; Ahn, W.S.; Stephanopoulos, G. The oxidative pentose phosphate pathway is the primary source of NADPH for lipid overproduction from glucose in *Yarrowia lipolytica*. *Metab. Eng.* **2015**, *30*, 27–39. [CrossRef]

applied sciences

MDPI

Article

Feasibility of Application of Near Infrared Reflectance (NIR) Spectroscopy for the Prediction of the Chemical Composition of Traditional Sausages

Eleni Kasapidou [1,*], Vasileios Papadopoulos [1] and Paraskevi Mitlianga [2]

[1] Department of Agriculture, School of Agricultural Sciences, University of Western Macedonia, 53100 Florina, Greece; vpapadopoulos@uowm.gr

[2] Department of Chemical Engineering, School of Engineering, University of Western Macedonia, 50100 Kozani, Greece; pmitliagka@uowm.gr

* Correspondence: ekasapidou@uowm.gr; Tel.: +30-69-7734-8326

Featured Application: A potential application is the prediction of the chemical composition of sausages with near infrared reflectance (NIR) spectroscopy devices in small scale production conditions.

Abstract: In the present study, the potential of application of near infrared reflectance (NIR) spectroscopy for the estimation of the chemical composition of traditional (village style) sausages was examined. The chemical composition (moisture, ash, protein and, fat content) was determined by standard reference methods. For the development of the calibration model, 39 samples of traditional fresh sausages were used, while for external validation, 10 samples of sausages were used. The correlation coefficients of calibration (RMSEC) and standard errors (SEC) were 0.92 and 1.58 (moisture), 0.77 and 0.18 (ash), 0.87 and 0.89 (protein) and 0.93 and 1.73 (fat). The cross-validation correlation coefficients (RMSECV) and standard errors (SECV) were 0.86 and 2.13 (moisture), 0.56 and 0.26 (ash), 0.78 and 1.17 (protein), and 0.88 and 2.17 (fat). The results of the calibration model showed that NIR spectroscopy can be applied to estimate with very good precision the fat content of traditional village-style sausages, whereas moisture and protein content can be estimated with good accuracy. The external validation confirmed the ability of NIR spectroscopy to predict the chemical composition of sausages.

Keywords: traditional sausages; chemical composition; near infrared reflectance (NIR) spectroscopy; calibration; validation

Citation: Kasapidou, E.; Papadopoulos, V.; Mitlianga, P. Feasibility of Application of Near Infrared Reflectance (NIR) Spectroscopy for the Prediction of the Chemical Composition of Traditional Sausages. *Appl. Sci.* **2021**, *11*, 11282. https://doi.org/10.3390/app112311282

Academic Editor: Agata Górska

Received: 22 October 2021
Accepted: 26 November 2021
Published: 29 November 2021

Publisher's Note: MDPI stays neutral with regard to jurisdictional claims in published maps and institutional affiliations.

1. Introduction

Traditional (village style) Greek sausages are the most commonly consumed meat products in Greece. In the past, sausages were prepared at home shortly before Christmas. Nowadays, sausages are produced throughout the year in the adjacent premises of butcher shops and sold, non-package, in bulk. Sausage ingredients such as seasonings vary between different regions influenced by culture, history, climate, and agriculture, but the production technology is the same. In general, pork meat and fat are coarsely chopped and thoroughly mixed with salt, pepper, and other seasonings, and the sausage batter is stuffed in natural casings. The sausages are partially dried and kept chilled until consumption [1].

According to the Greek Code of Foodstuffs, Beverages, and Objects of Common Use [2], as amended in 2014, traditional sausages are characterized as raw sausages, prepared only from lean meat and fat, salt and seasonings, that should be consumed only after heat processing. The sausage batter is stuffed into edible cases, and the sausages are air dried, and sometimes they are subjected to smoking. The addition of nitrites or nitrates and poultry meat is not allowed. Muscle protein content should not be lower than 12%

and fat content should not be higher than 35%. The chemical composition of traditional sausages has been studied in the past [3–5] and the results showed significant variability even between different production batches from the same butcher shop [6]. However, Regulation 1169/2011 of the European Union [7], on the provision of food information to the consumers, does not apply to traditional sausages because they are produced in small quantities. Consequently, consumers do not have the option to choose products with improved chemical composition and nutritional value during purchase. In addition to that, the chemical composition of meat products also affects products' eating quality [8].

Near infrared reflectance (NIR) spectroscopy is a valuable, cost-effective, and reliable tool for the estimation of various quality parameters of meat and meat products such as pH(u), color, eating quality characteristics, and chemical composition [9]. With regard to chemical composition, NIR spectroscopy has been used successfully in the industry to determine the content of the principal components (moisture, fat, and protein) in meat products [10]. However, in the vast majority of the studies, NIR spectroscopy has been used for the estimation of the chemical composition of intact or minced muscle, and in only a few studies has NIR spectroscopy has been applied for the estimation of the chemical composition of sausages that were produced in the laboratory following standard procedures commonly applied in the meat industry [11,12]. Furthermore, to the best of our knowledge, there are no studies on the applicability of NIR spectroscopy for the prediction of the composition of retail meat products produced in small quantities, such as traditional sausages that are prepared in the premises of butcher shops. Finally, the development of portable handheld NIR spectroscopy devices that will allow the in situ accurate estimations of meat and meat products' chemical composition is considered a research priority [10,13], and it is attracting attention for application in the production of traditional products [14].

The purpose of this preliminary study was to develop a calibration model to evaluate the potential of NIR spectroscopy to predict the chemical composition of commercial samples of traditional sausages produced in small quantities in butcher shops.

2. Materials and Methods

2.1. Sampling

2.1.1. Samples for the Development of the Calibration Model

Sausage samples (39) used for the development of the calibration models were occasionally analyzed in the laboratory in the period 2016–2018. In detail, 24 samples came from the same six different butcher shops as part of another trial that monitored the variability in basic composition per production batch [6]. The remaining samples (n = 15) came from other butcher shops also located in the Region of Western Macedonia, Greece. In any case, all samples used in the study came from different production batches, independently if they were purchased from the same shop. All butcher shops produced the sausages on their premises, in accordance with the Greek legislation [15]. Samples were purchased one to two days after their production, according to information provided by the seller. The drying period following sausage production was chosen to be similar in an attempt to standardize the moisture content of the samples and take into consideration that weight loss during drying is approximately 30% [1]. The samples used in the present study were collected from the same region in order to avoid matrix differences since ingredients such as spices, condiments, and seasonings vary between different regions.

2.1.2. Samples for Validation of the Calibration Model

The calibration model was validated with ten samples of traditional sausages that were purchased from different butcher shops also located in the Region of Western Macedonia in June 2018, after one or two days from their production. Samples from these shops were not used in the development of the calibration model, to achieve an independent set of data for the validation [16]. The ratio between calibration samples and validation samples was approximately 4:1. The same analogy between calibration samples and validation samples was reported by Ortiz-Somovilla et al. [11] for pork sausages whereas in the study of Ritthiruangdej et al. [17] the ratio between calibration samples and validation samples was 3:1. Additionally, according to Anderson [18] at least ten samples are required for external validation of a calibration model for meat and meat products.

2.2. Preparation of Samples for Proximate Analysis

Analyses were performed in samples of 1 kg each, consisting of 2–4 sausages, to ensure product homogeneity because the sausage batter often is mixed manually. Sausages were minced in a domestic type food chopper, taking care to produce a uniform finely ground sample to avoid physical variation between the samples reducing calibration error sources [19] and improve NIR spectroscopy predictability [20]. The homogenized samples were stored in plastic containers with tight-fitting lids to protect them from moisture interchange. Care was taken to minimize the air gap between the sample and the lid of the container to avoid sample deterioration during storage at 0 °C [21]. The homogenized sausage samples were thoroughly mixed by hand, with a spatula, before the determination of the chemical composition either by the standard reference methods or by NIR spectroscopy.

2.3. Proximate Analysis (Reference Methods)

The chemical composition of the sausages was determined according to the recommended standard methods described in AOAC [22]. Moisture content was determined by the 950.46 method after drying of the homogenized sample in a convection chamber (ED-115, Binder GmbH, Tuttlingen, Germany) at 102 °C until a constant weight was obtained. Dishes were placed in a desiccator and weighed when they obtained room temperature. Moisture was reported as weight loss from pre- and post-drying weights. Ash content was determined using the 920.153 method. Samples were incinerated at 550 °C for 12 h in a muffle furnace (model LM 412.07, Linn High Therm GmbH, Eschenfelden, Germany) until light grey ash was obtained. Ashing dishes were placed in a desiccator and weighted after reaching room temperature. Ash content was calculated by weight loss before and after incineration. Protein content was determined according to the method 928.08 by using nitrogen digestion (Turbotherm type TT/12M) and distillation (Vapodest type 40) apparatuses (Gerhardt Apparate GmbH & Co. KG, Königswinter, Germany) and converted to crude protein by multiplying the nitrogen content by 6.25. Fat content was determined according to the 991.36 method by extraction with petroleum ether using a Soxtherm/Multistat type SE-416 macro automated system (Gerhardt Apparate GmbH & Co. KG, Königswinter, Germany). Fat content was calculated as the proportional difference between the sample's weight before and after solvent extraction. All analyses were conducted in duplicate, and they were completed within one week following sample collection. Summations of the constituents falling within the range of 97–103% of the analytical sample weight were generally considered as acceptable [23].

2.4. NIR Spectroscopy
2.4.1. Spectra Acquisition

The homogenized sausage samples were subjected to NIR spectroscopy using a Spectra Star 2400-D spectrophotometer (Unity Scientific, Milford, MA, USA). The samples were placed in a rotating circular sample cup, set in stepped mode at 24 scans, and scanned at 2 nm intervals over the 1200–2400 nm spectral range. The spectrum was automatically recorded as log (1/Transmittance). The samples were prepared in duplicate, and two spectra were collected per individual sample, i.e., a total of four spectra per sausage.

2.4.2. Spectra Analysis

The calibration models for each parameter (moisture, ash, protein, and fat contents) were developed using Ucal^TM software suite (version 3.0.0.23; Unity Scientific, Brookfield, CT, USA) which supports the Partial Least Squares PLS optimized chemometric models and by applying "leave-one-out" cross-validation. Before the development of the calibration models, the chemical composition data, obtained by the application of the reference methods, were checked to exclude any erroneous results. The calibration data set was screened for anomalous spectra, and the outliers were excluded from the calibration process. Standard calibration error (SEC), root mean square standard calibration error (RMSEC), standard cross-validation error (SECV) and root mean square cross-validation error (RMSECV) were calculated for each examined variable.

2.4.3. Regression Analysis

Linear regression analysis of reference values and predicted values for each analyzed component was performed using Microsoft Excel for Office 365 software (version 2107). Pearson correlation coefficients for the reference values and predicted values for each analyzed component were performed using SPSS software (version 26.0).

3. Results and Discussion

3.1. Development of the Calibration Model

Table 1 presents the chemical composition (mean, standard deviation, and range of values) of the traditional sausages, as determined by application of the reference methods and by application of NIR spectroscopy for the calibration set of samples. The basic composition was similar to the one reported in the market surveys of Papadima et al. [3], Ambrosiadis et al. [4] and Konstantinidis et al. [5].

Moisture content varied to almost 30% between samples even though sausages were purchased one to two days after their production in an attempt to standardize their moisture content. The differences are related to the fact that some producers use drying chambers whereas others prefer to air dry the sausages. Variations in ash content are related to differences in spices, condiments, and seasonings used by the producers in their attempt to create product differentiation. The average protein content was higher than 12% and in accordance with the requirements of the Greek Code of Foodstuffs, Beverages, and Objects of Common Use [2]. There was one sample that the protein content was lower than 12%. Finally, the average fat content did not exceed the highest permitted value of 35%, by the Greek Code of Foodstuffs, Beverages, and Objects of Common Use [2]. However, in 23% of the examined samples, this value was exceeded. The findings in relation to protein and fat content enhance the need for in situ monitoring of the chemical composition of the sausage batter to enable compliance with the legislation.

The contents of all components were very variable, but according to González-Mohino et al. [14], this is expectable due to the limited degree of mechanization and control of the final product, causing a higher degree of heterogeneity. This finding is corroborated by the findings of the study of Parasolglou et al. [6] that reported batch variability in the range of 2.40–18.05%, 7.55–17.66%, 2.34–11.11%, and 9.52–26.71% for moisture, ash, protein and fat contents respectively, in sausages produced by the same butcher shop.

Table 1. Chemical composition according to the reference measurements and NIR spectroscopy for the calibration set of samples (n = 39).

Parameter	Mean Values	Standard Deviation	Range
Reference Methods			
Moisture (%)	47.73	5.215	31.87–59.24
Ash (%)	2.49	0.369	1.85–3.28
Protein (%)	17.61	2.222	11.08–22.41
Fat (%)	31.41	6.888	20.72–53.76
NIR Spectroscopy			
Moisture (%)	47.21	4.865	34.65–59.94
Ash (%)	2.41	0.305	1.62–3.09
Protein (%)	17.82	2.412	10.65–21.88
Fat (%)	31.33	6.320	19.48–49.32

Typical spectra of traditional sausages with NIR spectroscopy, are presented in Figure 1. The ability of NIR spectroscopy to estimate the chemical composition of traditional sausages is presented in Table 2.

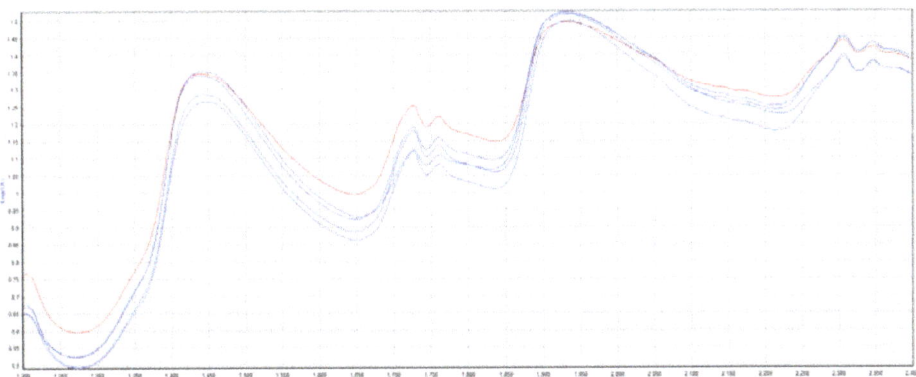

Figure 1. Near infrared spectra of traditional sausages.

Table 2. Calibration statistics for estimating the chemical composition of traditional sausages for the calibration samples.

Parameter	SEC	RMSEC	SECV	RMSECV
Moisture (%)	1.58	0.92	2.13	0.86
Ash (%)	0.18	0.77	0.26	0.56
Protein (%)	0.89	0.87	1.17	0.78
Fat (%)	1.73	0.93	2.17	0.88

SEC = standard calibration error; RMSEC = root mean square standard calibration error; SECV = standard cross-validation error; RMSECV = root mean square standard cross-validation error.

The calibration correlation coefficients (RMSEC) in all components were greater than 0.80, indicating that NIR spectroscopy can reliably estimate the chemical composition of meat and meat products [24]. The highest correlation coefficient for the calibration (RMSEC) was observed for the prediction of the fat content and the lowest to the protein content. Ortiz-Somovilla et al. [11] and Gaitán-Jurado et al. [12] observed comparable results for pork sausages. As reported in the review studies of Prieto et al. [10,25] and Prevolnik et al. [24], ash content is not frequently determined in trials examining the applicability of NIR spectroscopy for the prediction of the chemical composition of meat and meat products. The ash and protein cross-validation correlation coefficients (RMSECV) were less than 0.80, suggesting uncertainty in estimating the contents of the above components. Figure 2 presents the prediction plots for moisture, ash, protein, and fat contents for the calibration set of sausages. The majority of data points for all measured components were clustered along the regression lines.

The accuracy of the estimation is defined as good when the cross-validation correlation coefficient (RMSECV) lies in the range between 0.70–0.89, whereas coefficients higher than 0.90 are related to an excellent prediction of the chemical composition [26]. Values of both standard calibration errors (SEC) and standard cross-validation errors (SECV), for all examined variables, were within the range of those reported in the bibliography [10,24,25,27] on the use of NIR spectroscopy to assess the chemical composition of meat and meat products. The highest standard calibration (SEC) and cross-validation (SECV) errors were observed in the estimation of the moisture and fat contents, and the lowest values were observed in the estimation of the protein content. Reference data showed that the range was wider for moisture and fat (31.87–59.94 and 20.72–53.76 respectively) contents and narrower for protein (11.08–22.41) content, and this fact could account for differences in the accuracy of the estimation since variability can positively affect the NIR spectroscopy predictability [28]. The limited ability of NIR spectroscopy to estimate the protein content of traditional sausages was also confirmed in the studies by Prieto et al. [10,25] and Alomar et al. [29] and is due to the relatively small range of values for the protein content that hampers the detection of differences between samples. NIR spectroscopy is not usually applied for the estimation of the ash content in meat and meat products because inorganic compounds, in ionic forms, show no absorption in the near infrared spectrum and, therefore, such a calibration model would have minimal use [10,30]. Furthermore, Ritthiruangdej et al. [17] also reported poor accuracy in the prediction of ash content in Thai steamed pork sausages. The latter workers related the poor accuracy to very small range of ash content. Collell et al. [31] reported lower accuracy in the prediction of NaCl content, a major component of ash, in fermented pork sausages.

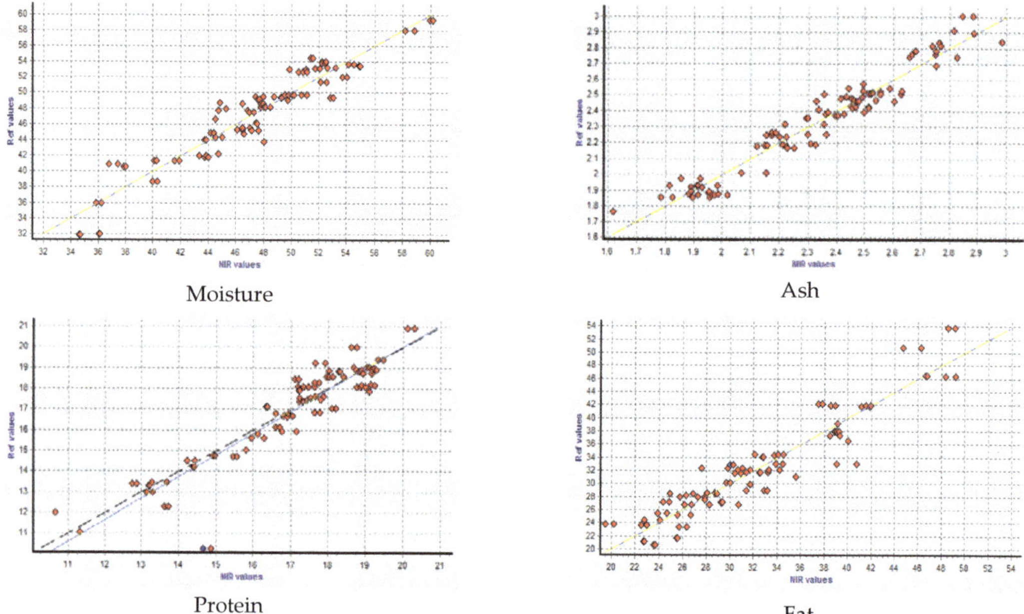

Figure 2. Plots for measured (Ref) versus estimated (NIR) values for moisture, ash, protein, and fat content of traditional sausages. Dashed lines represent the reference lines for the expected behavior. Solid lines are the actual regression lines.

Due to endogenous variances in the meat and fat used for the production of sausages, a considerable number of samples is required for the development of an adequate calibration model. These variations are related to the type of meat, which includes different animal species and anatomical parts of the carcass [10]. However, Prevolnik et al. [24] reported that between 30 and 150 samples are necessary to construct a calibration model with sufficient range for the individual components. The samples used for the development of the model must have significant variability in their chemical composition because high variability results in a robust calibration model [18]. This is feasible to a certain extent since retail samples such as the ones used in the present study should comply with the various regulations for food production and meet consumer criteria for their acceptability.

Finally, in foods produced from muscle tissue (meat and fish), the primary source of variability arises from their compact and heterogeneous nature, making it difficult to collect high-quality spectra. Scattering effects are primarily related to textural inhomogeneity due to variations in protein fiber arrangements and to the presence of intramuscular fat and connective tissue [30].

In this study, 39 samples were used for the development of the calibration model and many of these samples were produced in the same butcher shops. However, the great variability between production batches [6] allowed the use of these samples in the development of the calibration model. Additionally, the calibration model was initially constructed with 33 samples, as reported in the study of Tzemou et al. [32]. The additional 6 samples, used in the construction of this model were purchased from completely different butcher shops than the ones used for the first calibration model. It was observed that there were little differences in the calibration statistics between the two models indicating that the possibility to further improve the model was limited. This finding is attributed to the fact that the same level of variation was maintained. In any case, analysis of a greater number of samples results in better calibration statistics. However, the acceptable level of accuracy defines the number of analyzed samples since it is also associated with increased analytical costs.

3.2. Validation of the Calibration Model

Table 3 shows the chemical composition (mean, standard deviation, and range of values) of traditional sausages, according to the reference methods and NIR spectroscopy, for the validation set of samples. The composition was also very variable and similar to the composition of the samples used for the development of the calibration model. With regard to fat content, failure to comply with the legislation was also observed in 20% of the samples where the content was higher than 35%.

The highest deviation between the standard reference methods values and the NIR spectroscopy values was observed in the moisture content, and this can be attributed to changes occurring during the time elapsing between the two types of analyses. Whereas in the present study, the homogenized sausage samples' storage period did not exceed one week, at laboratory conditions, for routine analysis, changes in the moisture content during storage are dependent on the laboratory capacity to process the analyses and available equipment for sample storage such vacuum packaging. Thus, changes in the moisture content during analyses cannot be excluded.

Concerning nutrients such as protein and fat, whose content is declared in food labels, deviations in their contents fell within the allowances (tolerance) for variation between nutrient values declared on a label and values established by the monitoring authorities. The limits of permissible deviations are $\pm20\%$ when either the protein or the fat content ranges from 10–40 g/100 g of product [33]. In the present study, NIR spectroscopy values and reference methods values deviated approximately $\pm3\%$ for protein content and approximately -5% to 10% for fat content. It is also noted that NIR spectroscopy was successful in detecting the samples that failed to comply with the legislation with regard to the fat content. In other words, determination of the chemical composition of traditional sausages with NIR spectroscopy will not significantly differ from the chemical composition determined with standard reference methods, enabling the application of NIR spectroscopy for fast routine determination of the sausage composition.

Regression values for measured (Ref) and estimated (NIR) values for moisture, ash, protein, and fat content of traditional sausages used for the external validation are presented in Figure 3. The highest R^2 (0.9671) value was observed between the reference values and the NIR values for moisture content, whereas the lowest R^2 value was observed for ash content (0.3424). There was a good agreement between reference values and NIR values for fat content (0.8156), whereas an average agreement was found between reference values and NIR values for protein content (0.5708).

Table 4 shows the Pearson correlation coefficients between the measured and the estimated (NIR) values for moisture, ash, protein, and fat contents. Significant positive correlations ($p < 0.001$–0.01) were observed for all examined parameters. The highest coefficient is observed for moisture content, followed by the coefficients for fat and protein contents.

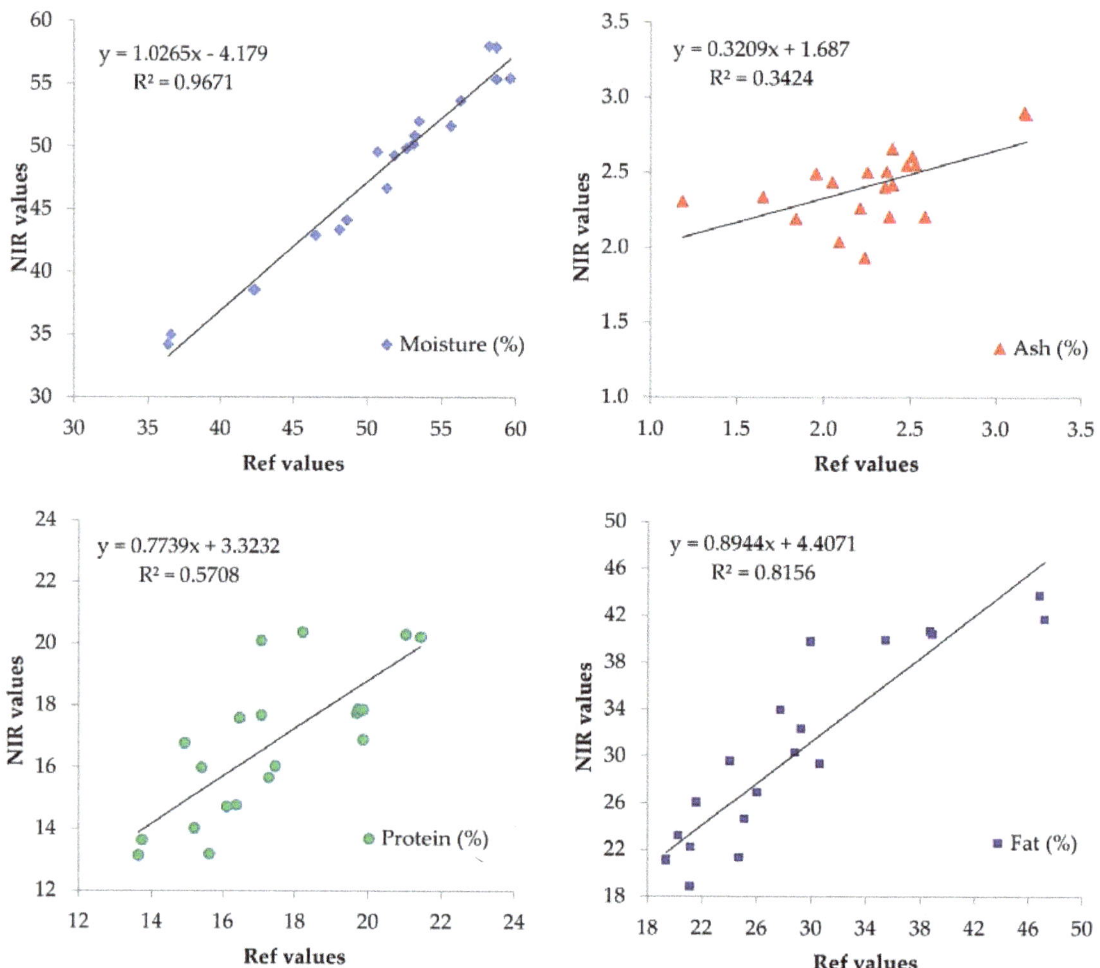

Figure 3. Regression values for measured (Ref) and estimated (NIR) values for moisture, ash, protein, and fat content of traditional sausages used for the external validation.

It is noted that this calibration model is valid when the samples' characteristics are the same as those used in the external validation. Therefore, if there are changes in the degree of drying or in the regulations for either fat or protein content, the application of the calibration model may lead to invalid results. However, it is also worth reporting that the main advantage for the application of NIR spectroscopy in the estimation of the composition of food products is the fact the database can be constantly expanded with the addition of the compositional data from new samples, leading to the development of a new calibration model that incorporates changes in the product composition.

Table 3. Chemical composition according to reference measurements and NIR spectroscopy for external validation samples and deviations between NIR spectroscopy values and reference methods values (n = 10).

Parameter	Mean Values	Standard Deviation	Range
		Reference Methods	
Moisture (%)	50.73	6.934	36.46–59.62
Ash (%)	2.29	0.452	1.19–3.18
Protein (%)	17.32	2.321	13.64–21.45
Fat (%)	28.83	8.513	19.38–47.18
		NIR Spectroscopy	
Moisture (%)	47.90	7.238	34.21–58.06
Ash (%)	2.42	0.248	1.94–2.91
Protein (%)	16.73	2.377	13.14–20.37
Fat (%)	30.20	8.431	17.84–43.72
	Deviations between NIR spectroscopy values and reference methods values		
Moisture (%)	−2.83	1.326	−4.73−−0.15
Ash (%)	0.13	0.367	−0.38–1.12
Protein (%)	−0.59	1.643	−2.99–3.04
Fat (%)	1.36	3.730	−5.48–9.87

Table 4. Pearson correlation coefficients for measured (Ref) and estimated (NIR) values for moisture, ash, protein, and fat content of traditional sausages were used for the external validation.

Parameter	Correlation Coefficient	Significance
Moisture (%)	0.983	***
Ash (%)	0.585	**
Protein (%)	0.756	***
Fat (%)	0.903	***

** = $p < 0.01$; *** = $p < 0.001$.

4. Conclusions

The present study results showed that NIR spectroscopy can be applied as a screening technique when a high but not an absolute level of accuracy is required. NIR spectroscopy can estimate moisture and fat contents of traditional sausages with high accuracy, whereas prediction of the protein content can also be achieved with reasonable accuracy. However, the development of a model for the estimation of the chemical composition of sausages, representing variations in sausage ingredients and/or production conditions, requires analysis of a greater number of samples. Future research should focus on developing calibration models for the next generation of low-cost, simple-operation portable NIR spectroscopy instruments that would help butcher shop owners produce healthier sausages that are in compliance with the specifications of the Greek Food Legislation.

Author Contributions: Conceptualization, E.K. and P.M.; methodology, P.M.; investigation E.K. and V.P.; formal analysis, E.K.; resources, P.M.; data curation, E.K. and V.P.; writing—original draft preparation, E.K.; writing—review and editing, E.K. and P.M. All authors have read and agreed to the published version of the manuscript.

Funding: This research received no external funding.

Institutional Review Board Statement: Not applicable.

Informed Consent Statement: Not applicable.

Data Availability Statement: Data is contained within the article.

Acknowledgments: The authors would like to thank the company Unity, Ltd., Cholargos, Athens, Greece, for the development of the calibration model.

Conflicts of Interest: The authors declare no conflict of interest.

References

1. Papagianni, M.; Ambrosiadis, I.; Filiousis, G. Mould growth on traditional Greek sausages and penicillin production by Penicillium isolates. *Meat Sci.* **2007**, *76*, 653–657. [CrossRef] [PubMed]
2. Greek state. Code of Foodstuffs, Beverages and Objects of Common Use—Part A: Foodstuffs and beverages (Classification, specifications and definitions of meat products and meat preparations). *Gov. Gaz* **2014**, 91. (In Greek)
3. Papadima, S.N.; Arvanitoyannis, I.; Bloukas, J.G.; Fournitzis, G.C. Chemometric model for describing Greek traditional sausages. *Meat Sci.* **1999**, *51*, 271–277. [CrossRef]
4. Ambrosiadis, J.; Soultos, N.; Abrahim, A.; Bloukas, J.G. Physicochemical, microbiological and sensory attributes for the characterization of Greek traditional sausages. *Meat Sci.* **2004**, *66*, 279–287. [CrossRef]
5. Konstandinidis, K.; Kasapidou, E.; Govaris, A.; Mitlianga, P. Evaluation of the chemical composition and the nutritional quality of traditional (village) style Greek Sausages. In Proceedings of the 59th International Congress of Meat Science and Technology, Izmir, Turkey, 18–23 August 2013; p. 129.
6. Parasoglou, V.-K.; Tzemou, M.; Mitlianga, P.; Kasapidou, E. Variation in the chemical composition of traditional sausages per production batch. In Proceedings of the 4th Panhellenic Conference of Production Animals and Food Hygiene, Volos, Greece, 12–14 May 2017.
7. Regulation (EU) No 1169/2011 of the European Parliament and of the Council of 25 October 2011 on the Provision of Food Information to Consumers, Amending Regulations (EC) No 1924/2006 and (EC) No 1925/2006 of the European Parliament and of the Council, and Repealing Commission Directive 87/250/EEC, Council Directive 90/496/EEC, Commission Directive 1999/10/EC, Directive 2000/13/EC of the European Parliament and of the Council, Commission Directives 2002/67/EC and 2008/5/EC and Commission Regulation (EC) No 608/2004. Available online: https://eur-lex.europa.eu/eli/reg/2011/1169/2018-01-01 (accessed on 25 January 2021).
8. Sheard, P.R.; Hope, E.; Hughes, S.I.; Baker, A.; Nute, G.R. Eating quality of UK-style sausages varying in price, meat content, fat level and salt content. *Meat Sci.* **2010**, *85*, 40–46. [CrossRef]
9. Dixit, Y.; Casado-Gavalda, M.P.; Cama-Moncunill, R.; Cama-Moncunill, X.; Markiewicz-Keszycka, M.; Cullen, P.J.; Sullivan, C. Developments and challenges in online NIR spectroscopy for meat processing. *Compr. Rev. Food Sci. Food Saf.* **2017**, *16*, 1172–1187. [CrossRef]
10. Prieto, N.; Roehe, R.; Lavín, P.; Batten, G.; Andrés, S. Application of near infrared reflectance spectroscopy to predict meat and meat products quality: A review. *Meat Sci.* **2009**, *83*, 175–186. [CrossRef] [PubMed]
11. Ortiz-Somovilla, V.; España-España, F.; Gaitán-Jurado, A.J.; Pérez-Aparicio, J.; de Pedro-Sanz, E.J. Proximate analysis of homogenized and minced mass of pork sausages by NIRS. *Food Chem.* **2007**, *101*, 1031–1040. [CrossRef]
12. Gaitán-Jurado, A.J.; Ortiz-Somovilla, V.; España-España, F.; Pérez-Aparicio, J.; De Pedro-Sanz, E.J. Quantitative analysis of pork dry-cured sausages to quality control by NIR spectroscopy. *Meat Sci.* **2008**, *78*, 391–399. [CrossRef]
13. Zamora-Rojas, E.; Pérez-Marín, D.; de Pedro-Sanz, E.; Guerrero-Ginel, J.E.; Garrido-Varo, A. Handheld NIRS analysis for routine meat quality control: Database transfer from at-line instruments. *Chemom. Intell. Lab. Syst.* **2012**, *114*, 30–35. [CrossRef]
14. González-Mohino, A.; Pérez-Palacios, T.; Antequera, T.; Ruiz-Carrascal, J.; Olegario, L.S.; Grassi, S. Monitoring the Processing of Dry Fermented Sausages with a Portable NIRS Device. *Foods* **2020**, *9*, 1294. [CrossRef] [PubMed]
15. Greek state. Regulation of operation subjects for the production of meat products in the premises of butcher shops. *Gov. Gaz* **2014**. (In Greek)
16. Harrington, P.D.B. Multiple versus single set validation of multivariate models to avoid mistakes. *Crit. Rev. Anal. Chem.* **2018**, *48*, 33–46. [CrossRef] [PubMed]
17. Ritthiruangdej, P.; Ritthiron, R.; Shinzawa, H.; Ozaki, Y. Non-destructive and rapid analysis of chemical compositions in thai steamed pork sausages by near-infrared spectroscopy. *Food Chem.* **2011**, *129*, 684–692. [CrossRef] [PubMed]
18. Anderson, S. Determination of fat, moisture, and protein in meat and meat products by using the FOSS FoodScan near-infrared spectrophotometer with FOSS artificial neural network calibration model and associated database: Collaborative study. *J. AOAC Int.* **2007**, *90*, 1073–1083. [CrossRef]
19. Workman, J.J. NIR spectroscopy calibration basics. In *Handbook of Near-Infrared Analysis*, 3rd ed.; Burns, D.A., Ciurczak, E.W., Eds.; CRC Press: Boca Raton, FL, USA, 2008; pp. 123–150.
20. Cozzolino, D.; Murray, I. Effect of sample presentation and animal muscle species on the analysis of meat by near infrared reflectance spectroscopy. *J. Near Infrared Spectrosc.* **2002**, *10*, 37–44. [CrossRef]
21. Williams, P. Sampling, sample preparation, and sample selection. In *Handbook of Near-Infrared Analysis*, 3rd ed.; Burns, D.A., Ciurczak, E.W., Eds.; CRC Press: Boca Raton, FL, USA, 2008; pp. 267–295.
22. Association of Analytical Communities International. *Official Methods of Analysis of AOAC International*, 17th ed.; AOAC International: Gaithersburg, MD, USA, 2003.
23. Greenfield, H.; Southgate, D.A.T. *Food Composition Data—Production, Management and Use*, 2nd ed.; Food and Agriculture Organization of the United Nations: Rome, Italy, 2003.

24. Prevolnik, M.; Candek-Potokar, M.; Skorjanc, D. Ability of NIR spectroscopy to predict meat chemical composition and quality: A review. *Czech J. Anim. Sci.* **2004**, *49*, 500–510. [CrossRef]

25. Prieto, N.; Pawluczyk, O.; Dugan, M.E.R.; Aalhus, J.L. A review of the principles and applications of near-infrared spectroscopy to characterize meat, fat, and meat products. *Appl. Spectrosc.* **2017**, *71*, 1403–1426. [CrossRef]

26. Shenk, J.S.; Westerhaus, M.O. Calibration the ISI Way. In *Near Infrared Spectroscopy: The Future Waves*; Davies, A.M.C., Williams, P.C., Eds.; NIR Publications: Chichester, UK, 1996; pp. 198–202.

27. Porep, J.U.; Kammerer, D.R.; Carle, R. On-line application of near infrared (NIR) spectroscopy in food production. *Trends Food Sci. Technol.* **2015**, *46*, 211–230. [CrossRef]

28. Su, H.; Sha, K.; Zhang, L.; Zhang, Q.; Xu, Y.; Zhang, R.; Li, H.; Sun, B. Development of near infrared reflectance spectroscopy to predict chemical composition with a wide range of variability in beef. *Meat Sci.* **2014**, *98*, 110–114. [CrossRef]

29. Alomar, D.; Gallo, C.; Castaneda, M.; Fuchslocher, R. Chemical and discriminant analysis of bovine meat by near infrared reflectance spectroscopy (NIRS). *Meat Sci.* **2003**, *63*, 441–450. [CrossRef]

30. Weeranantanaphan, J.; Downey, G.; Allen, P.; Sun, D.W. A review of near infrared spectroscopy in muscle food analysis: 2005–2010. *J. Near Infrared Spectrosc.* **2011**, *19*, 61–104. [CrossRef]

31. Collell, C.; Gou, P.; Picouet, P.; Arnau, J.; Comaposada, J. Feasibility of near-infrared spectroscopy to predict aw and moisture and NaCl contents of fermented pork sausages. *Meat Sci.* **2010**, *85*, 325–330. [CrossRef] [PubMed]

32. Tzemou, M.; Parasoglou, V.K.; Ioannidou, T.; Papadopoulos, V.; Mitlianga, P.; Kasapidou, E. Prediction of chemical composition of Greek traditional sausages by near-infrared reflectance spectroscopy. In Proceedings of the 3rd IMEKO Foods Conference: Metrology Promoting Harmonization and Standardization in Food and Nutrition, Thessaloniki, Greece, 1–4 October 2017; P27. pp. 236–239.

33. European Commission. Guidance Document for Competent Authorities for the Control of Compliance with EU Legislation on: Regulation (EU) No 1169/2011 of the European Parliament and of the Council of 25 October 2011 on the Provision of Food Information to Consumers, Amending Regulations (EC) No 1924/2006 and (EC) No 1925/2006 of the European Parliament and of the Council, and Repealing Commission Directive 87/250/EEC, Council Directive 90/496/EEC, Commission Directive 1999/10/EC, Directive 2000/13/EC of the European Parliament and of the Council, Commission Directives 2002/67/EC and 2008/5/EC and Commission Regulation (EC) No 608/2004 and Council Directive 90/496/EEC of 24 September 1990 on Nutrition Labelling of Foodstuffs and Directive 2002/46/EC of the European Parliament and of the Council of 10 June 2002 on the Approximation of the Laws of the Member States Relating to Food Supplements with Regard to the Setting of Tolerances for Nutrient Values Declared on a Label. 2012. Available online: https://ec.europa.eu/food/sites/food/files/safety/docs/labelling_nutrition-vitamins_minerals-guidance_tolerances_1212_en.pdf (accessed on 10 February 2021).

Article

Lipid Fraction Properties of Homemade Raw Cat Foods and Selected Commercial Cat Foods

Agnieszka Górska, Diana Mańko-Jurkowska *, Joanna Bryś and Agata Górska

Department of Chemistry, Institute of Food Sciences, Warsaw University of Life Sciences, 02-787 Warsaw, Poland; agnieszka_gorska@sggw.edu.pl (A.G.); joanna_brys@sggw.edu.pl (J.B.); agata_gorska@sggw.edu.pl (A.G.)
* Correspondence: diana_manko_jurkowska@sggw.edu.pl; Tel.: +48-22-593-76-29

Abstract: The purpose of the present study was to characterize lipid fraction extracted from five self-prepared and seven commercial cat foods using gas chromatography (GC) and pressurized differential scanning calorimetry (PDSC) techniques. Self-prepared food recipes were composed using *BARFny kalkulator*, software dedicated for balancing cat diets, and prepared on the basis of fresh raw meat and offal. Extracted fat fractions were compared qualitatively and quantitatively with literature data for the fat of whole prey items to check the main assumptions of the software used. The fatty acid (FA) composition and distribution were determined using GC. The PDSC method was used for the determination of the oxidative stability of extracted fats. The obtained results indicate that self-prepared cat foods contained a high level of essential fatty acids (EFA) but low oxidative stability, especially for those with significant amounts of polyunsaturated FA. The FA profile and oxidative stability were examined for four dry and three wet commercial cat foods. It was found that their omega-6 to omega-3 ratio was beneficial reaching 5.3:1 to 10.1:1, despite the low amount of EFA. The longer induction time was determined for fats extracted from commercial cat foods than for self-prepared ones, which indicate their higher oxidative stability.

Keywords: raw meat cat diet; essential fatty acids; fatty acids profile; fatty acids distribution; oxidative stability

Citation: Górska, A.; Mańko-Jurkowska, D.; Bryś, J.; Górska, A. Lipid Fraction Properties of Homemade Raw Cat Foods and Selected Commercial Cat Foods. *Appl. Sci.* **2021**, *11*, 10905. https://doi.org/10.3390/app112210905

Academic Editor: Anabela Raymundo

Received: 9 June 2021
Accepted: 16 November 2021
Published: 18 November 2021

Publisher's Note: MDPI stays neutral with regard to jurisdictional claims in published maps and institutional affiliations.

1. Introduction

Most cat owners feed their pets with commercial cat food because it is convenient and economical; however, over the last decade, the use of alternative diets, such as BARF (bone and raw food/biologically appropriate raw food), other raw meat-based diets, home-cooked diets and whole prey model diets have become more popular [1–3]. These raw food diets are often used by owners who have concerns about the wholesomeness and nutritional value of the ingredients used in commercial pet food. However, formulating a complete and balanced pet food needs knowledge about the nutritional requirements of cats.

In the literature, it is hard to find studies in which the nutrient composition of raw meat-based foods was examined by analytical techniques. It was only found that Kerr et al. [4,5] and Hamper et al. [6] examined the chemical composition of several raw meat-based diets for cats as well as the digestibility of these foods in cats and kittens. Most publications related to raw meat-based food for cats usually only present theoretical considerations on the impact of this type of food on the health of cats, along with its advantages and disadvantages [2,7–10]. Wilson et al. [10] performed computerized assessments of several types of homemade cat recipes, including some raw recipes, using the balancing software (*Balance IT*) available online. Laboratory analysis could confirm the correctness of the computerized assessment, but unfortunately, they were not performed. To the best of our knowledge, this is the first study dealing with the experimental research of the self-prepared cat food lipid fraction. The software used to compose the diets was *BARFny kalkulator*—a nutrition-balancing software, similar to the *Balance IT* used by Wilson et al. [10].

Fat typically provides most of the energy, increases the palatability of food [11,12], is a carrier of fat-soluble vitamins and is a source of essential fatty acids (EFA) for cats, which are recognized as obligate carnivores. The EFA for cats are linoleic (LA), alpha-linolenic (ALA) and arachidonic (AA) acids. The conditionally EFA are eicosapentaenoic (EPA) and docosahexaenoic (DHA) acids. They are responsible, among others, for skin and coat condition, kidney function and reproduction [13–15].

Triacylglycerols (TAG) are the predominant component of most dietary fats and oils. Identification and quantitative analysis of the composition of TAG is the first step in evaluating the quality of fats when considering their nutritional effects [16]. The distribution of FA in TAG molecules is particularly important in the process of their digestion and absorption in the gastrointestinal tract. The rate of hydrolysis at the *sn*-2 position of TAG is very slow and the FA in the *sn*-2 position remain intact as 2-monoacylglycerols during digestion and absorption. FA located in the *sn*-1 and *sn*-3 positions are preferentially hydrolyzed, leaving a 2-monoacylglycerol [17]. Free FA are long chain saturated if released; they form acid–calcium soaps that are insoluble in aqueous media at the pH of the intestine, which causes impaired absorption of both SFA and calcium [17,18]. Calcium loss can affect Ca:P ratio in the diet, which can be associated with kidney damage in cats [19,20].

One of the most important parameters in the characterization of fats and oils is oxidative stability. Lipid oxidation is a major challenge to pet food preservation [21]. It determines the final quality and nutritional properties of the food product because it is the main reaction responsible for their degradation [22]. In food chemistry, pressurized differential scanning calorimetry (PDSC) is a convenient, reproducible and fast method used to determine the oxidative stability of fats and oils [23,24]. Moreover, it is a nonchemical method without the time-consuming processing of test samples and it requires a small amount of sample (a few milligrams) for analysis [25,26]. This technique has already been used in the studies of fat extracted from dry pet foods [27,28].

The aim of our study is to characterize the lipid fraction properties of homemade cat foods and present it in light of FA composition and oxidative stability data obtained for selected dry and wet commercial cat foods.

2. Material and Methods

2.1. Chemicals

All the solvents and reagents were of chromatographic or analytical grade and purchased from Avantor Performance Materials Poland S.A. (Gliwice, Poland), except for the standard compounds, bile salts and lipase from porcine pancreas (type II), which were supplied by Sigma–Aldrich (Saint Louis, MO, USA).

2.2. Self-Prepared Cat Foods

Five self-prepared food recipes (Supplemental Materials) were composed using a software for balancing cat diets called the *BARFny kalkulator* (https://www.barfnyswiat.org/viewtopic.php?t=896—accessed on 28 February 2018). The tool was designed according to the authors' knowledge in this term and it is based on available data of the composition of raw skeletal muscles, internal organs, poultry bones, fish, as well as supplements. It also shows the cats' nutrient requirements. According to the author statement, *BARFny kalkulator* is mainly based on requirements from four sources [29–32].

The prepared diets differed primarily in the origin of the meat. They were based on poultry and/or beef (a source of easily digestible and nutritious protein, rich in taurine, iron, zinc and B vitamins, which contains tendons, veins and tallow). Two diets contained one meat ingredient: SP-2—turkey and SP-5—beef, and three of them contained two meat ingredients: SP-1—chicken and beef, SP-3—chicken and turkey and SP-4—chicken and duck (the complete composition of the diets can be found in the Supplementary File). The meat was purchased in a local grocery. The type of meat in the diet was selected taking into account its nutritional value, palatability for cats, popularity of use by cat owners and their easy access. Supplements (source of vitamins and minerals) were purchased online (https://www.lunderland.org/; http://pokusa.org/ accessed on 28 February 2018) and Tokovit E in a local pharmacy.

The fat content in the diet-balancing software used in these studies was set at 20–38% of dry matter, and the dry matter content was set as 25% for all samples. Authors suggest to maintain protein:fat ratio at about 2:1. Following these tips, the fat content in self-prepared foods (SP-1–SP-5) calculated by this software was around 30% (Table 1).

Table 1. Main animal ingredients: dry matter (DM), fat content and fatty acid composition (%) in fats extracted from tested self-prepared (SP) cat foods.

Product Designation	SP-1	SP-2	SP-3	SP-4	SP-5
Main Animal Ingredients	Chicken Meat, Beef Meat	Turkey Meat	Chicken Meat, Turkey Meat	Chicken Meat, Duck Meat	Beef Meat
DM, % †	25.13	25.14	25.12	25.10	25.12
DM, %	24.02 ± 0.94 b	22.80 ± 1.58 ab	23.53 ± 2.08 b	20.47 ± 1.81 a	27.73 ± 1.81 c
Fat in DM, % †	32.11	33.48	33.59	27.87	29.45
Fat in DM, %	35.48 ± 2.63 d	25.14 ± 0.97 a	28.58 ± 0.44 b	32.00 ± 1.11 c	36.30 ± 1.27 d
C10:0	0.04 ± 0.01 ab	0.13 ± 0.02 d	0.08 ± 0.02 c	0.03 ± 0.01 a	0.05 ± 0.01 bc
C12:0	0.33 ± 0.01 b	1.95 ± 0.17 d	1.34 ± 0.08 c	0.37 ± 0.03 b	0.08 ± 0.01 a
C14:0	2.13 ± 0.04 b	2.44 ± 0.16 c	2.10 ± 0.04 b	1.28 ± 0.04 a	3.00 ± 0.13 d
C14:1	0.59 ± 0.06 b	0.20 ± 0.02 a	0.17 ± 0.03 a	0.13 ± 0.02 a	0.79 ± 0.15 c
C15:0	0.25 ± 0.01 c	0.18 ± 0.01 b	0.17 ± 0.01 b	0.13 ± 0.01 a	0.60 ± 0.01 d
C16:0	20.39 ± 0.15 b	17.36 ± 0.52 a	17.43 ± 0.23 a	19.87 ± 0.03 b	24.24 ± 0.94 c
C16:1	4.48 ± 0.01 d	3.14 ± 0.11 a	3.28 ± 0.02 b	4.00 ± 0.03 c	3.24 ± 0.05 b
C17:0	0.69 ± 0.01 c	0.40 ± 0.01 b	0.39 ± 0.02 b	0.36 ± 0.01 a	1.29 ± 0.02 d
C17:1	0.56 ± 0.01 b	0.12 ± 0.02 a	0.12 ± 0.01 a	0.14 ± 0.01 a	0.77 ± 0.01 c
C18:0	10.81 ± 0.01 c	6.86 ± 0.15 ab	7.18 ± 0.09 b	6.68 ± 0.07 a	18.71 ± 0.48 d
C18:1 n-9	39.72 ± 0.01 e	27.86 ± 0.24 a	30.53 ± 0.39 b	36.41 ± 0.10 c	38.60 ± 1.04 d
C18:2 n-6	12.88 ± 0.02 b	29.14 ± 0.22 e	27.15 ± 0.15 d	22.65 ± 0.12 c	3.12 ± 0.07 a
C18:3 n-3	0.87 ± 0.02 b	3.40 ± 0.05 e	2.66 ± 0.07 d	1.30 ± 0.03 c	0.69 ± 0.02 a
C20:1 n-9	0.79 ± 0.05 bc	0.71 ± 0.11 b	0.85 ± 0.05 c	0.81 ± 0.01 bc	0.55 ± 0.08 a
C20:2 n-6	0.23 ± 0.01 b	0.33 ± 0.03 c	0.38 ± 0.02 d	0.31 ± 0.01 c	0.08 ± 0.05 a
C20:3 n-6	1.11 ± 0.01 b	1.66 ± 0.11 d	1.47 ± 0.09 c	1.51 ± 0.01 c	0.52 ± 0.03 a
C20:4 n-6	1.48 ± 0.02 b	1.41 ± 0.07 ab	1.66 ± 0.10 c	1.32 ± 0.02 a	1.31 ± 0.53 a
C20:5 n-3	0.12 ± 0.0 1b	0.14 ± 0.01 c	0.14 ± 0.02 c	0.16 ± 0.01 c	0.05 ± 0.01 a
C22:6 n-3	1.22 ± 0.05 bc	1.19 ± 0.12 abc	1.34 ± 0.18 c	1.13 ± 0.01 ab	1.02 ± 0.07 a
Σ other	1.37 ± 0.02 ab	1.39 ± 0.10 ab	1.57 ± 0.07 c	1.47 ± 0.01 bc	1.27 ± 0.13 a
MUFA/PUFA	2.58	0.86	1.00	1.46	6.47
MUFA/SFA	1.33	1.09	1.22	1.44	0.92
PUFA/SFA	0.52	1.27	1.21	0.99	0.14
Omega-6/omega-3	7.12	6.88	7.41	9.97	2.86

† Based on content calculated by the software. Obtained data for DM, fat in DM and fatty acids are expressed as means ± S.D. The different lower-case letters in the same row indicate significantly different values ($p < 0.05$). MUFA—monounsaturated fatty acids, PUFA—polyunsaturated fatty acids, SFA—saturated fatty acids.

Each diet was prepared three times as follows:

- The fresh meat was ground twice;
- To balance the food, additives such as hemoglobin, salmon oil, egg shell meal, etc. were added with water to ground meat in the amount determined by the *BARFny kalkulator* for a given mass of this meat (the addition of egg yolks, salmon oil and vitamin E was the same for all diets);
- All ingredients are thoroughly mixed with each other. Each food sample was made in greater scale, portioned and frozen.

Self-prepared cat foods were designed as SP (1–5). All tested foods are listed in Table 1 with an indication of their main animal ingredients and fat content in dry matter.

2.3. Commercial Cat Foods

Seven commercial dry and wet foods for cats were used in this study. Three packs of each food were purchased in different local specialized pet stores in order to avoid the same lots of food. Each food was a mid-priced product of a different popular brand, dedicated for adult cats and unexpired. Commercial foods were based on commonly used ingredients in the pet food industry, such as poultry and fish meal, dehydrated and hydrolyzed proteins, animal fat, meat and animal by-products and plants. The complete composition of each diet can be found in the Supplementary File.

Research was not sponsored by any of the commercial diet manufacturers. Commercial diets were identified by code letters to preserve the privacy of the manufacturers. Dry foods were designed as CD (1–4) and wet foods as CW (5–7). All tested foods are listed in Table 2.

Table 2. Main animal ingredients: dry matter (DM), fat content and fatty acid composition (%) in fats extracted from tested commercial dry (CD) and wet (CW) cat foods.

Product Designation	CD-1	CD-2	CD-3	CD-4	CW-5	CW-6	CW-7
DM, % †	90	90	92	90	20	18	19
DM, %	94.35 ± 0.37 d	94.10 ± 0.28 cd	94.28 ± 0.19 cd	93.46 ± 0.42 c	20.18 ± 0.63 b	18.93 ± 0.89 a	19.17 ± 0.13 a
Fat in DM, % †	22.22	20.00	17.39	17.77	12.50	25.00	14.21
Fat in DM, %	16.31 ± 0.57 cd	17.39 ± 2.15 d	13.16 ± 3.42 bc	14.78 ± 1.32 cd	10.11 ± 2.59 ab	10.38 ± 1.73 ab	8.89 ± 0.67 a
C10:0	0.08 ± 0.03 ab	0.02 ± 0.01 a	0.20 ± 0.01 d	0.16 ± 0.05 cd	0.12 ± 0.07 bc	0.06 ± 0.01 ab	0.08 ± 0.04 ab
C12:0	0.40 ± 0.10 ab	0.07 ± 0.01 a	1.85 ± 0.14 c	0.45 ± 0.05 ab	1.98 ± 1.03 c	1.07 ± 0.24 b	0.29 ± 0.17 a
C14:0	1.36 ± 0.21 b	0.73 ± 0.02 a	2.72 ± 0.09 d	2.19 ± 0.10 ad	2.09 ± 0.75 c	1.37 ± 0.24 b	1.22 ± 0.36 ab
C14:1	0.17 ± 0.02 a	0.19 ± 0.01 a	0.30 ± 0.02 b	0.23 ± 0.02 ab	0.25 ± 0.10 ab	0.20 ± 0.04 a	0.17 ± 0.59 a
C15:0	0.15 ± 0.02 ab	0.11 ± 0.01 a	0.21 ± 0.01 c	0.19 ± 0.01 bc	0.16 ± 0.05 bc	0.15 ± 0.02 ab	0.19 ± 0.04 bc
C16:0	20.08 ± 1.10 a	22.15 ± 0.23 ab	22.52 ± 0.03 ab	26.27 ± 0.67 c	22.41 ± 3.09 ab	23.03 ± 1.83 ab	23.91 ± 2.82 bc
C16:1	4.46 ± 0.35 ab	5.72 ± 0.10 d	5.15 ± 0.10 cd	4.89 ± 0.05 bc	3.95 ± 0.62 a	3.94 ± 0.35 a	3.95 ± 0.51 a
C17:0	0.37 ± 0.01 ab	0.34 ± 0.03 a	0.37 ± 0.01 ab	0.41 ± 0.01 bc	0.35 ± 0.04 ab	0.40 ± 0.04 abc	0.45 ± 0.07 c
C17:1	0.13 ± 0.01 bc	0.11 ± 0.01 ab	0.14 ± 0.01 c	0.18 ± 0.01 d	0.13 ± 0.02 abc	0.11 ± 0.01 a	0.15 ± 0.02 c
C18:0	6.52 ± 0.37 b	7.62 ± 0.07 c	5.42 ± 0.04 a	8.99 ± 0.04 d	6.58 ± 0.63 b	9.69 ± 0.18 e	8.89 ± 0.57 d
C18:1 n-9	38.05 ± 1.08 d	40.33 ± 0.09 d	29.08 ± 0.18 a	32.70 ± 0.76 bc	30.41 ± 2.95 ab	31.68 ± 1.76 ab	34.90 ± 2.34 c
C18:2 n-6	20.48 ± 0.23 c	18.38 ± 0.09 b	25.58 ± 0.24 e	17.12 ± 0.02 a	25.18 ± 1.06 e	21.83 ± 0.51 d	17.60 ± 0.46 ab
C18:3 n-3	2.56 ± 0.06 c	1.83 ± 0.04 a	2.31 ± 0.04 b	3.12 ± 0.09 d	2.51 ± 0.20 c	1.89 ± 0.06 a	2.59 ± 0.11 c
C20:1 n-9	1.19 ± 0.13 d	0.52 ± 0.03 ab	0.44 ± 0.02 ab	0.61 ± 0.07 bc	0.80 ± 0.22 c	0.34 ± 0.015 a	0.44 ± 0.11 ab
C20:2 n-6	0.39 ± 0.05 b	0.19 ± 0.01 a	0.20 ± 0.03 a	0.22 ± 0.06 a	0.26 ± 0.07 a	0.23 ± 0.03 a	0.43 ± 0.06 b
C20:3 n-6	0.34 ± 0.03 a	0.61 ± 0.04 bc	0.42 ± 0.01 ab	0.64 ± 0.01 bc	0.78 ± 0.17 c	2.30 ± 0.12 d	2.29 ± 0.34 d
C20:4 n-6	0.63 ± 0.05 d	0.11 ± 0.01 a	1.32 ± 0.06 e	0.23 ± 0.03 b	0.34 ± 0.06 c	0.13 ± 0.03 a	0.15 ± 0.04 ab
C20:5 n-3	0.08 ± 0.10 a	0.06 ± 0.04 a	0.08 ± 0.01 a	0.05 ± 0.02 a	0.08 ± 0.03 a	0.16 ± 0.02 b	0.07 ± 0.03 a
C22:6 n-3	1.03 ± 0.09 d	0.19 ± 0.01 a	0.77 ± 0.03 c	0.29 ± 0.03 a	0.53 ± 0.23 b	0.37 ± 0.07 ab	0.26 ± 0.07 a
Σ other	1.53 ± 0.08 c	0.71 ± 0.03 a	0.97 ± 0.03 b	1.12 ± 0.02 b	1.09 ± 0.20 b	1.11 ± 0.01 b	1.98 ± 0.21 d
MUFA/PUFA	1.73	2.19	1.14	1.78	1.20	1.35	1.69
MUFA/SFA	1.52	1.51	1.05	1.00	1.05	1.01	1.13
PUFA/SFA	0.88	0.69	0.92	0.56	0.88	0.75	0.67
Omega-6/omega-3	5.96	9.26	8.72	5.27	8.53	10.13	7.01

† Based on guaranteed analysis of producer. Obtained data for DM, fat in DM and fatty acids are expressed as means ± S.D. The different lower-case letters in the same row indicate significantly different values ($p < 0.05$).

2.4. Fat Extraction and Dry Matter Determination

Fat was extracted using Folch's method, according to the procedure described by Boselli et al. [33]. Extraction was done immediately after both methods by opening commercial foods and thawing self-prepared ones. Extraction was performed in triplicate for a given diet by conducting one for each representative sample.

The dry matter of self-prepared and commercial cat food samples was obtained by their oven-drying for 24 h at 105 °C in an electrically heated laboratory oven type SLN 115 STD (Pol-Eko-Aparatura) in triplicate.

2.5. GC Measurements

The determination of FA composition in the examined samples was carried out by a GC analysis of FA methyl esters. They were prepared through esterification with KOH/methanol (1M) according to ISO 5509:2001. An YL6100 GC chromatograph equipped with a flame ionization detector and BPX70 capillary column of 60 m (length) $\times$ 0.25 mm (internal diameter) $\times$ 0.25 μm (film thickness) was used. The oven temperature was programmed as follows: an initial temperature of 70 °C was maintained for 0.5 min, then it was increased by 15 °C min^{-1} to 160 °C, then from 160 °C to 200 °C it was increased by 1.1 °C min^{-1}, then kept at 200 °C for 6 min, then in the next step from 200 °C to 225 °C it was increased by 30 °C min^{-1} and kept at 225 °C another 1 min. The temperature of the injector was 225 °C (with a split ratio of 1:50) and the detector temperature was 250 °C. The carrier gas was nitrogen, at a flow rate of 1.2 mL min^{-1}. For each sample measurements were carried out three times. The results were expressed as relative percentages of each fatty acid. Reference fatty acid methyl esters from Sigma–Aldrich were used as standards for identification and quantitation purposes.

2.6. TAG Hydrolysis

In order to define the positional distribution of FA in the *sn*-2 and *sn*-1,3 positions of the TAG molecule, hydrolysis in the presence of pancreatic lipase was applied. Briefly, 0.1 g of fat sample was mixed with 1 mL of Tris buffer (1 mol L^{-1} and pH = 8), 0.1 mL of CaCl$_2$ (2.2%) and 0.25 mL aqueous solution of bile salts (0.05%). After 30 s of mixing, 20 mg of purified pancreatic lipase (porcine pancreatic lipase, crude type II) was added. The samples were placed in a water bath for 3 min at 40 °C. After incubation, the reaction was stopped by adding 1 mL of hydrochloric acid (6 mol L^{-1}) and 4 mL of diethyl ether. The products obtained as a result of hydrolysis were analyzed by extraction from the reaction mixture and separation by thin-layer chromatography with hexane/diethyl ether/acetic acid (50:50:1, *v:v:v*). The 2-monoacylglycerol band was removed from silica gel plates, extracted with diethyl ether, converted to methyl esters, as mentioned above, and quantified by GC. Under reaction conditions, there was little or no hydrolysis of the esters' secondary hydroxyl groups by pancreatic lipase [34].

2.7. PDSC Measurements

The oxidative stability of fats extracted from commercial and homemade cat foods was determined using a differential scanning calorimeter (DSC Q20, TA Instruments, New Castle, DE, USA) coupled with a high-pressure cell (Q20P). Weighed fat samples (3–4 mg) were placed on an open aluminum pan in the heating sample chamber of the PDSC cell. Experiments were performed under oxygen atmosphere with an initial pressure of 1380 kPa and with the 100 mL min^{-1} gas flow rate. The isothermal temperature for each sample was programmed at 100 °C. Obtained diagrams were analyzed using TA Universal Analysis 2000 software. For each sample, measurements were carried out three times, and the maximum oxidation time was determined based on the maximum rate of heat flow with an accuracy of 0.005.

The maximum PDSC oxidation and induction time (τ_{max}) was determined on the basis of the maximum rate of oxidation (maximum rate of heat flow) for all fats extracted from commercial and self-prepared cat foods.

2.8. Statistical Analysis

The statistical analysis was performed using the Stargraphics Plus, version 4.1 software. The collected data were statistically processed using the one–way analysis of variance ANOVA. Differences were considered to be significant at a p-value $\leq$ 0.05, according to Tukey's Multiple Range Test. Three samples of each food were used for the analysis. All analyses were conducted in triplicate.

3. Results

3.1. Fat Content and Fatty Acids Composition

The fat content in the dry matter of tested foods (Tables 1 and 2) was calculated on the basis of extracted fat mass and obtained dry matter as well as taking into account the producers' declaration (guaranteed analysis) regarding moisture and fat in products for commercial cat foods and *BARFny kalkulator* data for raw ones.

The results of the determination of FA composition of fat extracted (crude fat) from twelve commercial and self-prepared cat foods are presented in Tables 1 and 2.

The percentages of saturated fatty acids (SFA), monounsaturated fatty acids (MUFA) and polyunsaturated fatty acids (PUFA) in total FA for all foods are presented in Figure 1. In the fraction of SFA, palmitic and stearic acids have the highest percentage for all foods. In turn, oleic acid has the most significant share in the MUFA fraction and linoleic acid has the highest percentage among PUFA in all foods.

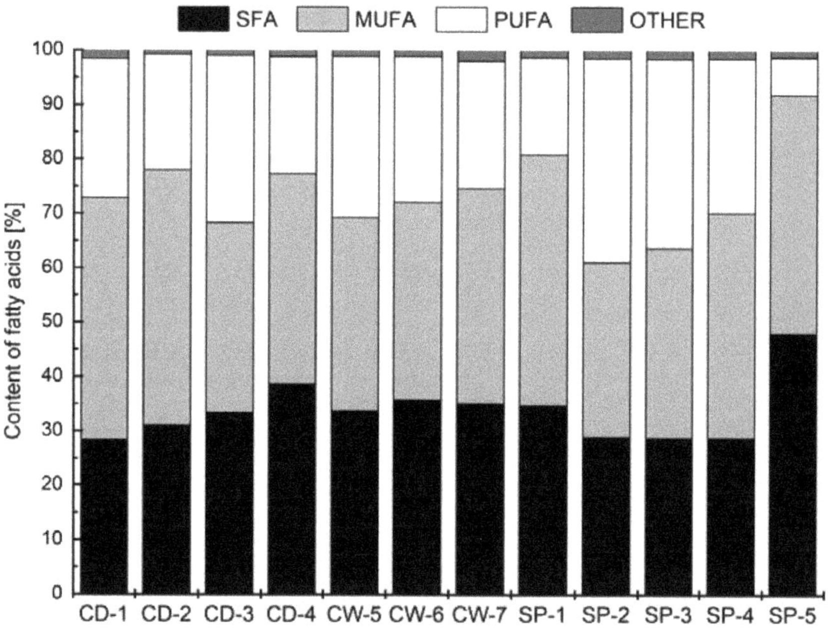

Figure 1. The content of saturated (SFA), monounsaturated (MUFA) and polyunsaturated fatty acids (PUFA) for fats extracted from commercial wet (CW), dry (CD) and self-prepared (SP) cat foods.

The MUFA to PUFA, MUFA to SFA and PUFA to SFA ratios for all foods were also compared (Tables 1 and 2). The MUFA to PUFA ratio is the lowest for SP-2 (below 1) and the highest for SP-5 (more than 6), while in most cases it is between 1 and 2. Additionally, the omega-6 to omega-3 ratio was also calculated and presented in Tables 1 and 2 (the percentage of omega-3, omega-6 and omega-9 in all foods are presented in Supplemental Materials Figure S1). For most of tested foods omega-6/omega-3 was in the range of 5:1 to 10:1.

The results obtained for EFA for cats (Supplemental Materials Figure S2) were re-calculated using determined dry matter content and GC data and presented as grams of each EFA per 1000 g of dry matter for self-prepared food (Table 3). The highest content of ALA and LA was detected for SP-2 and SP-3 and the lowest for SP-5. Comparing the FA composition of all self-prepared cat foods, it can be noticed that those based on poultry (SP-2–SP-4) are characterized by a higher content of LA and ALA than in SP-1 and SP-5, containing beef. The highest content of AA and EPA+DHA was found in the case of SP-1 and the lowest for SP-2.

3.2. Fatty Acid Distribution

The content of FA (%) of the TAGs and of isolated *sn*-2 monoacylglycerols was used to calculate the composition of FA in the *sn*-1 and *sn*-3 positions (Supplemental Materials Tables S1 and S2). The percentage of selected FA (C16:0, C18:0, C18:1 and C18:2) in the *sn*-2 position of TAGs of fats extracted from self-prepared cat foods was presented in Figure 2. In Figure 2 a dotted line indicates the even distribution of FA between the internal and external positions of TAGs. Taking these calculations into account, over 33.3% of palmitic acid was located at the *sn*-2 position of TAGs for SP-1 only; for other foods it was below 33.3% (in the range from 19% to 22%). In turn, stearic acid in all tested foods was located at the *sn*-2 position in less and at linoleic acid in more than 33.3% of TAGs from extracted fats. The oleic acid presented the most varied distribution between the external and internal positions of TAG. In some cases this distribution was even (for SP-2 and SP-3) but it was also directed towards *sn*-2 (for SP-5) as well as towards *sn*-1,3 (for SP-1).

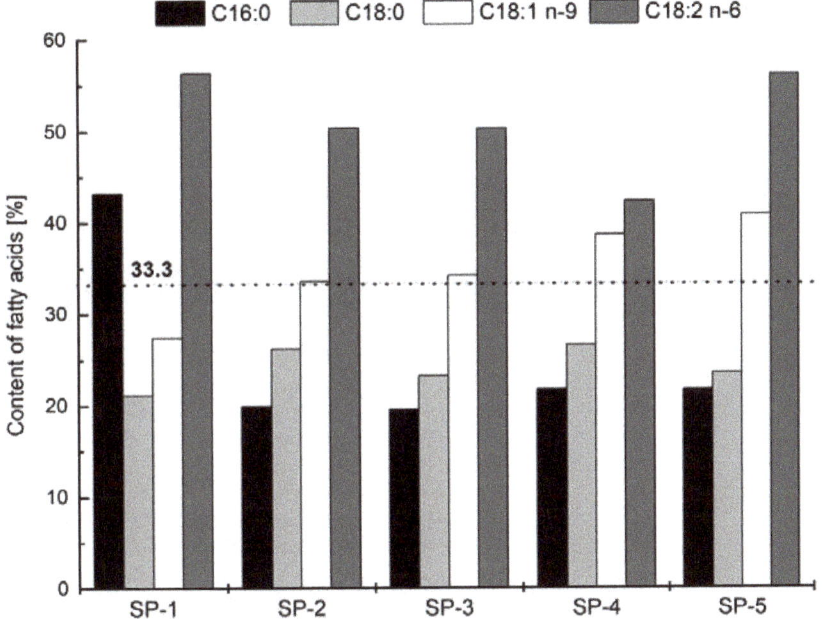

Figure 2. The percentage of a given fatty acid in *sn*-2 position of triacylglycerols (TAGs) of fats extracted from SP cat foods, expressed as the relative fatty acid [(sn-2 fatty acid $\times$ 100%)/(3 $\times$ total fatty acid in TAGs)]. The line indicates the statistical (even) distribution of FA between three TAG positions (33%).

3.3. Oxidative Stability

The oxidation induction time for each extracted fat, determined under isothermal conditions (at 100 °C), is presented in Figure 3. The longest τ_{max}, thus the highest oxidative stability, was revealed for CD-2 and CD-3 (46.05 and 53.59 min, respectively) and the shortest τ_{max} for SP-2 and SP-3 (7.14 and 3.32 min, respectively).

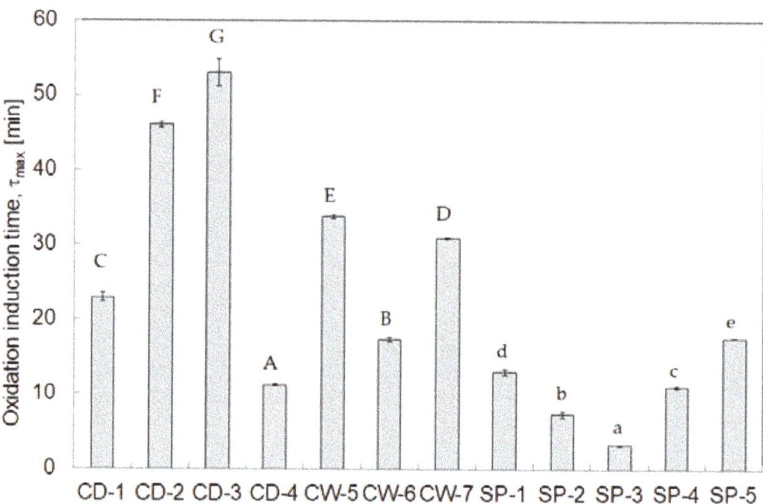

Figure 3. The oxidation induction time at 100 °C obtained for the lipid fraction extracted from commercial and self-prepared cat foods. Different letters indicate that the samples are significantly different at $p < 0.05$ (the upper cases (A–G) for commercial cat foods and the lower cases (a–e) for self-prepared ones).

4. Discussion

4.1. Lipid Fraction Properties of Self-Prepared Cat Food

The fat content in self-prepared foods oscillated around 30% (Table 1), indicating an effective extraction. The differences may be associated with variations in the composition of individual meat cuts, especially when the food is produced in very small batches.

The SP foods should have a composition similar to what cats eat in nature. Plantinga et al. [35] estimated the dietary profile of free-roaming feral cats based on a composition of the consumed preys living in the wild. The fat content varied between preys, from 9% to 31%, and the mean fat content in the diet of cats was 22.8% in dry matter. Pet food regulatory bodies, notably the Association of American Feed Control Officials in the United States and the European Pet Food Industry Federation in Europe, recommend a minimum of 9 g of fat per 100 g of dry matter, which is a significantly lower content than what a cat eats in nature. The fat content of all raw foods is significantly higher than the minimum dietary levels for cats (Table 3), but differs from the mean value for preys set by Plantinga [35].

The obtained results for LA and ALA are consistent with the literature data, suggesting that LA is found mostly in vegetable oils and only in some animal fats (e.g., chicken fat); beef is not a rich source of omega-6 fatty acids [36–38]. Moreover, very high contents of oleic acid as well as high contents of palmitic and stearic acids in SP-1 and SP-5 are associated with high amounts of these acids in beef [38].

The amount of individual EFA in a cat's diet affects its health, growth and reproduction. The results of EFA for tested diets, whole prey items (available data only for captive preys) [39] and recommendations are presented as grams per kg of dry matter (Table 3). The concentrations of total FA as well as EFA for all foods are above minimal recommendations for all life stages. The LA and AA content is highly variable, but it can be noticed that ALA content is considerably lower in whole prey items than in self-prepared foods. The EPA+DHA content is several times higher in self-prepared food than in whole prey items. A higher amount of EPA+DHA is beneficial for many health aspects and essential at the stage of growth and reproduction. The low EPA+DHA content in whole prey items is probably connected with their diet during the farming period (data for free-living preys are unavailable). This content in the wild preys is supposed to be higher because of differences in their food. The higher EFA content in SP foods is also due to the overall higher fat content of raw diets. The fat content of the prey is closer to the model fat content established by Plantinga [35]. Another reason for the differences is the different content and ratio of meat, bones and internal organs used in the raw diets and their other content in the whole animal. In addition, not all internal organs are easily available on the market and they are replaced in self-prepared diets with those easily available.

The distribution of selected FA in TAG was analyzed (Figure 2) and it can be noticed that in most cases SFA (C16:0 and C18:0) occupy the external positions of TAGs. The SFA in *sn*-1 and *sn*-3 positions will be poorly absorbed in the digestive tract because they will cleave off and react with free calcium ions to form insoluble calcium soaps [17,18]. This will cause both calcium and FA losses in the feces. In turn, LA, which is EFA for mammals, mainly occupied the *sn*-2 position of TAGs and subsequently will be easily absorbed [17].

From the nutritional point of view, the distribution of FA in TAG is very important, but even more important, is their composition in the fat. The EFA for cats are LA and AA from the omega-6 group as well as ALA, EPA and DHA belonging to the omega-3 group [40,41]. The role of omega-6 and omega-3 fatty acids in maintaining health is affected by their ratio. Most fats and oils in cats' foods are richer in omega-6 than omega-3 fatty acids, but it is important for pets to contain an appropriate balance of both. Determining the optimal dietary omega-6 to omega-3 ratio for pet food is still under investigation, but some recommendations indicate ratios of 10:1 to 5:1 [42]. In the case of tested foods, omega-6/omega-3 was in this range (except SP-5, with the ratio below 5:1). Salmon oil has been added to all self-prepared cat food to improve this ratio. Such supplementation seems necessary because the raw diet is intended for cats at any stage of their life, especially during growth and reproduction when they have an increased need for EFA. The addition of salmon oil as well as cod-liver oil and algae meal affects the higher content of EPA, DHA and AA in self-prepared cat food. The added salmon oil was the oil of wild Atlantic salmon, which according to the producer analysis, compared to oil of farmed salmon, has a higher content of EPA and DHA as well as higher ratio of omega-3 to omega-6 (farmed salmon—1:1 and wild salmon—10:1) [43].

FA composition is one of the factors that determine the oxidative stability of fats and oils [24]. As a rule, samples with longer τ_{max} are more stable than those with shorter τ_{max} determined at the same temperature. Extremely short τ_{max}, especially for SP-2 and SP-3 (Figure 3), is probably associated with the greatest amount of PUFA in their composition (Figure 1) and a higher PUFA to SFA ratio than in the other tested foods (Table 1). The longest τ_{max} among self-prepared foods was observed for SP-5, for which the content of SFA or MUFA was the highest and PUFA was the lowest. Moreover, it should be mentioned that only SP-5 fat was solid, which was due to the high content of SFA.

4.2. The FA Composition and Oxidative Stability of Commercial Cat Food

In light of the obtained results for raw foods, the FA profiles of selected dry and wet commercial foods were determined and presented. The same extraction method (Folch method) was used. The Folch method for fat extraction from dry pet food was also tested by other authors [27,28]. The determined extracted fat content for commercial cat foods was lower (especially for CW-6 and CW-7) than the producers' statements (Table 2), which means the extraction was not effective. Moreover, in commercial wet foods the percentage of fat in dry matter was lower than in the case of the dry ones.

The FA composition is presented in Table 2. The differences in FA profiles of commercial and self-prepared foods may be due to the type of meat and the ingredients used (human grade meat, animal by-products, meat meal or mechanically separated meat). Among all commercial dry cat foods CD-3 had the highest content of LA. This is probably due to the presence of plant origin oils (soya and borage oils) in its composition. These oils are rich in PUFA, especially in high content of LA [44,45]. In the case of wet foods, CW-7 has the lowest LA content and it is the only commercial food that does not contain plants.

It should be emphasized that the raw diet concept assumes that it is suitable for cats at every stage of life and that tested commercial foods are for adult cats. The American Feed Control Officials and the European Pet Food Industry Federation do not provide recommendations for the minimum levels of some of the EFA (ALA, EPA and DHA) in the diet of adult cats. In the case of CD-1, in which salmon meal (source of omega-3 [39]) is the main component, an increased content of EPA+DHA in relation to other tested commercial foods was observed (Table 2). For other foods the required omega-6 to omega-3 ratio is not due to the high content of EPA and DHA, but because of the relatively high levels of ALA. Plants are a rich source of ALA [14]. The ingredients of plants origin are added to commercial diets but not to self-prepared ones (Supplemental Materials). Moreover, such ingredients do not contain AA (present only in fat of animal origin [15]). The AA content in commercial foods is relatively low, excluding CD-3 (Table 2). The low level of AA balances the ratio of omega-6 to omega-3 in commercial diets.

The oxidative stability of commercial cat foods was determined by using the PDSC technique, the same as the self-prepared food. As predicted, τ_{max} for commercial cat foods is much longer than for self-prepared ones. Commercial cat foods contain different antioxidants and preservatives, which increase their resistance to oxidation. Moreover, technological procedures (e.g., baking or extrusion) used during their production reduce the free fat available for oxidation [46]. Self-prepared food contained only natural antioxidants in the form of vitamin E as an essential nutrient and PUFA stabilizer [37]. The BARF diet is stored frozen, inter alia, to provide additional protection against oxidation during storage.

Table 3. Concentrations of crude fat (%), essential and conditionally essential fatty acids (EFA) (g/kg DM) in self-prepared foods, whole prey items, American Feed Control Officials (AAFCO) and United States and the European Pet Food Industry Federation (FEDIAF) recommendations.

	Self-Prepared Raw Meat Food					Whole Prey Items ‡				Recommendations			
										AAFCO		FEDIAF	
Fat in DM, %	SP-1	SP-2	SP-3	SP-4	SP-5	Mouse	Rat	Rabbit	Quail	Adult	GR †	Adult	GR †
	35.48	25.14	28.58	32.00	36.30	29.92	26.35	22.05	19.24	9	9	9	9
LA	45.67	73.26	77.60	72.47	11.31	27.03	33.47	10.28	18.76	6	6	5	5.5
ALA	3.09	8.55	7.60	4.14	2.52	0.46	1.13	0.47	0.71	-	0.2	-	0.2
AA	5.25	3.54	4.73	4.21	4.76	2.18	6.86	0.34	2.75	0.2	0.2	0.06	0.2
EPA	0.43	0.35	0.40	0.51	0.18	0.02	0.22	0.01	0.04	-	-	-	-
DHA	4.31	2.99	3.83	3.62	3.69	0.79	2.28	0.03	0.72	-	-	-	-
EPA + DHA	4.74	3.34	4.23	4.13	3.87	0.81	2.5	0.04	0.76	-	0.12	-	0.1

† Growth and reproduction period. ‡ According to Kerr et al. [47].

5. Conclusions

- The self-prepared cat foods contain high levels of EFA, which is beneficial for adult cats and essential for the cats' growth and reproduction stage.
- For the first time the distribution of FA in TAG for cat food has been determined.
- The commercial and self-prepared cat foods have a similar ratio of omega-6 to omega-3 but self-prepared diets provide more EPA and DHA.
- It is shown that the type of meat in self-prepared cat food affects the composition and distribution of fatty acids.
- Fat samples from self-prepared cat foods containing beef are characterized by the highest oxidative stability, which can be due to a lower ratio of PUFA/SFA.

Supplementary Materials: The following are available online at https://www.mdpi.com/article/10.3390/app112210905/s1, Figure S1: The content of omega-3, omega-6 and omega-9 fatty acids for fats extracted from commercial and self-prepared cat food. Figure S2: The content of essential omega-3 and omega-6 fatty acids for fats extracted from commercial and self-prepared cat food. Table S1: The composition of selected fatty acids (%) in the internal (sn-2) and external (sn-1,3) positions of TAG in fats extracted from commercial cat food. Table S2: The composition of selected fatty acids (%) in the internal (*sn*-2) and external positions (*sn*-1,3) of TAG in fats extracted from self-prepared cat food.

Author Contributions: Conceptualization, A.G. (Agnieszka Górska) and D.M.-J.; methodology, A.G. (Agnieszka Górska), J.B. and A.G. (Agata Górska); software, J.B.; validation, J.B., A.G. (Agata Górska), A.G. (Agnieszka Górska) and D.M.-J.; formal analysis, A.G. (Agnieszka Górska), D.M.-J. and J.B.; investigation, A.G. (Agnieszka Górska) and D.M.-J.; resources, A.G. (Agnieszka Górska) and D.M.-J.; writing—original draft preparation, A.G. (Agnieszka Górska), D.M.-J.; writing—review and editing, A.G. (Agnieszka Górska), D.M.-J., J.B. and A.G. (Agata Górska); supervision, A.G. (Agata Górska) All authors have read and agreed to the published version of the manuscript.

Funding: The study was financially supported by sources of the Polish Ministry of Education and Science within funds of the Institute of Food Sciences of Warsaw University of Life Sciences (WULS), for scientific research.

Institutional Review Board Statement: Not applicable.

Informed Consent Statement: Not applicable.

Data Availability Statement: The data generated or analysed during this study are available from the corresponding author on reasonable request.

Conflicts of Interest: The authors declare no conflict of interest.

References

1. Michel, K.E. Unconventional Diets for Dogs and Cats. *Vet. Clin. N. Am.* **2006**, *36*, 1269–1281. [CrossRef] [PubMed]
2. Remillard, R.L. Homemade Diets: Attributes, Pitfalls, and a Call for Action. *Top. Companion Anim. Med.* **2008**, *23*, 137–142. [CrossRef]
3. Parr, J.M.; Remillard, R.L. Handling Alternative Dietary Requests from Pet Owners. *Vet. Clin. N. Am.* **2014**, *44*, 667–688. [CrossRef] [PubMed]
4. Kerr, K.R.; Vester Boler, B.M.; Morris, C.L.; Liu, K.J.; Swanson, K.S. Apparent total tract energy and macronutrient digestibility and fecal fermentative end-product concentrations of domestic cats fed extruded, raw beef-based, and cooked beef-based diets. *J. Anim. Sci.* **2012**, *90*, 515–522. [CrossRef]
5. Kerr, K.R.; Beloshapka, A.N.; Morris, C.L.; Parsons, C.M.; Burke, S.L.; Utterback, P.L.; Swanson, K.S. Evaluation of four raw meat diets using domestic cats, captive exotic felids, and cecectomized roosters. *J. Anim. Sci.* **2013**, *91*, 225–237. [CrossRef] [PubMed]
6. Hamper, B.A.; Kirk, C.A.; Bartges, J.W. Apparent nutrient digestibility of two raw diets in domestic kittens. *J. Feline Med. Surg.* **2016**, *18*, 991–996. [CrossRef]
7. Schlesinger, D.P.; Joffe, D.J. Raw food diets in companion animals: A critical review. *Can. Vet. J.* **2011**, *52*, 50–54.
8. Freeman, L.M.; Chandler, M.L.; Hamper, B.A.; Weeth, L.P. Current knowledge about the risks and benefits of raw meat–based diets for dogs and cats. *J. Am. Vet. Med. Assoc.* **2013**, *243*, 1549–1558. [CrossRef]
9. Fredriksson-Ahomaa, M.; Heikkilä, T.; Pernu, N.; Kovanen, S.; Hielm-Björkman, A.; Kivistö, R. Raw Meat-Based Diets in Dogs and Cats. *Vet. Sci.* **2017**, *4*, 33. [CrossRef]
10. Wilson, S.A.; Villaverde, C.; Fascetti, A.J.; Larsen, J.A. Evaluation of the nutritional adequacy of recipes for home-prepared maintenance diets for cats. *J. Am. Vet. Med. Assoc.* **2019**, *254*, 1172–1179. [CrossRef]
11. Macdonald, M.L.; Rogers, Q.R.; Morris, J.G. Role of Linoleate as an Essential Fatty Acid for the Cat Independent of Arachidonate Synthesis. *J. Nutr.* **1983**, *113*, 1422–1433. [CrossRef] [PubMed]
12. Morris, J.G. Idiosyncratic nutrient requirements of cats appear to be diet-induced evolutionary adaptations. *Nutr. Res. Rev.* **2002**, *15*, 153–168. [CrossRef]
13. Rivers, J.P.W. Essential fatty acids in cats. *J. Small Anim. Pract.* **1982**, *23*, 563–576. [CrossRef]
14. Lenox, C.E.; Bauer, J.E. Potential Adverse Effects of Omega-3 Fatty Acids in Dogs and Cats. *J. Vet. Intern. Med.* **2013**, *27*, 217–226. [CrossRef] [PubMed]
15. Bauer, J.J.E. Facilitative and functional fats in diets of cats and dogs. *J. Am. Vet. Med. Assoc.* **2006**, *229*, 680–684. [CrossRef]
16. Mu, H.; Høy, C.E. The digestion of dietary triacylglycerols. *Prog. Lipid Res.* **2004**, *43*, 105–133. [CrossRef]
17. Kubow, S. The influence of positional distribution of fatty acids in native, interesterified and structure-specific lipids on lipoprotein metabolism and atherogenesis. *J. Nutr. Biochem.* **1996**, *7*, 530–541. [CrossRef]
18. Stroebinger, N.; Rutherfurd, S.M.; Henare, S.J.; Hernandez, J.F.P.; Moughan, P.J. Fatty Acids from Different Fat Sources and Dietary Calcium Concentration Differentially Affect Fecal Soap Formation in Growing Pigs. *J. Nutr.* **2021**, *151*, 1102–1110. [CrossRef] [PubMed]
19. Summers, S.C.; Stockman, J.; Larsen, J.A.; Zhang, L.; Rodriguez, A.S. Evaluation of phosphorus, calcium, and magnesium content in commercially available foods formulated for healthy cats. *J. Vet. Intern. Med.* **2019**, *34*, 266–273. [CrossRef]
20. Stockman, J.; Villaverde, C.; Corbee, R.J. Calcium, Phosphorus, and Vitamin D in Dogs and Cats. *Vet. Clin. N. Am.* **2021**, *51*, 623–634. [CrossRef]
21. Tran, Q.D.; Hendriks, W.H.; van der Poel, A.F.B. Effects of extrusion processing on nutrients in dry pet food. *J. Sci. Food Agric.* **2008**, *88*, 1487–1493. [CrossRef]

22. Saldaña, M.D.A.; Martínez-Monteagudo, S.I. Oxidative Stability of Fats and Oils Measured by Differential Scanning Calorimetry for Food and Industrial Applications. In *Applications of Calorimetry in a Wide Context—Differential Scanning Calorimetry, Isothermal Titration Calorimetry and Microcalorimetry*; Elkordy, A.A., Ed.; IntechOpen: London, UK, 2013. [CrossRef]

23. Symoniuk, E.; Ratusz, K.; Krygier, K. Comparison of the oxidative stability of linseed (Linum usitatissimum L.) oil by pressure differential scanning calorimetry and Rancimat measurements. *J. Food Sci. Technol.* **2016**, *53*, 3986–3995. [CrossRef] [PubMed]

24. Maszewska, M.; Florowska, A.; Dłużewska, E.; Wroniak, M.; Marciniak-Lukasiak, K.; Żbikowska, A. Oxidative Stability of Selected Edible Oils. *Molecules* **2018**, *23*, 1746. [CrossRef] [PubMed]

25. Danthine, S.; De Clercq, N.; Dewettinck, K.; Gibon, V. Monitoring batch lipase catalyzed interesterification of palm oil and fractions by differential scanning calorimetry. *J. Therm. Anal. Calorim.* **2014**, *115*, 2219–2229. [CrossRef]

26. Gray, J.I. Measurement of lipid oxidation: A review. *J. Am. Oil Chem. Soc.* **1978**, *55*, 539–546. [CrossRef]

27. Ciemniewska-Żytkiewicz, H.; Ratusz, K.; Bryś, J.; Reder, M.; Koczoń, P. Determination of the oxidative stability of hazelnut oils by PDSC and Rancimat methods. *J. Therm. Anal. Calorim.* **2014**, *118*, 875–881. [CrossRef]

28. Hołda, K.; Głogowski, R. Selected quality properties of lipid fraction and oxidative stability of dry dog foods under typical storage conditions. *J. Therm. Anal. Calorim.* **2016**, *126*, 91–96. [CrossRef]

29. Anderson, R.S.; Meyer, H. *Ernährung und Verhalten von Hund und Katze*; Schlütersche Verlag: Hannover, Germany, 1984; ISBN 978-3877060865.

30. Horzinek, M.C.; Lutz, H.; Schmidt, V. *Krankheiten der Katze*; Enke: Stuttgart, Germany, 1999; ISBN 978-3830410492.

31. Strombeck, D.R. *Home-Prepared Dog & Cat Diets: The Healthful Alternative*; Wiley-Blackwell: Hoboken, NJ, USA, 1999; ISBN 978-0813821498.

32. Wanner, M. *Die Ernährung der Katze. Skript für Studierende der Veterinärmedizin an der Universität Zürich*; University of Zurich: Zurich, Switzerland, 2004.

33. Boselli, E.; Velazco, V.; Caboni, M.F.; Lercker, G. Pressurized liquid extraction of lipids for the determination of oxysterols in egg-containing food. *J. Chromatogr. A* **2001**, *917*, 239–244. [CrossRef]

34. Mattson, F.H.; Lutton, E.S. The Specific Distribution of Fatty Acids in the Glycerides of Animal and Vegetable Fats. *J. Biol. Chem.* **1958**, *233*, 868–871. [CrossRef]

35. Plantinga, E.A.; Bosch, G.; Hendriks, W.H. Estimation of the dietary nutrient profile of free-roaming feral cats: Possible implications for nutrition of domestic cats. *Br. J. Nutr.* **2011**, *106*, S35–S48. [CrossRef]

36. Li, D.; Ng, A.; Mann, N.J.; Sinclair, A.J. Contribution of meat fat to dietary arachidonic acid. *Lipids* **1998**, *33*, 437–440. [CrossRef]

37. Wood, J.D.; Enser, M.; Fisher, A.V.; Nute, G.R.; Sheard, P.R.; Richardson, R.I.; Hughes, S.I.; Whittington, F.M. Fat deposition, fatty acid composition and meat quality: A review. *Meat Sci.* **2008**, *78*, 343–358. [CrossRef] [PubMed]

38. Hwang, Y.-H.; Joo, S.-T. Fatty Acid Profiles, Meat Quality, and Sensory Palatability of Grain-fed and Grass-fed Beef from Hanwoo, American, and Australian Crossbred Cattle. *J. Food Sci. Anim. Resour.* **2017**, *37*, 153–161. [CrossRef]

39. Wall, R.; Ross, R.P.; Fitzgerald, G.F.; Stanton, C. Fatty acids from fish: The anti-inflammatory potential of long-chain omega-3 fatty acids. *Nutr. Rev.* **2010**, *68*, 280–289. [CrossRef]

40. Biagi, G.; Mordenti, A.L.; Cocchi, M. The role of dietary omega-3 and omega-6 essential fatty acids in the nutrition of dogs and cats: A review. *Prog. Nutr.* **2004**, *6*, 97–107.

41. Tallima, H.; El Ridi, R. Arachidonic acid: Physiological roles and potential health benefits—A review. *J. Adv. Res.* **2018**, *11*, 33–41. [CrossRef] [PubMed]

42. Bunglavan, S.J.; Pratheesh, M.D.; Anoopraj, R.; Harish, C.; Davis, J. Therapeutic uses of omega fatty acids in cats. *Indian Pet J.* **2011**, *12*, 1–2.

43. Filburn, C.R.; Griffin, D. Effects of supplementation with a docosahexaenoic acid-enriched salmon oil on total plasma and plasma phospholipid fatty acid composition in the cat. *Int. J. Appl. Res. Vet. Med.* **2005**, *3*, 116–123.

44. Jokic, S.; Sudar, R.; Svilović, S.; Vidovic, S.; Bilić, M.; Velic, D.; Jurkovič, V. Fatty acid composition of oil obtained from soybeans by extraction with supercritical carbon dioxide. *Czech J. Food Sci.* **2013**, *31*, 116–125. [CrossRef]

45. Tasset-Cuevas, I.; Fernández-Bedmar, Z.; Lozano-Baena, M.D.; Campos-Sánchez, J.; De Haro-Bailón, A.; Muñoz-Serrano, A.; Alonso-Moraga, Á. Protective Effect of Borage Seed Oil and Gamma Linolenic Acid on DNA: In Vivo and In Vitro Studies. *PLoS ONE* **2013**, *8*, e56986. [CrossRef]

46. Gibson, M.; Alavi, S. Pet Food Processing—Understanding Transformations in Starch during Extrusion and Baking. *Cereal Foods World* **2013**, *58*, 232–236. [CrossRef]

47. Kerr, K.R.; Kappen, K.L.; Garner, L.M.; Swanson, K.S. Commercially Available Avian and Mammalian Whole Prey Diet Items Targeted for Consumption by Managed Exotic and Domestic Pet Felines: Macronutrient, Mineral, and Long-Chain Fatty Acid Composition. *Zoo Biol.* **2014**, *33*, 327–335. [CrossRef] [PubMed]

applied
sciences

Article

Application of Different Compositions of Apple Puree Gels and Drying Methods to Fabricate Snacks of Modified Structure, Storage Stability and Hygroscopicity

Ewa Jakubczyk [1,*], Anna Kamińska-Dwórznicka [1], Ewa Ostrowska-Ligęza [2], Agata Górska [2], Magdalena Wirkowska-Wojdyła [2], Diana Mańko-Jurkowska [2], Agnieszka Górska [2] and Joanna Bryś [2,*]

[1] Department of Food Engineering and Process Management, Institute of Food Sciences, Warsaw University of Life Sciences, 02-776 Warsaw, Poland; anna_kaminska1@sggw.edu.pl
[2] Department of Chemistry, Institute of Food Sciences, Warsaw University of Life Sciences, 02-787 Warsaw, Poland; ewa_ostrowska_ligeza@sggw.edu.pl (E.O.-L.); agata_gorska@sggw.edu.pl (A.G.); magdalena_wirkowska@sggw.edu.pl (M.W.-W.); diana_manko_jurkowska@sggw.edu.pl (D.M.-J.); agnieszka_gorska@sggw.edu.pl (A.G.)
* Correspondence: ewa_jakubczyk@sggw.edu.pl (E.J.); joanna_brys@sggw.edu.pl (J.B.)

Citation: Jakubczyk, E.; Kamińska-Dwórznicka, A.; Ostrowska-Ligęza, E.; Górska, A.; Wirkowska-Wojdyła, M.; Mańko-Jurkowska, D.; Górska, A.; Bryś, J. Application of Different Compositions of Apple Puree Gels and Drying Methods to Fabricate Snacks of Modified Structure, Storage Stability and Hygroscopicity. *Appl. Sci.* **2021**, *11*, 10286. https://doi.org/10.3390/app112110286

Academic Editor: Anna Lante

Received: 7 October 2021
Accepted: 31 October 2021
Published: 2 November 2021

Publisher's Note: MDPI stays neutral with regard to jurisdictional claims in published maps and institutional affiliations.

Abstract: The aim of this study was to determine the effect of incorporation of apple puree and maltodextrin to agar sol on the sorption properties and structure of the dried gel. The effect of different drying methods on the sorption behaviour of aerated apple puree gels was also observed. The gels with the addition of 25% and 40% concentration of apple puree and with or without maltodextrin were prepared and dried. The foamed agar gel was subjected to freeze-drying, air-drying and vacuum-drying. The sorption properties of dried gels (adsorption isotherms, water uptake in time) were investigated. The relations between the glass transition temperature, water activity and water content were also obtained for some apple snacks. The increase in apple puree in freeze-dried gels increased the hygroscopicity and decreased the glass transition temperature (T_g). The water content at given activity and hygroscopicity were reduced by the addition of maltodextrin, which also caused the increase in T_g. The application of different drying methods enabled obtaining different structures of material. The open-pore, fragile materials were obtained by freeze-drying, the expanded matrix with big holes was characteristic for vacuum-dried gels, but the closed pores with thick walls were created during the air-drying.

Keywords: fruit gel; aeration; drying; sorption isotherms; glass transition; maltodextrin

1. Introduction

Nowadays, the fast pace of life promotes the growth of snack consumption, which has become an important part of the daily diet [1]. Many consumers expect tasty and healthy food products based on fresh or processed fruits and vegetables as an alternative to the energy-dense, nutrient-poor snacks [2]. Additionally, these types of products should have an attractive appearance and texture. The application of a wide range of drying methods for fruits materials enables creating snacks with different structure and texture [3–7].

Drying is one of the oldest and common unit operations used to reduce the water moisture content to the level that guarantees the stability of food products during long-term storage. The application of this technology leads to slowing down enzymatic and non-enzymatic reactions as well as preventing microbial contamination of food. The quality of dehydrated products varies depending on the methods and conditions of drying [8,9]. A simple and relatively inexpensive dehydration technology, widely used in the food industry, is air-drying. During convective drying, the material is subjected to a continuous flow of hot air, which transports the evaporated water. The dehydration by the air-drying may cause undesirable changes in the quality of the material such as shrinkage, loss of volatile compounds, colour and nutrients [10,11].

Drying at low temperature and pressure (e.g., during freeze-drying or vacuum-drying) requires longer times of dehydration in comparison to other drying methods, but the obtained products are characterised by the high quality [8]. Porosity, colour, rehydration and retention of many nutrients (anthocyanins, phenolics, vitamin C) and flavours of fruit products obtained with the application of freeze-drying were more favourable than observed if other drying methods were used [12]. In the case of freeze-drying, the colour of apples, pears and oranges was preserved [13]. Freeze-drying contributed minimal volume reduction of different fruits (from 5 to 15%), while convective-drying caused excessive changes in volume (around 80%) [11]. The application of freeze-drying enabled 63% vitamin C retention in guava in comparison to only 25% when the other drying methods were used [14]. Freeze-drying of sea buckthorn pulp caused loss of vitamins E, C and phenolics reaching about 35%, 20% and 4%, respectively. However, it was possible, using freeze-drying, to retain more nutrients than by drying with a heat pump [15]. The preservation of lycopene, carotenoids, unsaturated fatty acids in oils and lipid-based oxidisable compounds in freeze-dried fruits can be limited due to the acceleration of oxidative reactions at low water activities [16]. The main disadvantage of freeze-drying is a high cost of the process [13,16]. The chemical, mechanical and thermal pre-treatments of fruits (immersion in the alkaline, peeling, blanching) were applied to enhance water transport during the freeze-drying and reduce the duration of the process as well as its costs. However, the use of some pre-treatment methods induced the loss of nutrients and vitamins. The application of infrared radiation, microwave energy and ultrasounds power in the freeze-drying of plant based foods was developed for intensification of the dehydration process. The limited examples of snack products with the addition of fruit materials (pulp and puree) were mainly fabricated with the application of the freeze-drying method or with a microwave-assisted freeze-drying technique [3,5,7]. Additionally, these products were obtained with the addition of foaming and stabilising agents (pectin, maltodextrin, potato protein) [17] as well as with carriers (maltodextrin, gum Arabic, bamboo fibre, whey protein isolate) [6,7,17]. Martínez-Navarrete et al. [6] stated that the freeze-drying method can provide different food textures such as crunchy fruit snacks with good acceptance of consumers. However, the drawback of the freeze-drying process can be a collapse of structure, which was observed for sugar-rich foods due to the glass transition of the amorphous matrix [18,19].

Fruit-based products on the market such as juices and purees are rich in low molecular weight soluble solids (sugars and organic acids). The rapid removal of water during drying of the material containing these kinds of compounds results in the formation of the amorphous state of the solid matrix. The amorphous solids show changes in structure, physical and chemical properties during the time that affect the stability of the material and, consequently, reduce the shelf life of dried products. The amorphous matrix may transform its structure from a stable solid glass to liquid-like rubber. This phenomenon occurs with an increase in water content in the amorphous food system at the temperature defined as the glass transition temperature T_g [20–23]. The significant physical defects, changes of physicochemical properties of dehydrated foods, and the worsening of their quality such as structure collapse, caking, stickiness, and change of colour can be observed at storage temperatures above T_g. To overcome this problem, high molecular weight biopolymers (maltodextrin, gums) with high T_g can be incorporated into fruit materials before drying [24,25].

As T_g depends on water content, knowledge of the relation between water activity and water content of the product described by the moisture sorption isotherms is necessary to predict the quality and stability of products during storage [22,26–28]. Many mathematical models can be used to describe the water sorption isotherms: these include models that refer to multilayer sorption (GAB model, Lewicki model), kinetics model with monolayer sorption (BET model) and semi-empirical models (Peleg model, Hasley model) [29].

There are many studies concerning the sorption behaviour and glass transition of dried fruit products [20,25,26,30,31]. However, the current literature about the effects of fruit material addition on these properties in the case of the dried gel is still limited. Drying of gels enables obtaining novel structures and textures of products [3,32–34]. Novel gelled products can be created with different hydrocolloid agents and fruit materials. Among many hydrocolloids used in the production of gels, agar-agar is an effective gelling agent, which forms a thermo-reversible but stable gel over a wide range of temperatures [35]. Agar is a mixture of agarose and agaropectin fractions, but gelation is caused by the presence of agarose, which creates the physical gel. The structure of this gel is only formed by the polymer's molecules, which are linked by hydrogen bonds [36]. The formation of agar gels occurs after the cooling of sol solutions below 40 °C.

This study aimed to determine the impact of adding apple puree and maltodextrin to agar sol on the sorption properties and structure of the freeze-dried gel. The scope of research also involved the analysis of the effect of different drying methods on the investigated properties of aerated apple puree gels. The relations between the glass transition temperature, water activity, and water content were also investigated for selected snacks.

2. Materials and Methods

2.1. Materials and Preparation of Gels

Apple snacks were produced by the freeze-drying of 2% agar gel with the addition of apple puree (25%, 40%) and with or without maltodextrin. Additionally, one type of gel was obtained by foaming of agar sol solution with apple puree, maltodextrin and methylcellulose (foaming agent). The composition of gels (concentration of ingredients) and type of added gelling agent and carrier were selected based on preliminary studies. This variant of the gel was subjected to different drying methods: air-drying, vacuum-drying, and freeze-drying. The pure agar-agar gel was also freeze-dried. Table 1 presents the composition of the investigated gels. The ingredients used in the formulation of the fruit gel were: Idared apple puree (16°Brix, thermally preserved, provided by a local manufacturer), agar-agar powder (Hortimex Sp. z o.o., Konin, Poland), methylcellulose (Methocel 65HG, Sigma Aldrich Co., Louis, MO, USA), maltodextrin DE 7-13 (Pepees S.A., Łomża, Poland).

Table 1. Composition of gels.

Type of Gel	Apple Puree	Agar-Agar Powder	Maltodextrin	Methylcellulose	Water
			g/100 g		
P0%	0.0	2.0	0.0	0.0	98.0
P25%	25.0	2.0	0.0	0.0	73.0
P25%DE	25.0	2.0	2.0	0.0	71.0
P40%DE	40.0	2.0	3.2	0.0	54.8
P40%DE-MC	40.0	2.0	3.2	0.3	54.5

Before further procedures, puree was rubbed through the sieve with a mesh size of 0.6 mm. The agar-agar powder with or without maltodextrin was dispersed in distilled water, heated to 90 °C and continuously agitated at a speed of 60 rpm using a propeller stirrer R1342 (IKA® Labortechnik, Staufen im Breisgau, Germany). The obtained solution was cooled in the water bath until 60 °C and the apple puree was added and stirred at a speed of 60 rpm for 1 min. In the case of the aerated variant of the gel one more preparation step was used. The mixture of agar-agar sol, maltodextrin, and apple puree with the addition of methylcellulose was foamed using a kitchen mixer (Severin, Sundern, Germany) at a speed of 3500 rpm for 5 min. All variants of the gel product were poured into Petri dishes and allowed to set at temperature 4 °C for 12 h. After storage, the gels were cut by using a cork borer, and cylinders with a diameter of 13.5 mm and height of 13.7 mm were obtained.

2.2. Drying of Gels

2.2.1. Freeze-Drying

The gel samples (all variants) were placed on an aluminium tray and frozen at $-40\,^{\circ}\mathrm{C}$ for 4 h using a shock freezer (Irinox, Corbanese, Italy). In the next step, the samples were freeze-dried for 24 h with an application of Gamma 1-16LSC freeze-dryer (Martin Christ Gefriertrocknungsanlagen GmbH, Osterode am Harz, Germany) under the pressure of 63 Pa and at a shelf temperature of $20\,^{\circ}\mathrm{C}$.

2.2.2. Vacuum-Drying

The gel cylinders (only variant P40%DE-MC) were also dehydrated on a tray in a cabin-vacuum-drier (Conbest, Cracow, Poland) at the temperature of $50\,^{\circ}\mathrm{C}$ and under the pressure of 10 kPa with the tray load of 3.0 kg·m^{-2}. The dryer was equipped with a weighing system (Mensor, Warsaw, Poland), which enabled the control of the mass of the dried material. The material was dried to obtain the constant mass of the product.

2.2.3. Air-Drying

The gel samples (only variant P40%DE-MC) were air-dried with the application of a prototype laboratory convection dryer. The airstreams at a temperature of $70\,^{\circ}\mathrm{C}$ flowed parallel to the material layer with a speed of 1.5 m/s. The tray load was the same as for the vacuum-drying. The gels were dried to equilibrium moisture content.

2.3. Sorption Isotherms

The obtained dried gels were stored in desiccators containing calcium chloride for 1 month to equilibrate the water content in the material. The different saturated salt solutions (LiCl, CH$_3$COOK, MgCl$_2$, K$_2$CO$_3$, Mg(NO$_3$)$_2$, NaNO$_2$, NaCl, (NH$_4$)$_2$SO$_4$) and anhydrous calcium chloride (CaCl$_2$) were prepared to obtain the water activity in the range of 0–0.810. To avoid the microbial growth in dried samples, a small amount of thymol was placed in desiccators. The triplicate samples of the same variant of dried gels with known weight were placed in open weighing dishes in glass sorption jars with different water activity and stored for 3 months at a temperature of $25\,^{\circ}\mathrm{C}$. The mass of the samples was controlled every month to obtain the equilibrium value. The water content of the samples after storage was calculated based on the change in the measured initial and final (at equilibrium state) weight of the dried gels.

Table 2 presents the applied models for sorption isotherms used to predict the water sorption characteristics of dried gels. The regression analysis was performed with the application of the Table Curve 2D programme (Systat Software Inc., San Jose, CA, USA). The determination coefficient (R^2), the root mean square error (RMS), and parameter P were used to estimate the compliance of the model with empirical data (Table 2).

Table 2. List of sorption isotherm models.

Model	Equation		Description
GAB [37]	$u = \dfrac{u_m C k a_w}{[(1 - k a_w)[1 - k a_w + C k a_w]}$	(1)	u—equilibrium water content g/g $_{d.m.}$, u_m—monolayer moisture content (g/g $_{d.m.}$); C, k—constants, a_w—water activity
BET [38]	$u = \dfrac{u_m c a_w}{(1 - a_w)[1 + (c - 1)a_w]}$	(2)	c—constant

Table 2. *Cont.*

Model	Equation		Description
Lewicki [39]	$u = \dfrac{F}{(1 - a_w)^G} - \dfrac{F}{1 + a_w{}^H}$	(3)	F, G, H—constants
Peleg [40]	$u = A a_w{}^D + B a_w{}^E$	(4)	A, B, D, E—constants
Hasley [41]	$a_w = exp\left(\dfrac{-g}{u^n}\right)$	(5)	g, n—constants
Statistical parameters			
'RMS	$RMS = \sqrt{\dfrac{\sum_{i=1}^{N}\left(\dfrac{u_e - u_p}{u_e}\right)^2}{N}} \, 100\%$	(6)	
P	$P = \dfrac{\sum_{i=1}^{N}\left\lvert\dfrac{u_e - u_p}{u_e}\right\rvert}{N} \, 100\%$	(7)	u_e—an experimental value of water content, u_p—the predicted value of water content, N—number of observations

2.4. Water Sorption Kinetics

The water sorption kinetics of all kinds of dried samples were determined according to the procedure described by Jakubczyk, et al. [42]. The sample was weighed automatically (PW-Win software, Radwag, Warsaw, Poland) every minute up to 48 h during storage in a chamber with a relative humidity of 75.3% (saturated NaCl solution). The water uptake (g/gd.m.) by samples was calculated and presented during the time of the sorption experiment as the sorption kinetics curves.

2.5. Glass Transition Temperature

Modulated DSC experiments were carried out using TA Instruments Q200 differential scanning calorimeter (New Castle, DE, USA). Modulated differential scanning calorimetry (MDSC) was used to determine the glass transition temperature of the selected samples (freeze-dried gels: P0%, P25%, P25%DE and air-dried as well as vacuum-dried gels: P40%De-MC) in the range of water activity 0–0.75. The cell was purged with dry nitrogen (at 50 mL/min) and temperature calibration was performed on an empty oven with an application of standard pure indium and distilled water. The sapphire standard was tested for calibration of specific heat capacity. An empty hermetically sealed aluminium pan was used as a reference. The samples of mass 10–15 mg were hermetically sealed in aluminium pans and cooled from room temperature to $-90\,°C$ at $5\,°C/min$, then equilibrated for 5 min. During MDSC procedures, the samples were scanned from -90 to $150\,°C$ at parameters: a constant heating rate of $2\,°C/min$, an amplitude of $\pm1\,°C$, period of modulation -60 s. The obtained data were analysed with respect to the total, reversible and non-reversible heat flow [28,42,43]. TA Instruments Universal Analysis software was applied to determine the onset, midpoint and endpoint of transition. The midpoint of the glass transition was selected as the transition temperature. The mean value and standard deviation of the glass transition were calculated based on three replicates of measurements.

The Gordon-Taylor model [44] and Roos model [45] represented by Equations (8) and (9) were used to describe changes of T_g with water content and water activity. The goodness fit was estimated based on R^2, RMS and parameter P.

$$T_g = \frac{(1 - x_w)T_{gs} + kx_w T_{gw}}{(1 - x_w) + kx_w} \tag{8}$$

where: T_{gs}, T_{gw}, T_g—glass transition temperatures of solids, water, and their mixture, respectively; x_w—mass fraction of water; k—the Gordon–Taylor model parameter.

$$T_g = Aa_w + B \tag{9}$$

where: a_w—water activity; A, B—parameters of the Roos model.

The critical water content x_{wc} and critical water activity a_{wc} were defined as the values of these parameters at which the glass transition temperature was equal to the storage temperature of 25 °C.

A one-way ANOVA test (Tukey's method) was applied to determine the significance of differences among samples at the 95% significance level for constants.

2.6. Microstructure

The inner microstructure of the gel samples after drying (a_w~0.1–0.2) was observed using an environmental scanning electron microscope Quanta 200 ESM (FEI Company, Hillsboro, OR, USA) at an accelerating voltage of 30 kV in a high-vacuum mode. The cylindrical samples were cut lengthwise using a new razor blade each time. The images were taken at magnification in the range from 30 to 220×.

3. Results and Discussion

3.1. Sorption Properties of Dried Apple Puree Gels

The sorption isotherms characterise the relation between water content and the water activity of the product. The selection of the best model describing the sorption isotherms can be a challenge due to the different compositions and complex structure of food materials [46].

Table 3 presents the parameters of analysed models of the sorption isotherms of different dried apple puree snacks. The goodness of fit for models was estimated based on the determination coefficient (R^2), the root mean square error (RMS), and P parameter. The lower values of RMS and P values and the higher R^2, the better goodness of fit [26]. The GAB and Peleg models were described by high values of R^2, which ranged from 0.993 to 0.998 for all dried samples (Table 3). The RMS values varied from 5.99 to 27.59% for GAB model and from 5.79 to 17.64% for Peleg model (Table 3), which can be considered as satisfying. These models were also characterised by P values lower than 12.24%. Lee and Lee [47] as well as Filho et al. [48] stated that the sorption model can be acceptable when the parameter P is lower than 10%. It can be noticed that parameter P of the investigated models was very close to this limit. On the other hand, the Hasley model provided the worst representation of data, with P values ranging from 10.21 to 39.55%.

Table 3. Parameters of sorption isotherms models with goodness fit coefficients (R^2, RMS, P) of freeze-dried (FD), air-dried (AD) and vacuum-dried (VD) gels.

Model	Parameter	FD-P0%	FD-P25%	FD-P25%De	FD-P40%De	FD-P40%DE-MC	AD-P40%DE-MC	VD-P40%DE-MC
GAB	u_m	0.096 ± 0.009	0.074 ± 0.006	0.065 ± 0.005	0.066 ± 0.005	0.079 ± 0.008	0.112 ± 0.021	0.106 ± 0.035
	C	5.351 ± 0.941	2.904 ± 0.625	2.422 ± 0.308	2.493 ± 0.266	2.306 ± 0.412	0.689 ± 0.098	1.224 ± 0.197
	k	0.747 ± 0.022	0.999 ± 0.015	0.989 ± 0.011	1.020 ± 0.008	0.995 ± 0.019	0.943 ± 0.029	0.925 ± 0.018
	R^2	0.993	0.995	0.996	0.995	0.995	0.997	0.996
	RMS	5.99	15.57	14.48	13.02	10.06	27.59	11.26
	P	4.48	10.34	9.97	9.46	6.72	12.24	7.35
BET	u_m	0.068 ± 0.004	0.084 ± 0.009	0.080 ± 0.006	0.073 ± 0.009	0.105 ± 0.018	0.135 ± 0.022	0.072 ± 0.005
	c	7.138 ± 0.780	2.108 ± 0.613	1.514 ± 0.390	2.107 ± 0.711	1.202 ± 0.357	0.462 ± 0.105	2.997 ± 0.464
	R^2	0.986	0.961	0.973	0.948	0.974	0.993	0.978
	RMS	6.29	15.48	13.43	24.23	15.29	25.64	26.06
	P	4.92	12.29	10.33	21.33	12.18	14.20	19.59
Lewicki	F	1.598 ± 0.234	1.732 ± 0.828	1.302 ± 0.538	1.034 ± 0.286	1.535 ± 0.602	0.744 ± 0.094	0.350 ± 0.261
	G	0.089 ± 1.541	0.092 ± 0.041	0.101 ± 0.039	0.130 ± 0.032	0.102 ± 0.102	0.135 ± 0.080	0.341 ± 0.132
	H	282.20 ± 5.03	14.596 ± 2.171	14.087 ± 1.859	10.516 ± 1.386	13.458 ± 1.812	6.905 ± 0.483	6.632 ± 1.171
	R^2	0.852	0.995	0.996	0.995	0.995	0.994	0.992
	RMS	28.61	12.95	13.82	12.89	11.45	43.52	17.31
	P	22.88	8.59	9.17	9.14	7.91	18.16	10.97
Peleg	A	0.160 ± 0.011	0.995 ± 0.124	1.122 ± 0.111	0.845 ± 0.029	1.672 ± 0.031	0.741 ± 0.071	0.574 ± 0.021
	B	2.774 ± 0.086	8.642 ± 0.975	10.897 ± 1.412	6.813 ± 0.980	11.999 ± 1.076	7.089 ± 0.881	4.559 ± 0.467
	D	0.142 ± 0.021	0.258 ± 0.025	0.256 ± 0.018	0.184 ± 0.039	0.325 ± 0.034	0.226 ± 0.035	0.135 ± 0.029
	E	0.620 ± 0.015	1.226 ± 0.095	1.420 ± 0.101	1.063 ± 0.147	1.499 ± 0.132	1.569 ± 0.154	0885 ± 0.147
	R^2	0.994	0.995	0.996	0.995	0.995	0.998	0.998
	RMS	5.79	11.68	12.50	12.62	17.64	14.73	15.53
	P	4.37	8.18	8.64	9.25	11.77	8.19	7.93
Hasley	g	0.019 ± 0.001	0.078 ± 0.002	0.067 ± 0.002	0.082 ± 0.002	0.083 ± 0.002	0.085 ± 0.009	0.076 ± 0.003
	n	1.602 ± 0.035	0.973 ± 0.022	0.965 ± 0.023	0.904 ± 0.019	0.949 ± 0.027	0.829 ± 0.021	0.954 ± 0.022
	R^2	0.975	0.993	0.993	0.994	0.992	0.992	0.991
	RMS	17.60	34.27	36.99	27.04	20.30	94.24	37.23
	P	10.21	16.79	17.58	14.78	12.63	39.55	16.55

The GAB equation has been frequently applied to describe the sorption isotherms of many products due to the physical meaning of the model, which facilitates the interpretation of obtained results [38,49,50]. However, the constant C of the GAB model should be in the range of $5.67 \leq C \leq \infty$ because outside of this region the monolayer capacity is estimated with an error larger than 15.5% [51]. Table 3 shows that the values of C constant for GAB model did not fulfil this requirement in the case of investigated dried gels. For this reason, the Peleg model was selected to describe the sorption isotherms of apple dried snacks, since it presented the highest R^2 and the lowest RMS and P values. The Peleg model was also the most appropriate for the representation of sorption behaviour of freeze-dried avocado with maltodextrin and inulin [52], freeze-dried vegetable soups [53], as well as apple puree powders [42].

Figure 1 shows the sorption isotherms plots as experimental data and the Peleg fitted model for freeze-dried apple gels with different compositions (Figure 1a) and P40%DE-MC snacks obtained by the application of different drying methods (Figure 1b). The sorption curves obtained for the freeze-dried agar gel without other ingredients (P0%) and the vacuum-dried foamed gel (P40%De-MC) were sigmoidal in shape, which is characteristic for type II isotherms. The application of the procedure described by Blahovec and Yanniotis [54] based on nonlinear regression analysis of the data and using the plot of a_w/w vs. a_w (plots are not shown) enabled confirming that these isotherms represented type II. Additionally, the values of the c constant of the BET model obtained for these dried samples were higher than 2, which also indicated the isotherm type II. However, the BET model can be only applied to predict sorption behaviour at the water activity range lower than 0.5 [55]. In the case of isotherms type II, the interactions between water (sorbate) and food material (sorbent) are stronger than between sorbate–sorbate molecules. The freeze-dried agar gels showed a higher hygroscopicity at low water activity, which may affect the softer texture of the material. The material seemed to behave as a sponge.

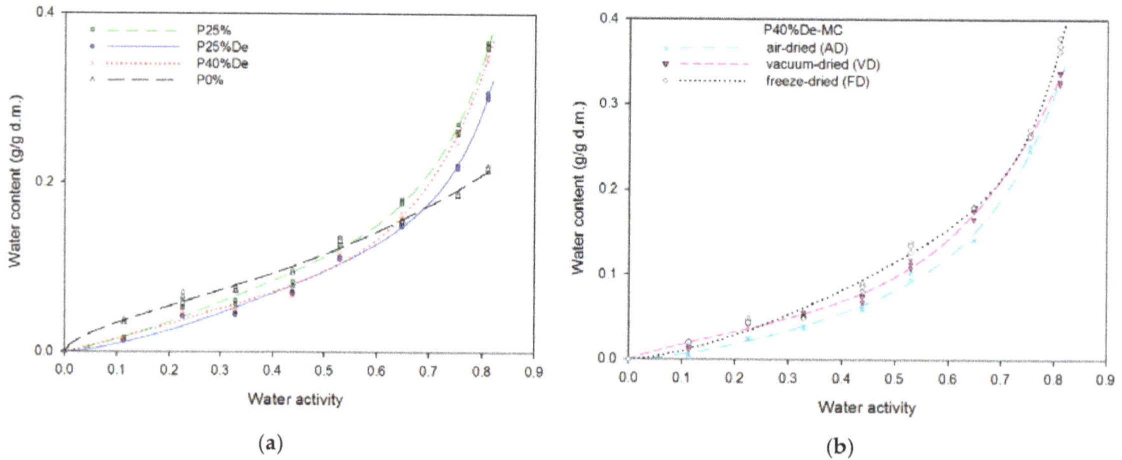

Figure 1. Water sorption isotherms of dried gels: (**a**) freeze-dried (non-foamed) gels with different compositions; (**b**) P40%DE-MC gels dried by AD, VD and FD techniques.

The foamed gel dried by the vacuum-drying method adsorbed a higher amount of moisture than the freeze-dried and the air-dried gels with the same composition at low water activities ($a_w < 0.2$) (Figure 1b). The vacuum-dried gels were probably characterised by more developed surface area as a result of foaming of material as well as expansion of matrix during the dehydration. However, this material (AD-P40%DE-MC) adsorbed less amount of water than freeze-dried gel (FD-P40%DE-MC) in the range of water activity from 0.20 to 0.65. It means that vacuum-dried gels were more stable products with a lower sorption capacity during the storage (in the analysed water activity range) than lyophilised material.

The other dried gels showed type III behaviour according to the Brunauer–Emmet–Teller (BET) classification [56]. These isotherms presented a convex shape without an inflection point (Figure 1). This behaviour is typical for products with high sugar content and for most fruits. The water content increased linearly with water activity in the range of low and intermediate a_w, this is the so-called multilayer sorption region. The capillary condensation region is characteristic at high water activities, then the rapid increase in water content with a_w can be observed due to sugar dissolution and solute–solvent interactions [57]. The slight sigmoid shape of the isotherm at low water activity is caused by the water sorption by biopolymers, but the sharp increase in water content at high water activities is caused by the presence of sugars [30].

The drying method affected the sorption properties of foamed gel with the same composition (Figure 1b). The isotherms type III were observed for the freeze-dried and the air-dried samples. Air-dried gel was characterised by the lowest moisture content in the investigated range of a_w. It may indicate the presence of less porous structure with closed pores of material obtained by convective drying. This structure of air-dried gels may affect the reduced water sorption by material. Tsami et al. [30] and Lee and Lee [47] concluded the freeze-drying created a porous structure and little shrinkage of products with a higher adsorptive capacity than other drying techniques. However, the addition of different amounts of apple puree and maltodextrin as well as the foaming process modified the sorption properties of dried gels. The foamed gel with the same composition absorbed a higher amount of moisture than non-foamed samples at the intermediate water activities ranging from 0.4 to 0.7. More porous structure of foamed and dried gels may enhance their hygroscopicity.

The composition, structure and size of molecules of food products may affect their sorption behaviour. The addition of maltodextrin (P25%, P25% De) caused the decrease in adsorbed water for the entire range of a_w. Water content of the material significantly decreased during water sorption when the crystallisation of solutes occurred [58]. The freeze-dried apple puree gels with the addition of maltodextrin were probably in amorphous state during the moisture sorption. Additionally, the presence of maltodextrin increased the glass transition temperature and maintained lower hygroscopicity of dried gels.

The increase in the apple addition to freeze-dried agar gels from 25 to 40% for samples with maltodextrin (P25%De, P40%De) caused the significant increase in water content at a given water activity (Figure 1a). It can be linked with the higher concentration of low molecular substances (sugars such as sucrose, fructose and glucose and organic acids) in gels that contained a larger amount of apple puree.

Figure 2 shows the water vapour sorption kinetics of the dried apple puree snacks. Higher water sorption was observed for apple puree dried gel prepared without maltodextrin. A similar effect of maltodextrin addition was noted for lemon juice powder [31] and apple leather [59]. The air-dried and vacuum-dried apple puree gels absorbed less water than freeze-dried samples. The shapes of kinetics curves for all freeze-dried gels were similar, and the samples reached an equilibrium of water uptake after 800 min. However, the amount of absorbed water differed with the composition of gels. Higher addition of apple puree caused the increase in water uptake of freeze-dried samples. Air-dried and vacuum-dried gels were less hygroscopic than freeze-dried gels with apple puree, which is in agreement with the results obtained for water sorption isotherms. The highest rate of water uptake was noted for freeze-dried pure agar gels because the material reached the equilibrium water content after 250 min. The kinetics of water uptake explains the changes of chemical properties during the time of storage with consideration of the rate of change. This type of measurement provides useful information during packaging, which enables estimating the time before the texture and colour of the product become unacceptable at a given relative humidity [59]. It is possible to observe the sorption behaviour of products after opening the packaging.

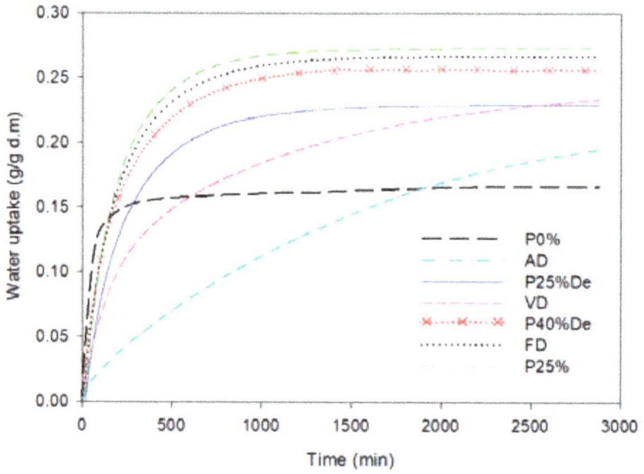

Figure 2. Effect of composition of gels and drying method on the water kinetics uptake of apple puree snacks.

3.2. Effect of Water Activity on the Glass Transition Temperature of Dried Gels

The water activity and glass transition temperature are crucial parameters in the evaluation of changes occurring in food products. Sugar recrystallisation is the main problem of stored products that contain low-molecular sugars in an amorphous state. Dried fruit products are commonly in an amorphous state. These materials are not stable due to a lack of thermodynamic equilibrium [24].

Values of glass transition temperature T_g obtained for selected dried apple puree gels and freeze-dried agar gel as a function of equilibrium water content are shown in Figure 3a. The experimental data and the Gordon–Taylor fitted model were presented. The glass transition temperature decreased with increasing water content due to the plasticising effect of water. Additionally, the addition of maltodextrin to apple puree gels caused the increase in T_g of the freeze-dried sample. A similar effect of a carrier on the T_g of different dried products was observed for mango powder [60], apple puree powder [42] and date syrup powder [61]. The highest glass transition temperature at the investigated range of water content was observed for freeze-dried agar gel without the addition of apple puree. The application of the air-dried method enabled obtaining apple snacks with slightly higher values of T_g in comparison with other dried apple puree gels. It means that the air-dried product will be more stable during the storage because the phase transitions occur at higher water activity at room conditions than for other dried gels. In addition, based on the sorption behaviour, it can be stated that the air-dried gel was less hygroscopic than other materials.

The parameter T_{gs} of the Gordon–Taylor model (Equation (8)) showed that the drying method of gels with the same composition (P40%DE) did not affect the T_{gs} values, but the k parameter obtained for the vacuum-dried sample was significantly higher than for the air-dried gel (Table 4). The k parameter is related to the plasticising effect of water. Higher value of k can be linked with a decrease in the glass transition temperature of the sample [60,62]. Lower susceptibility to water sorption of air-dried gels can be linked to the higher glass transition temperature. The increase in pulp addition caused a slight increase in T_{gs} (Table 4). Fongin et al. [60] stated that the glass transition temperature of anhydrous sample T_{gs} should be independent of pulp content in freeze-dried mango powder because the solid matrix of pulp is intrinsically crystalline. However, the increase in apple pulp addition in freeze-dried gels caused the increase in low molecular sugars and their presence affected a slight decrease in glass transition temperature. The addition of maltodextrin and a lower amount of incorporated pulp affected a fabrication of dried gels with higher stability (the possibility of structure collapse and the changes in texture of the product can be limited).

Table 4. Parameter of Ross and Gordon–Taylor models and critical values of moisture content (x_{wc}) and water activity a_{wc} at a temperature of 25 °C.

	Parameters of Model	Variants of Dried Gels				
		(FD) P0%	(FD) P25%	(FD) P25%De	(VD) P40%DE-MC	(AD) P40%DE-MC
Gordon-Taylor model	T_{gs}	132.32 [d],*	54.39 [a]	61.81 [b]	64.24[c]	64.58 [c]
	k	2.97 [a]	5.76 [c]	4.92 [a]	5.71 [c]	4.85 [a]
	R^2	0.989	0.997	0.995	0.992	0.992
	RMS	4.57	13.13	10.98	16.16	13.76
	P	3.48	7.58	8.79	11.50	4.85
Ross model	A	123.82 [d]	52.21 [a]	63.82 [b]	64.72 [b]	74.19 [c]
	B	−119.96 [d]	−146.48 [b]	−132.97 [c]	−152.50 [a]	−143.26 [b]
	R^2	0.965	0.985	0.993	0.988	0.991
	RMS	8.53	38.67	16.27	29.71	18.63
	P	6.83	20.12	11.74	19.54	14.54
T_g = 25 °C	x_{wc} (g/g$_{dm}$)	0.225 [a]	0.032 [b]	0.047 [c]	0.042 [c]	0.045 [c]
	a_{wc}	0.837 [e]	0.181 [a]	0.301 [c]	0.260 [b]	0.353 [d]

* the different letters in the rows indicates the significant difference between the obtained values for samples, $p \leq 0.05$.

The relation between the water activity and glass transition temperature can be described by Roos model (Figure 3b). However, the calculated RMS, R^2 and P values showed that the experimental data obtained in this study did not fit well for the Roos equation (Equation (9)).

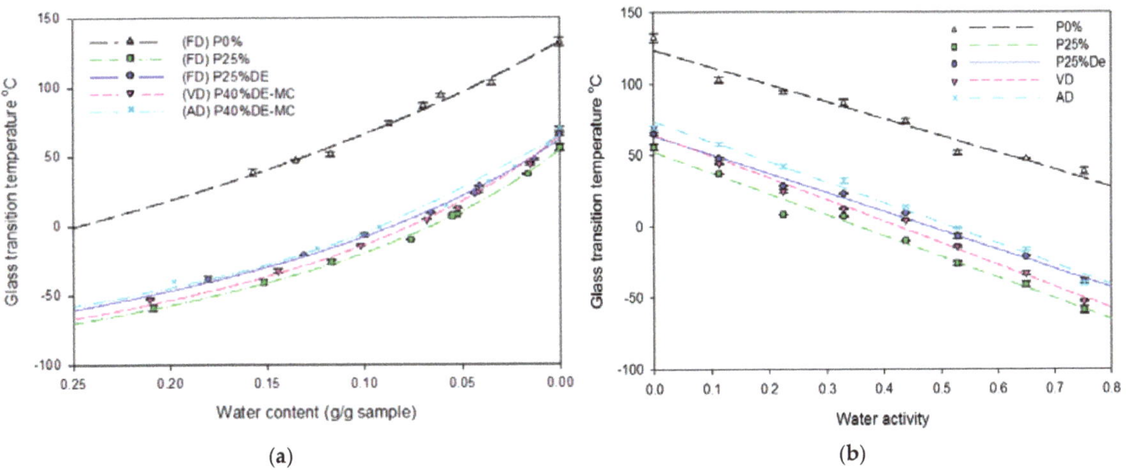

Figure 3. Changes of the glass transition temperature of selected dried gels to (**a**) water content; (**b**) water activity.

A higher value of critical water content and critical water activity at the glass transition temperature of 25 °C obtained from the relations of water content-water activity and water content-glass transition temperature may indicate the greater stability of physical properties of products during storage. Dried apple puree gels with maltodextrin were characterised by a significantly higher a_{wc} than dried gel without carrier addition. A similar tendency has been observed for grapefruit and mango powders [20,60]. In the case of the air-dried apple puree, higher critical values of water activity than for other dried gel with apple puree addition were observed. The results indicate that the freeze-dried agar gel (without apple puree addition) stored at room temperature will change its state from glassy to rubbery at a high-water activity of 0.837. The air-dried sample will be stable at water activity lower than 0.353.

3.3. Microstructure of Dried Gels

Figure 4 presents the structure of dried samples of different compositions or produced with the application of different drying techniques.

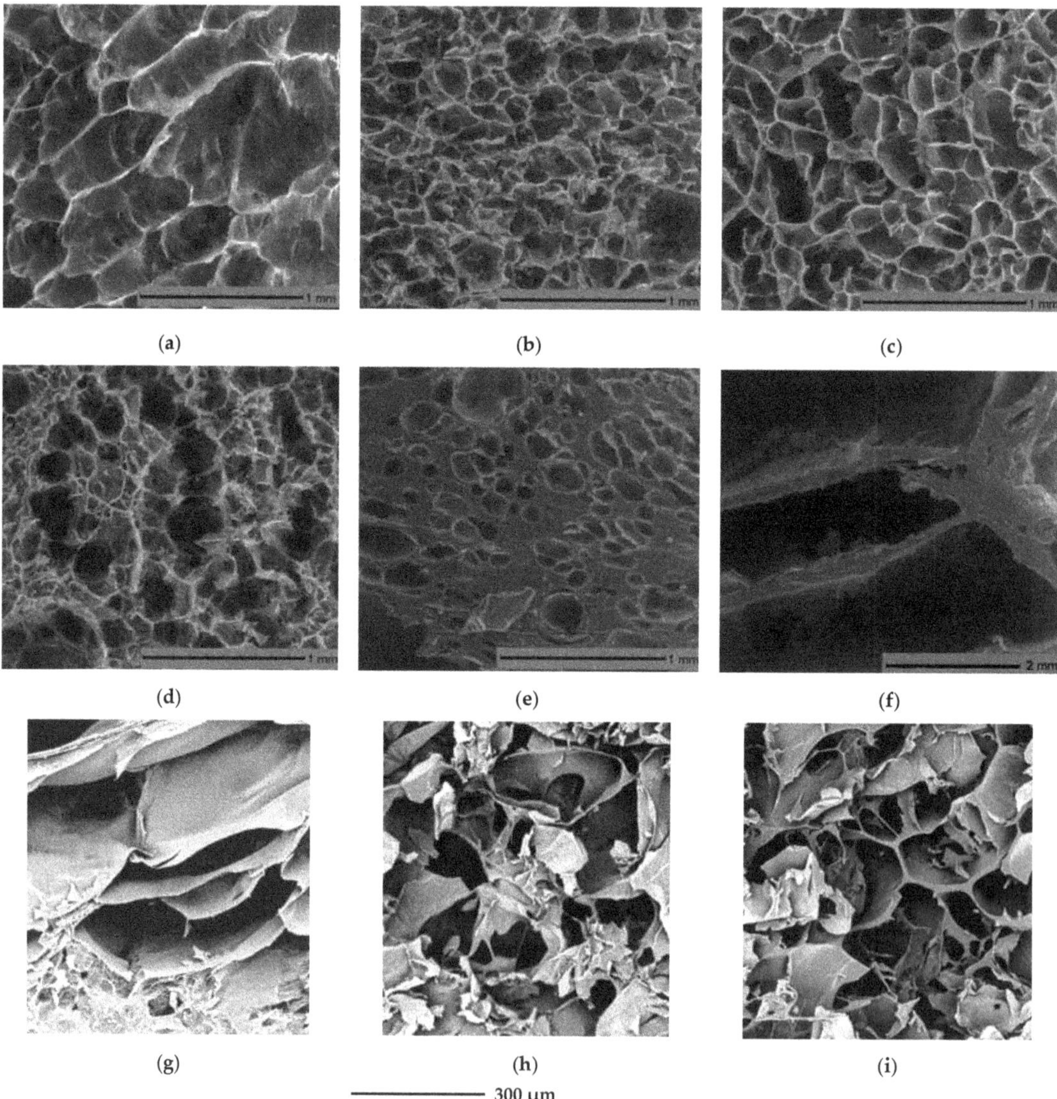

Figure 4. SEM views of different freeze-dried gels; (**a,g**) FD-P0%; (**b,h**) FD-P25%; (**c,i**)-FD- P40%DE; (**d**)-FD-P40%DE-MC; (**e**)-AD-P40%DE-MC; (**f**) VD-P40%DE-MC.

The addition of apple puree to agar gels changed the structure of freeze-dried products (Figure 4a,b). Long pores (Figure 4a,g) with thin walls can be observed for dried agar gels (without other ingredients). Freeze-dried samples with 25% addition of apple puree (Figure 4b) contained more round pores. An increase in puree concentration in gel led to obtaining freeze-dried samples with larger regular pores, but higher concentration of solids contributed to producing the matrix walls with a higher thickness (Figure 4c,i). Aeration of agar sol with the addition of apple puree affected the structure after the freeze-drying of the samples (Figure 4d). These dried gels were very porous and the thickness of pores' walls was reduced. Ciurzyńska and Lenart [63] also obtained the freeze-dried hydrocolloid gel, which was aerated. The structure of materials was very fragile and porous. Based on Figure 4e, it can be stated that air-dried apple puree gel was less porous with more visible shrinkage and contained more closed pores. Application of vacuum-drying of apple gel caused larger holes inside the material with thick walls of pores (Figure 4f). Sundaram and Durance [64] also noticed that the air-drying and the vacuum-drying of locust bean gum-pectin-starch composite gel resulted in samples with big holes. However, after the vacuum-drying, the volume of the sample was increased due to puffing. In our study, the same phenomenon was also observed for the vacuum-dried apple puree gel.

4. Conclusions

The nonlinear regression analysis showed that the Peleg model was adequate to describe the water sorption isotherms of all dried apple puree products as well as freeze-dried agar gel. The water adsorption isotherms of most dried gels represented type III sorption behaviour, which is common for a high sugar food. It can be linked with an increase in sugar concentration in gels due to the addition of apple puree. The different shapes of the sorption isotherm (represented type II) of the freeze-dried agar gel (without apple puree) was a result of the low concentration of solids in the material. The higher amount of adsorbed water at lower water activity was observed for the vacuum-dried apple puree gel, which can be related to expanded structure with large pores. However, the thick walls of pores created during the vacuum-drying impeded the hygroscopicity of the material. Low porosity and thick walls of pores in air-dried gels affected the lower moisture sorption and the increase in glass transition temperature. The open-pore structure and the presence of many pores with thin walls in freeze-dried apple gels led to the higher hygroscopicity of samples. The application of this drying method also caused a decrease in glass transition temperature and critical water activity of studied materials. Additionally, freeze-dried gel without maltodextrin was characterised by the low value of glass transitions, which may indicate the presence of the amorphous state. It may lead to the undesirable changes of texture (collapse of structure). The increase in apple puree addition gels from 25 to 40% caused the increase in hygroscopicity during the storage, which can be linked with the higher concentration of low molecular substances. This study shows that the composition of gel and the application of certain drying methods enabled obtaining products with tailored structure and sorption properties. The air-dried apple puree gels can be recommended as the snack's products with the lowest hygroscopicity and the highest stability at room conditions (the highest glass transition temperature and the critical water activity). In addition, a decrease in the amount of apple puree from 40 to 25% in dried gels improved their sorption properties.

Author Contributions: Conceptualisation, E.J.; methodology, E.J.; investigation, E.J., A.K.-D., E.O.-L., A.G. (Agata Górska), M.W.-W., D.M.-J., A.G. (Agnieszka Górska); formal analysis, E.J.; writing—original draft preparation, E.J.; writing—review and editing, E.J., A.G. (Agata Górska) and J.B. All authors have read and agreed to the published version of the manuscript.

Funding: This research received no external funding.

Institutional Review Board Statement: Not applicable.

Informed Consent Statement: Not applicable.

Data Availability Statement: The data generated or analysed during this study are available from the corresponding author on reasonable request.

Conflicts of Interest: The authors declare no conflict of interest.

References

1. Hess, J.M.; Jonnalagadda, S.S.; Slavin, J.L. What is a snack, why do we snack, and how can we choose better snacks? A review of the definitions of snacking, motivations to snack, contributions to dietary intake, and recommendations for improvement. *Adv. Nutr.* **2016**, *7*, 466–475. [CrossRef]
2. Bellisle, F. Meals and snacking, diet quality and energy balance. *Physiol. Behav.* **2014**, *134*, 38–43. [CrossRef]
3. Nussinovitch, A.; Jaffe, N.; Gillilov, M. Fractal pore-size distribution on freeze-dried agar-texturized fruit surfaces. *Food Hydrocoll.* **2004**, *18*, 825–835. [CrossRef]
4. Tiwari, S.; Ravi, R.; Bhattacharya, S. Dehumidifier assisted drying of a model fruit pulp-based gel and sensory attributes. *J. Food Sci.* **2012**, *77*, 263–273. [CrossRef] [PubMed]
5. Ozcelik, M.; Heigl, A.; Kulozik, U.; Ambros, S. Effect of hydrocolloid addition and microwave-assisted freeze drying on the characteristics of foamed raspberry puree. *Inn. Food Sci. Emerg.Technol.* **2019**, *56*, 102183. [CrossRef]
6. Martínez-Navarrete, N.; Salvador, A.; Oliva, C.; Camacho, M.M. Influence of biopolymers and freeze-drying shelf temperature on the quality of a mandarin snack. *LWT* **2019**, *99*, 57–61. [CrossRef]
7. Egas-Astudillo, L.A.; Martínez-Navarrete, N.; Camacho, M.M. Impact of biopolymers added to a grapefruit puree and freeze-drying shelf temperature on process time reduction and product quality. *Food Bioprod. Proc.* **2020**, *120*, 143–150. [CrossRef]
8. Llavata, B.; García-Pérez, J.V.; Simal, S.; Cárcel, J.A. Innovative pre-treatments to enhance food drying: A current review. *Curr. Opin. Food Sci.* **2020**, *35*, 20–26. [CrossRef]
9. Menon, A.; Stojceska, V.; Tassou, S.A. A systematic review on the recent advances of the energy efficiency improvements in non-conventional food drying technologies. *Trends Food Sci. Technol.* **2020**, *100*, 67–76. [CrossRef]
10. Drouzas, A.; Tsami, E.; Saravacos, G. Microwave/vacuum drying of model fruit gels. *J. Food Eng.* **1999**, *39*, 117–122. [CrossRef]
11. Ratti, C. Hot air and freeze-drying of high-value foods: A review. *J. Food Eng.* **2001**, *49*, 311–319. [CrossRef]
12. Harnkarnsujarit, N.; Kawai, K.; Suzuki, T. Impacts of freezing and molecular size on structure, mechanical properties and recrystallization of freeze-thawed polysaccharide gels. *LWT* **2016**, *68*, 190–201. [CrossRef]
13. Nowak, D.; Jakubczyk, E. The freeze-drying of foods-the characteristic of the process course and the effect of its parameters on the physical properties of food materials. *Foods* **2020**, *9*, 1488. [CrossRef]
14. Hawlader, M.N.A.; Perera, C.O.; Tian, M.; Yeo, K.L. Drying of guava and papaya: Impact of different drying methods. *Dry. Technol.* **2006**, *24*, 77–87. [CrossRef]
15. Araya-Farias, M.; Makhlouf, J.; Ratti, C. Drying of seabuckthorn (*Hippophae rhamnoides* L.) berry: Impact of dehydration methods on kinetics and quality. *Dry. Technol.* **2011**, *29*, 351–359. [CrossRef]
16. Bhatta, S.; Janezic, T.S.; Ratti, C. Freeze-drying of plant-based foods. *Foods* **2020**, *9*, 87. [CrossRef] [PubMed]
17. Ozcelik, M.; Ambros, S.; Morais, S.I.F.; Kulozik, U. Storage stability of dried raspberry foam as a snack product: Effect of foam structure and microwave-assisted freeze drying on the stability of plant bioactives and ascorbic acid. *J. Food Eng.* **2020**, *270*. [CrossRef]
18. Silva, M.A.; Sobral, P.J.A.; Kieckbusch, T.G. State diagrams of freeze-dried camu-camu (Myrciaria dubia (HBK) Mc Vaugh) pulp with and without maltodextrin addition. *J. Food Eng.* **2006**, *77*, 426–432. [CrossRef]
19. Silva-Espinoza, M.A.; Ayed, C.; Foster, T.; Camacho, M.D.M.; Martínez-Navarrete, N. The impact of freeze-drying conditions on the physico-chemical properties and bioactive compounds of a freeze-dried orange puree. *Foods* **2019**, *9*, 32. [CrossRef]
20. Telis, V.R.N.; Martínez-Navarrete, N. Collapse and color changes in grapefruit juice powder as affected by water activity, glass transition, and addition of carbohydrate polymers. *Food Biophys.* **2009**, *4*, 83–93. [CrossRef]
21. Roos, Y.H. Glass transition temperature and its relevance in food processing. *Annu. Rev. Food Sci. Technol.* **2010**, *1*, 469–496. [CrossRef] [PubMed]
22. Slade, L.; Levine, H. Beyond water activity—recent advances based on an alternative approach to the assessment of food quality and safety. *Crit. Rev. Food. Sci. Nutr.* **1991**, *30*, 115–360. [CrossRef]
23. Telis, V.R.N.; Sobral, P.J.A. Glass transitions and state diagram for freeze-dried pineapple. *LWT* **2001**, *34*, 199–205. [CrossRef]
24. Roos, Y.H.; Drusch, S. *Phase Transitions in Foods*; Academic Press: Oxford, UK, 2015; p. 380.
25. Gabas, A.L.; Telis, V.R.N.; Sobral, P.J.A.; Telis-Romero, J. Effect of maltodextrin and arabic gum in water vapor sorption thermodynamic properties of vacuum dried pineapple pulp powder. *J. Food Eng.* **2007**, *82*, 246–252. [CrossRef]
26. Mrad, N.D.; Bonazzi, C.; Boudhrioua, N.; Kechaou, N.; Courtois, F. Moisture sorption isotherms, thermodynamic properties, and glass transition of pears and apples. *Dry. Technol.* **2012**, *30*, 1397–1406. [CrossRef]
27. Witczak, T.; Witczak, M.; Socha, R.; Stępień, A.; Grzesik, M. Candied orange peel produced in solutions with various sugar compositions: Sugar composition and sorption properties of the product. *J. Food Proc. Eng.* **2017**, *40*, e12367. [CrossRef]
28. Ostrowska-Ligęza, E.; Jakubczyk, E.; Gorska, A.; Wirkowska, M.; Brys, J. The use of moisture sorption isotherms and glass transition temperature to assess the stability of powdered baby formulas. *J. Therm. Anal. Calorim.* **2014**, *118*, 911–918. [CrossRef]
29. Goula, A.M.; Karapantsios, T.D.; Achilias, D.S.; Adamopoulos, K.G. Water sorption isotherms and glass transition temperature of spray dried tomato pulp. *J. Food Eng.* **2008**, *85*, 73–83. [CrossRef]

30. Tsami, E.; Krokida, M.; Drouzas, A. Effect of drying method on the sorption characteristics of model fruit powders. *J. Food Eng.* **1998**, *38*, 381–392. [CrossRef]

31. Martinelli, L.; Gabas, A.L.; Telis-Romero, J. Thermodynamic and quality properties of lemon juice powder as affected by maltodextrin and arabic gum. *Dry. Technol.* **2007**, *25*, 2035–2045. [CrossRef]

32. Tiwari, S.; Chakkaravarthi, A.; Bhattacharya, S. Imaging and image analysis of freeze-dried cellular solids of gellan and agar gels. *J. Food Eng.* **2015**, *165*, 60–65. [CrossRef]

33. Cassanelli, M.; Norton, I.; Mills, T. Role of gellan gum microstructure in freeze drying and rehydration mechanisms. *Food Hydrocoll.* **2018**, *75*, 51–61. [CrossRef]

34. Ciurzyńska, A.; Jasiorowska, A.; Ostrowska-Ligęza, E.; Lenart, A. The influence of the structure on the sorption properties and phase transition temperatures of freeze-dried gels. *J. Food Eng.* **2019**, *252*, 18–27. [CrossRef]

35. Banerjee, S.; Ravi, R.; Bhattacharya, S. Textural characterisation of gellan and agar based fabricated gels with carrot juice. *LWT* **2013**, *53*, 255–261. [CrossRef]

36. Armisen, R.; Gaiatas, F. Agar. In *Handbook of hydrocolloids*; Phillips, G.O., Williams, P.A., Eds.; Elsevier: Cambridge, UK, 2009; pp. 82–107.

37. Bizot, H. Using the 'GAB' model to construct sorption isotherms. In *Physical Properties of Foods*; Jowitt, R., Escher, F., Hällström, B., Meffert, H.F.T., Spiess, W.E.L., Vos, G., Eds.; Applied Science Publishers: London, UK, 1983; pp. 43–54.

38. Timmermann, E.O.; Chirife, J.; Iglesias, H.A. Water sorption isotherms of foods and foodstuffs: BET or GAB parameters? *J. Food Eng.* **2001**, *48*, 19–31. [CrossRef]

39. Lewicki, P.P. A three parameter equation for food moisture sorption isotherms. *J. Food Process Eng.* **1998**, *21*, 127–144. [CrossRef]

40. Peleg, M. Assessment of a semi-empirical four parameter general model for sigmoid moisture sorption isotherms. *J. Food Eng.* **1993**, *16*, 21–27. [CrossRef]

41. Iglesias, H.A.; Chirife, J. A model for describing the water sorption behavior of foods. *J. Food Sci.* **1976**, *41*, 984–992. [CrossRef]

42. Jakubczyk, E.; Ostrowska-Ligęza, E.; Gondek, E. Moisture sorption characteristics and glass transition temperature of apple puree powder. *Int. J. Food Sci. Technol.* **2010**, *45*, 2515–2523. [CrossRef]

43. Rahman, M.S.; Al-Marhubi, I.M.; Al-Mahrouqi, A. Measurement of glass transition temperature by mechanical (DMTA), thermal (DSC and MDSC), water diffusion and density methods: A comparison study. *Chem. Phys. Lett.* **2007**, *440*, 372–377. [CrossRef]

44. Gordon, M.; Taylor, J.S. Ideal copolymers and the second-order transitions of synthetic rubbers. I. Non-crystalline copolymers. *J. Appl. Chem.* **1952**, *2*, 493–500. [CrossRef]

45. Roos, Y.H. Effect of moisture on the thermal behavior of strawberries studied using differential scanning calorimetry. *J. Food Sci.* **1987**, *52*, 146–149. [CrossRef]

46. Raharitsifa, N.; Ratti, C. Foam-mat freeze-drying of apple juice part 2: Stability of dry products during storage. *J. Food Proc. Eng.* **2010**, *33*, 341–364. [CrossRef]

47. Lee, J.H.; Lee, M.J. Effect of drying method on the moisture sorption isotherms for Inonotus obliquus mushroom. *LWT* **2008**, *41*, 1478–1484. [CrossRef]

48. Lopes Filho, J.; Romanelli, P.; Barboza, S.; Gabas, A.; Telis-Romero, J. Sorption isotherms of alligator's meat (*Caiman crocodilus yacare*). *J. Food Eng.* **2002**, *52*, 201–206. [CrossRef]

49. Frabetti, A.C.C.; de Moraes, J.O.; Porto, A.S.; da Silva Simão, R.; Laurindo, J.B. Strawberry-hydrocolloids dried by continuous cast-tape drying to produce leather and powder. *Food Hydrocoll.* **2021**, *121*, 107041. [CrossRef]

50. Rodríguez-Bernal, J.; Flores-Andrade, E.; Lizarazo-Morales, C.; Bonilla, E.; Pascual-Pineda, L.; Gutierréz-López, G.; Quintanilla-Carvajal, M.X. Moisture adsorption isotherms of the borojó fruit (*Borojoa patinoi. Cuatrecasas*) and gum arabic powders. *Food Bioprod. Process.* **2015**, *94*, 187–198. [CrossRef]

51. Lewicki, P.P. The applicability of the GAB model to food water sorption isotherms. *Int. J. Food Sci. Technol.* **1997**, *32*, 553–557. [CrossRef]

52. Stępień, A.; Witczak, M.; Witczak, T. Moisture sorption characteristics of food powders containing freeze dried avocado, maltodextrin and inulin. *Int. J. Biol. Macromol.* **2020**, *149*, 256–261. [CrossRef] [PubMed]

53. Jakubczyk, E.; Jaskulska, A. The effect of freeze-drying on the properties of Polish vegetable soups. *Appl. Sci.* **2021**, *11*, 654. [CrossRef]

54. Blahovec, J.; Yanniotis, S. Modified classification of sorption isotherms. *J. Food Eng.* **2009**, *91*, 72–77. [CrossRef]

55. Moraga, G.; Talens, P.; Moraga, M.; Martínez-Navarrete, N. Implication of water activity and glass transition on the mechanical and optical properties of freeze-dried apple and banana slices. *J. Food Eng.* **2011**, *106*, 212–219. [CrossRef]

56. Brunauer, S.; Deming, L.S.; Deming, W.E.; Teller, E. On a theory of the van der Waals adsorption of gases. *J. Am. Chem. Soc.* **1940**, *62*, 1723–1732. [CrossRef]

57. Djendoubi Mrad, N.; Bonazzi, C.; Boudhrioua, N.; Kechaou, N.; Courtois, F. Influence of sugar composition on water sorption isotherms and on glass transition in apricots. *J. Food Eng.* **2012**, *111*, 403–411. [CrossRef]

58. Fukami, K.; Kawai, K.; Takeuchi, S.; Harada, Y.; Hagura, Y. Effect of water content on the glass transition temperature of calcium maltobionate and its application to the characterization of non-arrhenius viscosity behavior. *Food Biophysics* **2016**, *11*, 410–416. [CrossRef]

59. Valenzuela, C.; Aguilera, J.M. Effects of maltodextrin on hygroscopicity and crispness of apple leathers. *J. Food Eng.* **2015**, *144*, 1–9. [CrossRef]

60. Fongin, S.; Alvino Granados, A.E.; Harnkarnsujarit, N.; Hagura, Y.; Kawai, K. Effects of maltodextrin and pulp on the water sorption, glass transition, and caking properties of freeze-dried mango powder. *J. Food Eng.* **2019**, *247*, 95–103. [CrossRef]
61. Farahnaky, A.; Mansoori, N.; Majzoobi, M.; Badii, F. Physicochemical and sorption isotherm properties of date syrup powder: Antiplasticizing effect of maltodextrin. *Food Bioprod. Proc.* **2016**, *98*, 133–141. [CrossRef]
62. Ahmed, J.; Rahman, M.S. Glass transition in foods. In *Engineering Properties of Foods*; Datta, A.K., Ahmed, J., Rao, M.A., Rizvi, S.S.H., Eds.; CRC Press: Boca Raton, FL, USA, 2014; pp. 93–120.
63. Ciurzyńska, A.; Lenart, A. Effect of the aerated structure on selected properties of freeze-dried hydrocolloid gels. *Int. Agrophys.* **2016**, *30*, 9–17. [CrossRef]
64. Sundaram, J.; Durance, T.D. Water sorption and physical properties of locust bean gum-pectin-starch composite gel dried using different drying methods. *Food Hydrocoll.* **2008**, *22*, 1352–1361. [CrossRef]

Review

Current Status of Optical Systems for Measuring Lycopene Content in Fruits: Review

Marcos-Jesús Villaseñor-Aguilar [1,2,3], José-Alfredo Padilla-Medina [1], José-Enrique Botello-Álvarez [1], Micael-Gerardo Bravo-Sánchez [1], Juan Prado-Olivares [1], Alejandro Espinosa-Calderon [1,4] and Alejandro-Israel Barranco-Gutiérrez [1,5,*]

[1] Doctorado en Ciencias de la Ingeniería, Tecnológico Nacional de México en Celaya, Celaya 38010, Mexico; mvillasenor@upgto.edu.mx (M.-J.V.-A.); alfredo.padilla@itcelaya.edu.mx (J.-A.P.-M.); enrique.botello@itcelaya.edu.mx (J.-E.B.-Á.); gerardo.bravo@itcelaya.edu.mx (M.-G.B.-S.); juan.prado@itcelaya.edu.mx (J.P.-O.); alejandro.espinosa@crodecelaya.edu.mx (A.E.-C.)
[2] Departamento de Ingeniería de Robótica y de Datos, Universidad Politécnica de Guanajuato, Cortazar 38496, Mexico
[3] Facultad de Ingeniería, Universidad de Celaya, Celaya 38080, Mexico
[4] CRODE, Tecnológico Nacional de México en Celaya, Celaya 38010, Mexico
[5] Cátedras-CONACyT, García Cubas esq. Av. Tecnológico, Celaya 38010, Mexico
* Correspondence: israel.barranco@itcelaya.edu.mx

Citation: Villaseñor-Aguilar, M.-J.; Padilla-Medina, J.-A.; Botello-Álvarez, J.-E.; Bravo-Sánchez, M.-G.; Prado-Olivares, J.; Espinosa-Calderon, A.; Barranco-Gutiérrez, A.-I. Current Status of Optical Systems for Measuring Lycopene Content in Fruits: Review. *Appl. Sci.* **2021**, *11*, 9332. https://doi.org/10.3390/app11199332

Academic Editor: Agata Górska

Received: 17 July 2021
Accepted: 24 September 2021
Published: 8 October 2021

Abstract: Optical systems are used for analysing the internal composition and the external properties in food. The measurement of the lycopene content in fruits and vegetables is important because of its benefits to human health. Lycopene prevents cardiovascular diseases, cataracts, cancer, osteoporosis, male infertility, and peritonitis. Among the optical systems focused on the estimation and identification of lycopene molecule are high-performance liquid chromatography (HPLC), the colorimeter, infrared near NIR spectroscopy, UV-VIS spectroscopy, Raman spectroscopy, and the systems of multispectral imaging (MSI) and hyperspectral imaging (HSI). The main objective of this paper is to present a review of the current state of optical systems used to measure lycopene in fruits. It also reports important factors to be considered in order to improve the design and implementation of those optical systems. Finally, it was observed that measurements with HPLC and spectrophotometry present the best results but use toxic solvents and require specialized personnel for their use. Moreover, another widely used technique is colorimetry, which correlates the lycopene content using color descriptors, typically those of CIELAB. Likewise, it was identified that spectroscopic techniques and multispectral images are gaining importance because they are fast and non-invasive.

Keywords: lycopene; optical system; colorimeter; spectroscopy; images; HPLC

1. Introduction

Optical systems in the agriculture and food sector are efficient tools that allow the external quality and characteristics of the internal composition of fruits to be determined, such as the shape, colors, maturity, vitamins, and soluble solids [1–4]. The operation principle of these systems is to associate the variable of interest of the vegetable or fruit matrix with the optical properties, such as the reflectance, transmittance, absorbance, fluorescence, and emission, which can be analyzed and associated with the radiation of light on the sample [5].

The study of the optical properties of each piece of fruit or vegetable is complex because of their non-uniform surface. The relevance of the knowledge of the structure allows identification of how light travels through food. In Figure 1, the light behavior schema on fruit radiated by a light beam is shown. Chen [5] reported that approximately 4% of the radiation incident on the fruit surface is reflected, on the outside, in the regular form of reflectance. The rest of the radiation is transmitted in all directions through the surface and the tissue of the fruit formed by the cellular structure, and the transmitted energy can be absorbed or reach the other surface of the fruit with the same or a different wavelength [6,7].

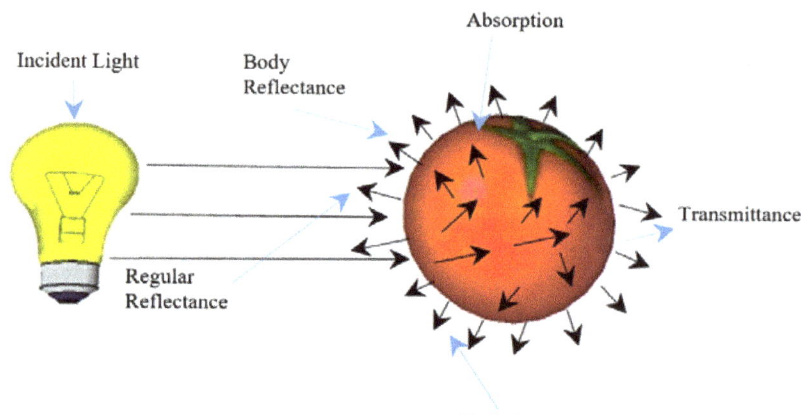

Figure 1. Schematic of the interaction between light and a fruit [5].

The optical properties of the lycopene molecule allow the absorption of ultraviolet and green electromagnetic regions because it has a structure of 40 carbons and eight isopropene units. These characteristics identify the family of carotenoids. The main functions attributed to carotenoids are fruit pigmentation, capture of light, and plant photoprotection [8–12]. The study of lycopene in fruits is very important because it has helpful antioxidant properties, which helps to avoid the generation of unstable molecules and free radicals that damage proteins, DNA, carbohydrates, and fats [13–15]. Some fruits that contain lycopene are tomato, watermelon, guava, grapefruit, papaya, and apricot. In Table 1, it can be observed that the fruits with the highest lycopene content are guava, tomato, watermelon, and papaya [16].

Table 1. Lycopene content of some fruits and vegetables [16].

Fruit	Tomato	Watermelon	Guava	Grapefruit	Papaya	Apricot
Lycopene µg/100 g wet weight	8.8–42	23–72	54	33.6	20–53	<0.1

By 2023, lycopene is expected to generate an economic impact of $133 million according to Industry ARC [17]. The aim of this work is to present a review of optical systems focused on the study of lycopene. For the analysis, an advantages and disadvantages comparison of the following optical systems is presented: the colorimeter, multispectral and hyperspectral imaging systems, high-performance liquid chromatography, and different spectroscopy types (IR, UV-VIS, Raman).

2. Lycopene Nutraceutical Properties and Its Effects on Human Health

Currently, the health sector focuses its research on foods that provide a benefit to humans due to their nutritional composition. Foods high in lycopene play an important role in the prevention of diseases. These include cardiovascular diseases, cataracts, cancer, osteoporosis, male infertility, and peritonitis [13,15,18]. The suggested dose of daily lycopene intake is 30 to 35 mg according to Rao et al. [19], similarly suggesting that daily consumption is 5 to 7 mg to maintain lycopene levels in blood and to combat oxidative stress, as well as the risks of chronic diseases. In the case of cardiovascular diseases, an intake of between 35 and 75 mg is recommended.

Figure 2 shows the mechanism for the prevention of diseases from [20]. The first phase of this mechanism considers the intake of lycopene, which is absorbed from 10 to 30% of the total content in the diet and the lifetime of this, in the blood, is 2 to 3 days [21–23]. Lycopene allows a reduction of reactive oxygen species (ROS), as it works as a mechanism to prevent chronic diseases. This is achieved with the increase of lycopene levels in the human body. Among the benefits provided are the regulation of gap genes, improvement of intercellular communication, hormone modulation, metabolism regulation and its improvement for the immune response, participation in carcinogenic metabolism, and in the metabolic pathway through the induction of enzymes.

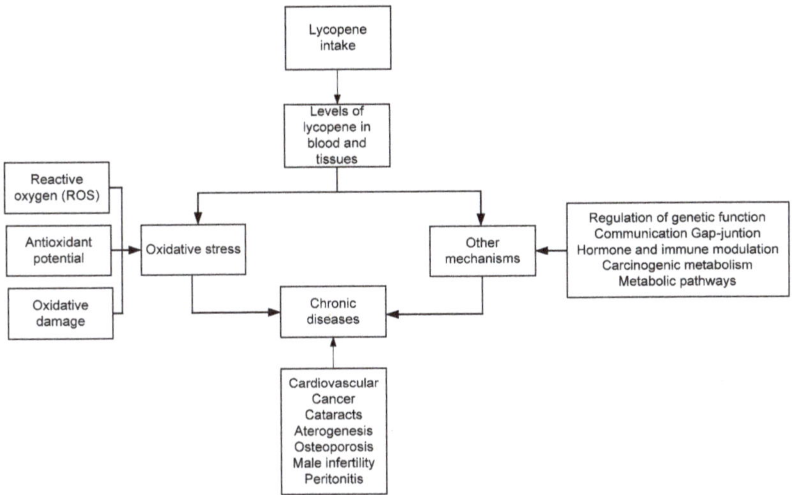

Figure 2. Role of lycopene in the prevention of chronic diseases [20].

3. Effects in the Biological and Physical-Chemical Properties of Lycopene by Electromagnetic Wave Radiation

The chemical structure of the lycopene molecule consists of 11 double bonds that allow the absorption of electromagnetic radiation between 200 and 490 nm [24]. Hashimoto et al. [25] reported that carotenoids absorb in the blue and green regions of the electromagnetic spectrum. The absorption in the blue region is due to the lycopene molecule, which has an outer electron that can move along the main carbon chain. It is conjugated and consists of alternating sequences of double (C=C) and simple carbon bonds (C-C) [26,27]. Radiation in the visible blue region of the lycopene molecule is a factor that supports carotenoid biosynthesis. This can affect the antioxidant content and the product quality [28–30]. Figure 3 shows the chemical structures and absorption spectra of lycopene, α-carotene, β-carotene, and lutein belonging to the same carotenoid's family. A characteristic highlighted of carotenoids is that their absorption spectra generally exhibit similar behaviors, and it has three peak values. In the case of lycopene, maximum absorption in acetone is found at 446, 474, and 504 nm [31,32].

When studying lycopene, it is necessary to consider the possible factors that can damage it. Amongst these are high temperatures, extreme pH values, and oxygen [33]. Radiation use allows inhibition and carotenoid stimulation. Also, the use of distant red-light radiation inhibited carotenoid generation in tomato and, with the red-light radiation, it was stimulated [30].

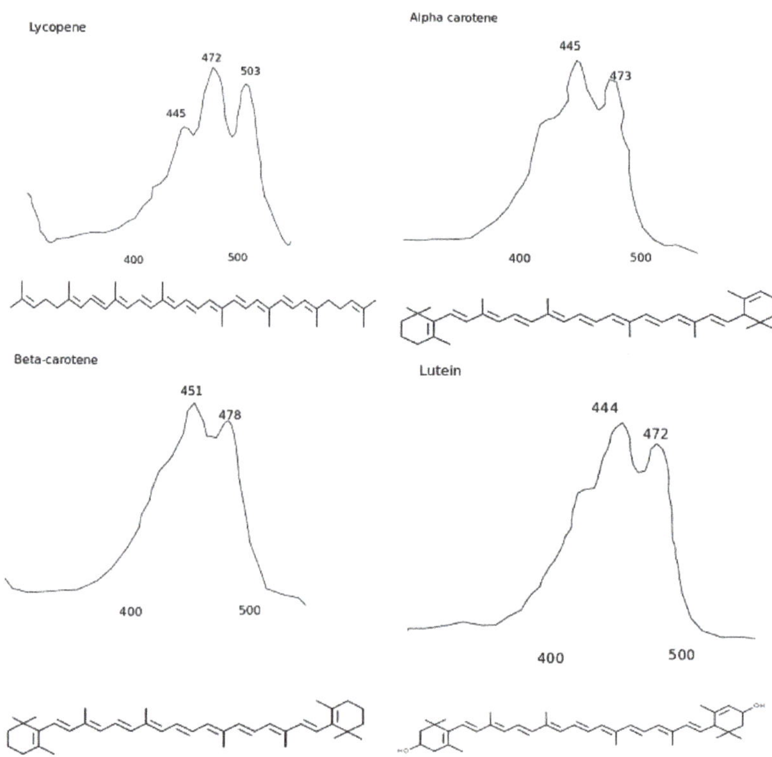

Figure 3. Chemical and molecular structures of selected carotenoids from Zielinska et al. [34].

Liu et al. [34] also established that the increase in carotenoid synthesis in tomatoes is logarithmic. In the study, 30 tomato samples in the breaking maturity stage with three variants of radiation for each analysis were used. The first treatment consisted of radiating the samples with red light (243 mw/cm^2). The second one irradiated the samples with far red light (488 mw/cm^2). The last treatment only isolated them in the dark. All samples were observed for 14 h per day, for a total of 8 days. Finally, they were kept in the dark. The results were: 692.5, 345.6, and 180.8 µg/g for the first, second, and third experiments, respectively.

3.1. Biological and Chemical Effects by Radiation in Lycopene

The food industry demands higher-quality food and durability and lower chemical waste content [35]. The quality of food can be affected by the following factors, such as the amount and intensity of the lighting that it is grown in, the temperatures to which the fruits are exposed, and the CO_2 content in the environment [36]. These are related to the chemical reactions that generate lycopene. This process is called biosynthesis. The optimal temperature for biosynthesis is 22 to 26 °C and the range under which it is affected is 30 to 35 °C [32,37–40]. Lycopene biosynthesis is shown in Figure 4, where the predecessor of this molecule can be seen. These include phytoene and ς-carotene. Molecules derived from lycopene also appear, including δ-carotene, α-carotene, γ-carotene, and β-carotene.

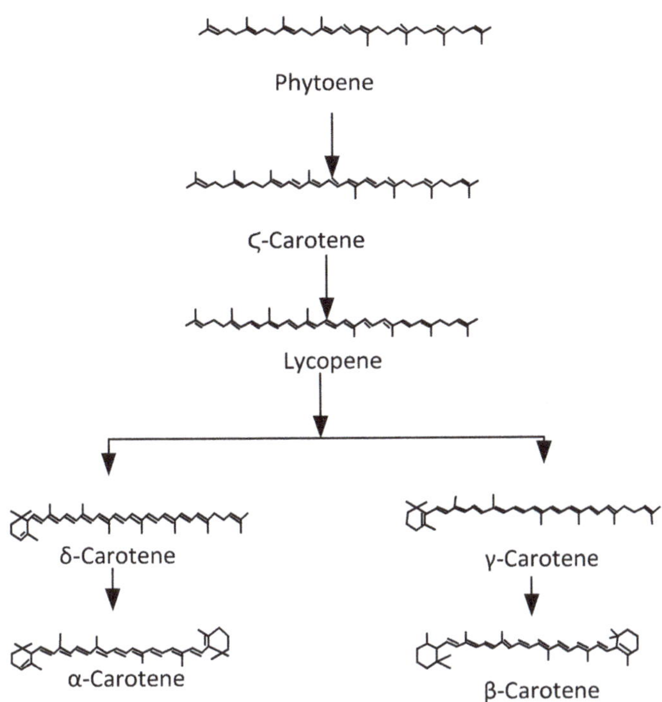

Figure 4. Simplified route of lycopene biosynthesis [39].

3.2. Fluorescent Lighting

Pesek et al. [41] explored the effects of radiation on vegetable juice with fluorescent light. For the investigation, they used two trials, and each trial had 2 g of juice. One sample was treated with fluorescent light at a constant intensity at 230 ft-c and 4 °C while the other sample remained isolated from illumination. The samples initially contained 3.0 µg/g of α-carotene, 8.2 µg/g of β-carotene, and 62.9 µg/g of lycopene. They were exposed to radiation for four days and showed a reduction in antioxidants: 25% for lycopene, and 75% for α-carotene and β-carotene. The authors concluded that the lycopene carbon rings at the ends of its chemical structure protect it from light degradation.

In another experiment, Lee et al. [42] reported lycopene's behavior at different temperatures (50, 100, and 150 °C) and radiated with fluorescent light at 25 °C for different periods. They observed that at 50 °C, isomerization predominated in the first 9 h and then degradation started. In contrast, for conditions between 100 and 150 °C, degradation proceeded faster than isomerization.

3.3. UV-C Radiation

The use of UV-C radiation generates positive biological effects, such as a reduction of food decomposition, and germicidal and antimicrobial effects [43,44]. Bhat [45] explored the impact of treating tomato juice by using ultraviolet radiation (UV-C) at different time intervals (0, 15, 30, and 60 min). In the study, the physicochemical properties, antioxidant activity, microbial load, and the color were evaluated. The researcher determined, by using color analysis, the value of the L* coordinate, which increased significantly, and for the coordinates a* and b*, the values decreased after the treatments with UV-C. Regarding the content of lycopene, no significant changes were perceived.

Liu et al. [34] used UV-C treatments and red and solar light on the tomato. They analyzed the behavior of the carotenoids, skin color, firmness of the tissue, and soluble reactive solids during the experiment. They found that sunlight and red light increased the lycopene content in tomato but UV light degraded it. They also mentioned that soluble solids are not affected by this type of radiation. Another similar study on tomato was proposed by Noga et al. [46]. They used handlings with red light and with short exposure to UV radiation, and achieved an increase of lycopene, ß-carotene, total flavonoids, and phenolic compounds. Table 2 shows the four treatments used and the concentration of lycopene obtained in periods of 5, 10, 15, and 20 days. The control treatment was kept in the dark. For the treatments dark + UV and red light + UV, the samples were irradiated with 4.98 kJ/m^2 of UV light for 30 min. In the treatments red light and red light + UV, tomatoes were irradiated with a special LED lighting module. The spectrum of the light was composed of 60% UV-B (280–320 nm with a dominant peak at 290 nm), 30% UV-A (320–400 nm), 4% UV-C (200–280 nm), and 6% visible light (400–700 nm). The red illumination was applied throughout the storage time, with a peak at 665 nm, equivalent to an available radiation photosynthesis (PAR) of 113 µmol/m^2 per day. The procedure that presented a significant change was the red and UV light, and this occurred between days 5 and 10.

Table 2. Lycopene content, per kg tomato irradiated by red light + UV [34].

Treatment	Exhibition Time	Lycopene Concentration mg/Kg				
		0 Days	5 Days	10 Days	15 Days	20 Days
Dark (Control)	24 h	42.07	44.22	131.92	961.37	871.16
Dark + UV	Only two UV radiation of 15 min daily	-	54.61	49.6 0	855.14	850.84
Red light	24 h	-	40.92	1049.08	1176.50	1507.17
Red light + UV	24 h of red light with two UV radiation of 15 min daily	-	54.65	1280.18	1324.43	1413.91

3.4. γ-Radiation

Kumar et al. [47] experimented with γ-rays and magnetic fields to establish a relationship with the behavior of the biochemical attributes that influences the maturation and quality of the fruits. Radiation with γ-rays showed reductions in maturation, in the synthesis of lycopene, and in the production of ethylene and reactive oxygen species (ROS). Silva-Sena et al. [48] studied the effect of γ-irradiation on the carotenoids and vitamin C contained in papaya. The lycopene of this fruit was not affected by the irradiation but the other carotenoids were. With this, the increase of carotenoids in the ripening of the papaya was delayed.

4. Current Systems for Estimating Lycopene

It is necessary to highlight that these optical systems have elements, such as multiple lenses, mirrors, prisms, and windows. In this sense, the measurement of lycopene is determined with the interaction of light and the optical properties of the samples [49]. This section deals with the optical systems for the identification, estimation, and measurement of lycopene. These can be grouped in high-performance liquid chromatography (HPLC), colorimetry, UV-Vis spectroscopy, IR spectroscopy, Raman spectroscopy, and multispectral and hyperspectral imaging systems. The analysis of these was focused on the mode of acquisition of lycopene information, the region of the electromagnetic spectrum, and the processing of the study sample.

4.1. High-Performance Liquid Chromatography (HPLC)

In the food sector, one of the main techniques that allows identification of the internal composition of food is high-performance liquid chromatography (HPLC). This lycopene measurement technique requires its extraction from the food. This is a process used to obtain an oil that contains the carotenoid. Conventionally, it is done with the use of solvents, heat, and agitation to separate the compounds from the sample. The HPLC technique employs a non-polar stationary phase and a mobile phase that acts as a carrier for the sample can be composed of various solvents. The components of the solution migrate according to the non-covalent interactions of the compounds with the column. These chemical interactions determine the separation of the contents in the sample [50]. Figure 5 shows the basic elements that integrate HPLC. These are the solvents of the mobile phase, the pumping system, the chromatograph, the injection system, the column for liquids, the detectors, waste, the control, and processing. The limit of detection (LOD) and limit of quantitation (LOQ) are important characteristics used to evaluate the efficiency of HPLC. Cámara et al. [51] reported that this method can determine concentrations of lycopene LOD = 0.6 µg. For the case of LOQ = 0.11, this parameter is related to the lowest concentrations with acceptable repeatability and accuracy.

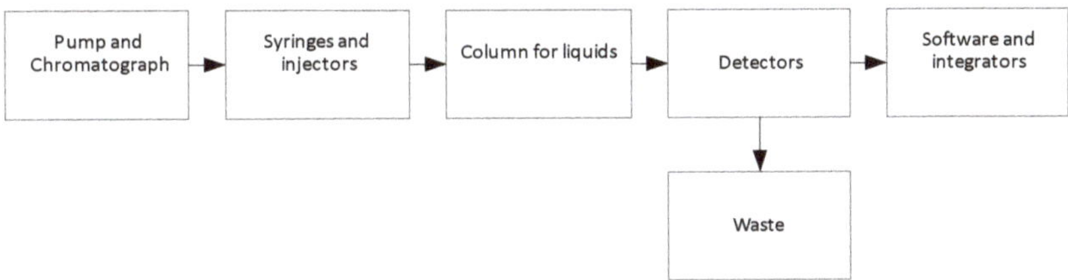

Figure 5. Elements that integrate a high-performance liquid chromatograph.

Figure 6 shows the optical system used by HPLC to identify different study molecules. It has seven elements, which are a UV-lamp (1), lamp mirror (2), flow cell (3), fold mirror (4), programmable or fixed slit (5), grating (6), and diode array detector (DAD) (7). Its operation uses the UV-lamp light, which is focused by the lamp mirror, towards the inlet of the flow cell. Subsequently, the light is guided by the optofluidic waveguides to the opposite end of the flow cell. This is focused on the fold mirror using the holographic grating. Subsequently, the light is scattered over the diode array detector (DAD) in order to have simultaneous access to all the information related to the wavelengths [37].

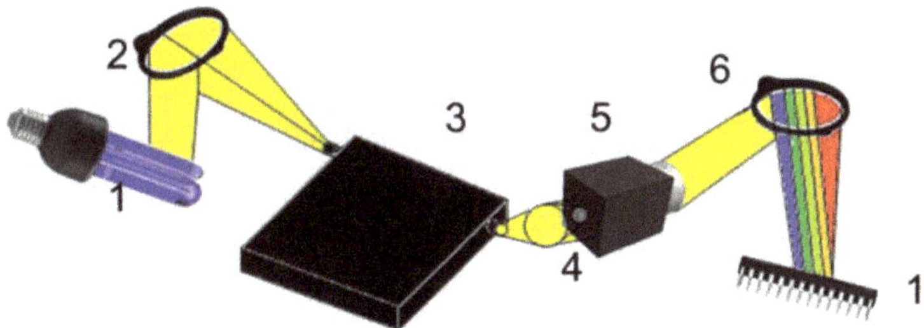

Figure 6. Elements that integrate the optical system of HPLC [37].

4.1.1. Lycopene Extraction Methods

The extraction of lycopene from food is relevant to facilitate the displacement of the same in the mobile phase, and this allows the optical detectors of the HPLC to identify and subsequently quantify the sample to be studied. Perkins-Veazie et al. [52] proposed a method that allows lycopene to be obtained from different varieties of watermelon. They used 2 g of watermelon tissue puree, 50 mL of hexane, 25 mL of acetone, 25 mL of ethanol, and 0.05% (*w/v*), butylated hydroxytoluene (BTH). The mixture was placed in a wrist-action shaker for 10 min at 5 °C and then stirred for another 5 min, adding double-distilled water. The sample settled for 15 min to achieve separation of the polar and non-polar layers. Subsequently, duplicates of 1 mL each of the hexane layer were removed for measurement. In the case of Skoog et al. [50], to extract lycopene from tomato, the samples of the fruit were homogenized by means of a blender for 3 min and the water content was measured by the constant weight method in a vacuum oven at 70 °C. The carotenoids were separated from the homogenate formed by a hexane, acetone, and ethanol mixture (50:25:25). It was stirred for 15 min during phase separation and the sample was settled. After this, the sample was settled. Subsequently, the polar and non-polar layers were filtered and separated [53]. Likewise, Cámara et al. [51] described a method of extracting beta-carotene and lycopene from fresh tomato, tomato sauce, ketchup, tomato juice, tomato puree, carrots, watermelon, peach, green chili, and medlar. The analytical method of extraction of the samples used the mixture of THF/ACN/methanol solvents (15:30:55 *v/v/v*). This extraction process required shaking the samples for 30 min. Subsequently, the lycopene was separated from the food matrix by layers.

4.1.2. Chromatographic Conditions

Cámara et al. and Barba et al. [51,54] handled the same characteristics in their chromatographer in order to quantify lycopene: a µBondapack C18 column (300 mm × 2 mm), 10 µm pore size, a µBondapack C18 precolumn (20 mm × 3.9 mm), 10 µm pore size, mobile phase 90:10 methanol/ACN (*v/v*) + 9 µM TEA, flow rate 0.9 mL/min, temperature of the column of 30 °C, and a detection wavelength λ = 475 nm. In the case of Arias et al. [1], they used a 4.6 × 250 mm, 5 µm polymer C18 column for the isolation of carotenoids, with a mobile phase isocratic system that was composed of methyl alcohol and methyl tert-butyl ether in a ratio of 3:7. The wavelength range used was 420–530 nm, and the length for the identification of the lycopene reported was 471 nm. On the other hand, Perkin et al. [53] used a Microsorb 277 MV-C-18 column with a 470 nm detection wavelength and solvent proportion of acetonitrile (40%), methanol (20%), methylene chloride (20%), and hexane (20%). Pedro et al. [55] determined lycopene and β-carotene (mg kg^{-1}) using a Shimatzu HPLC (Shimatzu, Co., Kyoto, Japan) equipped with a CTO-10A column furnace, a Sil-10A automatic injector, LC-10AD pumps, and a UV-visible SPD-10AV detector at 473 nm. The separation was achieved using a RP18 Zorbax ODS column (5 µm, 15 × 0.46 cm). The mobile phase was MetOH/THF/H$_2$O (67:27:6), Socratic at 1.0 mL/min.

4.2. Colorimeter

Color is one of the main indicators used to identify the maturity and physical and chemical composition of fruits and vegetables [56–59]. The colorimeter is an instrument that uses the CIELAB color space and provides unified measurements. It has a perception closer to that of humans [60–62]. The CIELAB color space consists of a coordinate space of three orthogonal components, which are L* (clarity), a* (redness), and b* (yellowness) [59,62,63].

Various investigations have used color for the measurement of lycopene. Arias et al. [56] analyzed the use of several relationships for the measurement of lycopene in tomato using the coordinates L*, a*, b*, c*, and h*, where the selection of these was in the stages of green maturation, yellowish with some pink regions, orange, soft red, red, intense red in firm fruit, and intense red with soft fruit. Color measurement of tomatoes in the equatorial region of the surface using a Minolta Chroma Meter CR-200 (Minolta) colorimeter (Camera Co. Ltd., Osaka, Japan) determined that the coordinate a* shows a linear correlation with the stages of maturation of the tomato. Splicing was observed in the groups of firm red and intense red, the luminosity factor L* decreased during the first five stages of maturation and then remained constant, and the b* value increased through the first four stages of maturity due to the synthesis of β-carotene and presented a low correlation with the maturity states.

Different models focus on the measurement of lycopene using the color space Cielab. Vazquez-Cruz et al. [1] proposed a model based on an artificial neural network (RNA). Its architecture has six entries, two hidden layers of 13 and eight neurons, respectively, and an output for the prediction of lycopene. The entries used were L*, a*, b*, a*/b*, and the leaf area index (LAI). The learning ratio was 0.4, with a time of 0.6 and the coefficient of the relationship was 0.98. Tilahun et al. [64] performed a prediction model of lycopene and β-carotene in tomatoes, using the color correlation coefficients, which were, a*, a*/b*, and $(a*/b*)^2$. Another investigation was that of Ye et al. [65], who used tomato of the Momotaro breed to create a model for lycopene estimation. They used a gram of tomato tissue, and the quantification was performed spectrophotometrically with a biophotometer (B PM-10 Bio, Taitec Corporation, Saitama, Japan). They used a standard solution of lycopene with concentrations of 0, 2, 5, 10, and 15 ppm and a colorimeter (NF333, Nippon Denshoku Industries Co., Ltd., Tokyo, Japan). The system was performed on an Android 4.2.2 tablet. Its operation consisted of capturing an image or images with the standard colors to calculate the values and reference chromaticity. Subsequently, they determined the color differences between the test fruit and each of the standard colors. The model for estimating lycopene used the color relationship (a*/b*) 2. Table 3 shows the factor with the highest correlation with the lycopene content, which is a*/b*, which was used by [1,54,64]. For the case of Ye et al. [65], they used the same squared correlation factor. Moreover, Hue and Chroma did not present an R^2 correlation lower than 0.7 with the lycopene content as reported by Arias et al. and Ye et al. [56,65].

Table 3. Comparison of lycopene measurement systems using color.

Factor	Arias et al. (2000) Linear Regression R^2	Vazquez-Cruz et al. (2013) Linear Regression R^2	Tilahun et al. (2018) Linear Regression R^2	Xujun Ye and Zhang (2018) Linear Regression R^2
L*	0.90	0.908	0.49	0.83
a*	0.87	0.909	0.92	0.78
a^2*	X	X	X	0.72
b*	0.09	0.02	0.52	0.50
b^2*	x	X	x	0.48
(a*/b*)	0.90	0.980	0.94	0.76
$(a*/b*)^2$	0.86	X	0.42	0.81
Hue Tan $-^1$ (b*/a*)	0.44	X	X	X
Chroma $C* = ((a*)^2 + (b*)^2)^{1/2}$	0.67	X	X	0.36
Lycopene mg/100 g wet weight	0.91	0.586	X	X

X = Descriptor not reported by the author.

4.3. Ultraviolet-Visible (UV-Vis) and Infrared (NIR) Spectroscopy

Spectroscopy is used to identify maturity states and the internal composition of some fruits and vegetables. It uses optical properties, such as the reflectance and absorbance, in a range of 780–2526 nm [66]. An important aspect in spectroscopy is that absorption causes fundamental vibrations. They are related to the functional groups -CH, -NH, -OH, and -SH [67,68]. The (NIR) spectroscopy consists of irradiating the study element with a light that can be reflected, absorbed, or transmitted. Once the measurement of the sample is made, the composition is established, as well as the structure, depending on the amount of light measured and the wavelength [69–71]. Figure 7 shows the elements integrating the optical system of the type of (NIR) spectroscopy: light source of the halogen or tungsten type, a monochromator, a sample holder, and detectors.

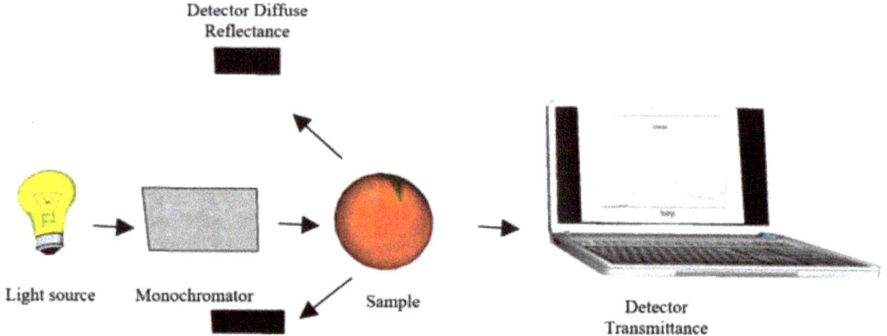

Figure 7. Elements that integrate an optical system of spectroscopy (NIR) [72].

Li et al. [57] identified the maturity stages of the tomato using spectroscopy (Vis-Nir). Through this, the reflectance spectra were obtained in the range 380 to 2500 nm; these present changes in the region of 400–700 nm and allow the fruits to be classified as shown in Figure 8. Similar studies allow the measurement of lycopene, as in the case of Tamburini et al. [73], who developed an online spectroscopy (NIR) system. For this, they selected the range of 900 to 1700 nm with the intention of measuring lycopene, β-carotene, and total soluble solids in watermelon (Citrullus lanatus). This system used the NIR On-Line RX-One (Buchi, Flawil, Switzerland), which is made up of a diode array detector (DAD) and a dual tungsten-halogen lamp. In 2013, they performed tests with watermelons on a conveyor belt without movement. From these, two spectra were obtained from two selected regions. By 2014, the four-sided spectra with the moving band were obtained. Three speeds were used for the band, which were 2100, 2400, and 2700 rpm, to obtain a total of 720 spectra. Finally, in 2015, they obtained 35 spectra randomly, using new fruits. The spectra obtained during the study were treated by the standard normal variance (SNV) and the first derivative for the reduction of noise and unwanted information. With the use of principal component analysis (PCA), they identified three groups of data that offer significant information. These groups are the year of sampling, the climate that is associated with the frequency of rains, extreme temperatures, and the physical-chemical characteristics. One aspect to highlight is that these factors together with the radiation modify the lycopene content in the fruits. The reported correlations for lycopene in the case of R^2_{cal} calibration were 0.877; for the validation R^2_{cv}, 0.756; and finally, for the external validation, R^2_{Ext}, the result obtained was 0.805.

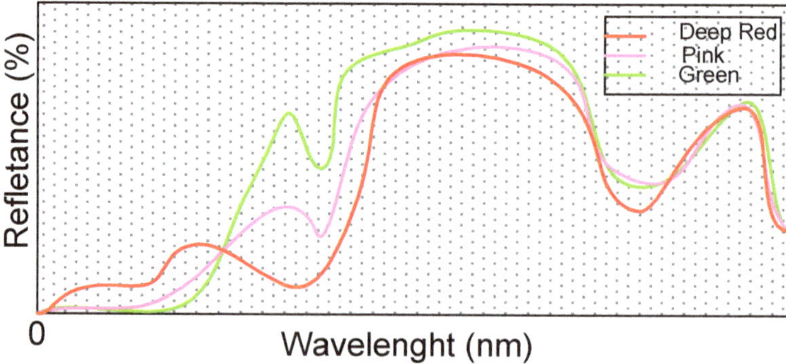

Figure 8. Identification of the maturity of tomato using spectroscopy (NIR) from Li et al. [57].

Tilahun et al. [66] performed a prediction model for lycopene and β-carotene in tomatoes, using a laptop and Vis-NIR spectra. In the development of the model, they used a range of 500–1100 nm and reported some relationship coefficients for the lycopene calibration of R^2_C = 0.89092 and in the case of β-carotene R^2_C = 0.88158. The correlation coefficients for the predictions were R^2_P = 0.85106 and R^2_P = 0.77353. The transmittance spectra of the intact tomatoes were obtained with a Vis-NIR spectrometer (Life & Tech, CO, Ltd., Yongin, Korea) using a halogen lamp as the Vis-NIR light source. The measurement on the fruit was done six times to reduce the noise, obtaining a total of 3500 data for each measurement with a spectrum resolution of 0.2 nm. Together with this, a total of 1160 spectra were obtained, which represent the breaking, pink, and red maturity state of the fruits in the study. For the calibration and cross-validation process, they used 50% of the spectra for each stage. The spectra obtained from the fruits were processed to eliminate unwanted information and noise. Original spectra were treated by a Hanning Window, standard normal variance (SNV), multiplicative dispersion correction (MSC), and the first derivative. Another study focused on lycopene measurement is that of Pedro et al. [55], who analyzed, in a non-destructive way, the soluble solids and carotenoids in tomato products. The tomato spectroscopy NIR was acquired after opening the samples. The aliquot of the fruit was placed at the bottom of a Petri dish (Schott 23 755 48 05), and the readings were performed on a Büchi NIRLab Spectrometer N-200 (Büchi Labortechnik AG, Postfach) equipped with a diffuse reflectance cell MSC-100. The method for the calibration models was that of partial least squares (PLS). For this, 42 samples were used during the calibration stage, with 126 spectra obtained. In the validation, 37 samples and 111 spectra were used. The proposed models used three regions that were 1000 to 1250 nm, 1250 to 1666.67 nm, and 1666.67 to 2500 nm. The original spectra were pre-processed by applying a medium smoother with a window width of 15 wave numbers. Among the proposed models, the main difference is the number of spectrum regions and the number of factors or main components they use. The best model reported by Pedro et al. [55] is the one that used the treatment with multiplicative signal correction (MSC) and the use of the spectrum from 1250 to 2500 nm using 5 factors. The correlation factor was 0.9996 and a mean squares error of prediction of 21.5779.

4.3.1. Raman Spectroscopy

Raman spectroscopy has been used to measure the lycopene content. The technique involves the use of high-energy monochromatic light, such as a laser, which disperses over the molecules and interacts with the photons of the sample. The study matrix does not require a complex preparation and they can be studied in either glass or polymer packaging [74–77]. Typical Raman measurements employ the highest intensity that is related to the longest wavelength and Stokes scattering [77]. The measurement is carried out using the vibrations of the stretches of the C-C bonds (v1) of the polyene chain. The other stretch of interest is the bond C=C and the deformation C-CH3, respectively known as (v2) and (v3) [78–83].

Withnall et al. [83] reported two peaks of greater magnitude for the estimation of carotenoids in tomato. The location of these is in the spectral regions of 1100–1200 and 1400–1600 cm^{-1}. In Figure 9, the study of the behavior of the maturity of the tomato is presented. It can be seen that the peaks coincide with those reported by Withnall et al. [83]. Measurement of the lycopene content was achieved using a portable Raman spectroscopy system and the maximum concentration of lycopene occurred in the state of deep red maturity [84].

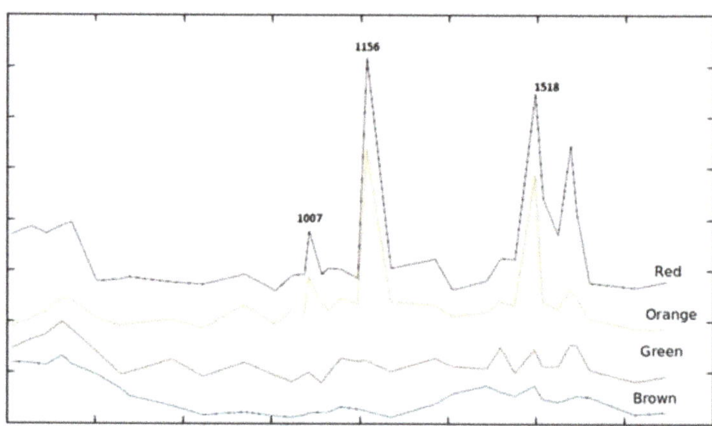

Figure 9. Spectra of tomato maturity behavior using Raman spectroscopy [84].

Qin et al. [85] developed a spatial compensated Raman spectroscopy system (SORS), for a non-destructive evaluation of the internal maturity of the tomatoes. The system shown in Figure 10 consists of a 785 nm laser in the spectra range 4000 to 50,000 nm, a 16-bit camera with 1024 × 256 pixels, and a Raman imaging spectrometer. This system accepts light by means of an input slot 5 mm long by 100 mm wide. The spectrometer works in the range of 779–1144 nm. The focusing unit consists of a band pass filter, a focus lens, and an optic fiber collimator. The biological materials' auto-flowering difficulty was solved by the curve adjustment method called modified polynomial means [86]. The method used to identify the lycopene was through the divergence of spectral information (SID) and relative entropy.

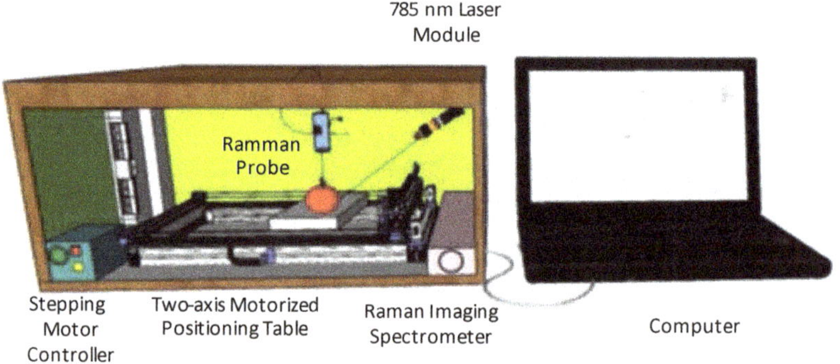

Figure 10. Raman spectroscopy system for lycopene measurement [86].

4.3.2. Multispectral (MSI) and Hyperspectral (HSI) Imaging Systems

The optical systems of multispectral and hyperspectral images are used to acquire both the spatial and spectral information of a product by combining traditional methods of imaging and spectroscopy. With them, internal and external information is obtained from the study matrix [86]. Qin et al. [87] performed the measurement of the optical properties of fruits and vegetables. The multispectral system shown in Figure 11 is made up of a light source and a hyperspectral image unit. The unit contains a focusing lens and a spectrograph. The broadband light was coupled to an optical fiber and a micro lens to generate a 1.0 mm diameter beam, which impinges on the sample at 15° from the vertical direction. The images were captured in online scan mode covering a spectral range between 200 and 1100 nm. The difference between the traditional UV-VIS and NIR spectroscopy techniques is based on the regional study of the matrix. In the case of multispectral and hyperspectral images, they evaluate the entire surface of the element of study, and provide spatial and spectral information [85].

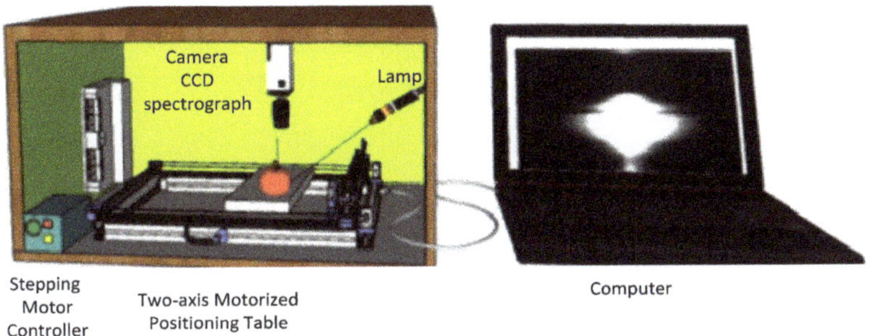

Figure 11. Multispectral system for lycopene measurement from Qin et al. [85].

Liu et al. [9] developed a system using multispectral images with chemo-metric methods to measure lycopene and the phenolic compounds in tomato. This system operated with 19 wavelengths from 405 to 970 nm. The total samples used for calibration and validation were 162 fruits in different stages of maturity. When calibrating, they used two-thirds of the total of fruits and the rest during validation. The proposed models to measure lycopene are partial least squares regression (PLS), the least squares support vector machine (LS-SVM), and the backpropagation neural network (BPNN). According to its results, BPNN obtained the best results for its R^2_C calibration of 0.957 and the R^2_p prediction of 0.938.

The research by Polder et al. [7] measured the superficial distribution of carotenes and chlorophyll in mature tomatoes using imaging spectrometry and PLS. The recorded range was 400 to 700 nm with a resolution of 1 nm. The images were recorded using two illuminators Dolan-Jenner PL900 (Andover St. Lawrence, Massachusetts, USA), with quartz halogen lamps of 150 W in the range of 380 and 2000 nm. The root mean square error of the lycopene concentration in one pixel was 0.95 and for the base tomato it was 0.96. Here, 500 mg samples of the fruit were milled with liquid nitrogen and 4 mL of acetone with 50 mg of $CaCO_3$. To the granules obtained after centrifugation, lycopene was extracted with 4 mL of acetone, 2 mL of hexane, and 5 mL of acetone: hexane (4:1). All solvents contained 0.1% (*w/v*) butylated hydroxytoluene (BHT).

Table 4 summarizes the application of different optical systems for the estimation of the lycopene content. The correlation coefficients are reported with the lycopene, instrumentation, wavelengths for the estimation of the same, treatment of the matrix, type of vegetable and fruit, acquisition method, and range of measurement. In the same table, the use of spectroscopy is highlighted, which allows a good correlation with tomato lycopene of R = 0.996 to be obtained. The proposed system uses the spectrum regions 4000 to 6000 cm^{-1} and 6000 to 8000 cm^{-1}, together with an MSC treatment. Another optical system that has a good correlation with the lycopene content in tomato is vegetable color measurement using the correlation a*/b*. It should be noted that high-performance liquid chromatography is commonly used for the validation of lycopene measurement systems. The relevance is that it allows quantification of the correlation between the states of maturity of the fruits when using the absorbance. The advantage of using this optical system is that it allows the creation of prediction models of lycopene content in tomato sauce, ketchup, tomato juice, tomato puree, carrots, watermelon, peach, green pepper, and medlar. HPLC measurements have the disadvantage that it requires a process of extraction of lycopene in the fruits. In this sense, the process is destructive and involves the use of toxic solvents. Furthermore, an important factor is the choice of a suitable mobile phase to achieve an adequate separation of this carotenoid. The use of spectral images and NIR spectroscopy presents a wide range of estimation of the lycopene content with 2.65–151.75 (mg/kg) [73] and in the case of spectral images it is 7.52–139 mg/g [7].

Table 4. Comparison of lycopene detection and measurement methods.

Food	Technique	Mode of Acquisition	Wavelength of Detection	Treatment of the Matrix	Rank of Measurement Wet Weight	Correlation	Analysis Photochemical of Lycopene	Reference
Watermelon (Citrullus lanatus (Thumb) Matsum & Nakai)	HPLC (190–950 nm)	Absorbance	475 nm	hydroxytoluene, butylated al 0.1% (p/v)	25 a 100 $3\ \mu g/g$	a^* chorma $1000\ a^*/(b^* + L)$	$R^2 = 0.723$ $R^2 = 0.468$ $R^2 = 0.357$ $R^2 = 0.123$	(Perkins-Veazie et al., 2001)
	Colorimeter (400–700 nm)	Absorbance		(BHT) in etanol (1:1, v/v)			$R^2 = 0.666$ $R^2 = 0.451$	
Lycopersicun esculentum Mill cv. Laura	HPLC (420–530 nm)	Absorbance	471 nm	hexane/ acetone/	10.37–29.25 mg/100 g 0.12–1.17 mg/100 g	a^* a^*/b^*	$R^2 = 0.82$ $R^2 = 0.96$	(Arias et al., 2000)
	Colorimeter	Absorbance		ethanol (50:25:25)	1.13–14.32 mg/100 g	$(a^*/b^*)^2$	$R^2 = 0.905$	
Ketchup, juice tomato, tomato pure, carrots, watermelon, green pepper and medlar	HPLC (190–800 nm)	Absorbance	502 y 446 nm	THF/ ACN/ metanol (15:30:55 $v/v/v$)	2–9 (g mL^{-1})	Absorbance values in 446 and 502 nm	$R_c = 0.99\ R_P = 0.99$	(Cámara et al., 2009)
Solanum lycopersicum	HPLC (190–950 nm)	Absorbance	450 nm	0.1% (w/v) butylated, hydroxytoluene (BHT) in ethanol (1:1, v/v)	6.456 105 (7.1×10^{-6}) mg/_L			(Vazquez-Cruz et al., 2013)
	Colorimeter (400–700 nm)	Absorbance				L*, a*, b*, a*/b*, and LAI	$R_P = 0.95$	
Solanum lycopersicum	Spectrophoto-metry	Transmittance	503 nm	acetone, ethanol, and hexan (25:25:50)	2.14–15.57 mg/Kg 2.23–16.67 mg/Kg 6.04–17.46 mg/Kg 2.41–16.19 mg/Kg 2.75–16.23 mg/Kg 6.10–16.40 mg/Kg	a^* a^*/b^* $(a^*/b^*)^2$ a^* a^*/b^* $(a^*/b^*)^2$	$R_c = 0.92$ $R_c = 0.94$ $R_c = 0.42$ $R_P = 0.90$ $R_P = 0.98$ $R_P = 0.52$	(Tilahun et al., 2018)
	Colorimeter (400–700 nm)	Absorbance						
	Spectroscopy VIS/NIR (500–1100 nm)						$R_c = 0.89$ $R_P = 0.85$	

Table 4. *Cont.*

Food	Technique	Mode of Acquisition	Wavelength of Detection	Treatment of the Matrix	Rank of Measurement Wet Weight	Correlation	Analysis Photochemical of Lycopene	Reference
Watermelon Citrullus lanatus	HPLC (195–650 nm)	Absorbance	450 nm	chloroform	2.65–151.75 (mg/kg)		$R_c = 0.756$ $R_P = 0.805$	(Tamburini et al., 2017)
	NIR Process Analyser 900–1700 nm	Reflectance						
Solanum lycopersicum	HPLC (190–600 nm)	Absorbance	473 nm	hexane/acetone/ethanol (50:25:25)	144–921 mg/Kg		$R_c = 0.9996$	(Pedro and Ferreira, 2005)
	Spectrometer 1000–2500 nm	Reflectance	4000–10,000 cm^{-1}.			MSC and segmentation of the spectrum region 4000 to 6000 cm^{-1} and 6000 to 8000 cm^{-1}		
Solanum lycopersicum	Spectrometer of pictures Raman 779–1144 nm	light scattering	1151 y 1513 cm^{-1}	chloroform				(Qin and Lu, 2008; Qin et al., 2001)
Solanum lycopersicum	Spectrograph 396–736 nm	Reflectance	396–736 nm	Nitrogen, Acetone, CaCO$_3$, hexane, (p/v) hydroxytoluene, butylated (BHT)	7.52–139.0 µg/g	Correlation PCA		(Polder et al., 2004)
Solanum lycopersicum	Spectroscopy VIS/NIR	Absorbance	503 nm	hexane/ethanol/acetone				(Liu et al., 2010)
	Videometer 405, 435, 450, 470, 505, 525, 570, 590, 630, 645, 660, 700, 780, 850, 870, 890, 910, 940 and 970 nm	Reflectance	Video-meter wave-lengths	(2:1:1), containing 2.5% BHT (butylated hydroxy toluene)	0.840–36.791 mg/kg	Neural network with 19 wavelengths	$R_c = 0.957$ $R_P = 0.938$	

5. Discussion

According to the literature review, it can be noted that in the food industry, different variants of optical systems are used to mathematically describe the lycopene content in fruits. It can be identified that optical systems that measure the lycopene content in fruits use reflectance, absorbance, light scattering, and transmission as a quantitative variable that detects and quantifies this molecule. This is possible because the frequencies used in this measurement are in the infrared, visible, and/or ultraviolet spectrums. Such is the HPLC case, as it uses the optical property of light absorption. Several methods [1,51,52,55,56,73] use this in focus, where the photodiode array detector of the HPLC operates in the range of 450–503 nm. This range is used to identify the maximum absorption peak of this carotenoid. Another proposal is [51], as they reported the use of two absorption peaks with wavelengths of 446 and 502 nm. In many studies, this equipment (HPLC) has achieved extremely reliable results of an R^2 = 0.99.

The frequency spectrum usage is an important characteristic to optimize lycopene measurement methods. For example, Perkins et al. [52] used an HPLC with a spectrum between 190 and 950 nm and in its detection only used the wavelength around 475 nm. Arias et al. [56] worked with HPLC in the spectrum between 420 and 530 nm and its detection in the wavelength around 471 nm. Cámara et al. [51], Pedro et al. [55], Vazquez-Cruz [1], Tamburini et al. [73], and Tilahun [64] also worked with wide electromagnetic spectrum systems in lycopene detection. It is understood that equipment is designed for quantification at different wavelengths, but by designing equipment in a smaller range, this reduces component costs. Polder et al. [7] used a range of the electromagnetic spectrum between 396–736 nm to recognize lycopene. However, other references make it possible to identify a narrower useful zone between 446 and 503 nm. This allows us to label this space as one range where lycopene can be acknowledged, as shown in Figure 12.

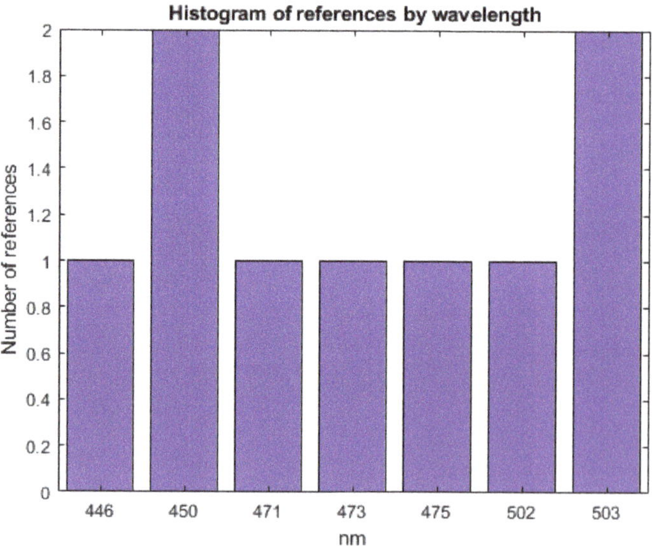

Figure 12. Range of the electromagnetic spectrum most used to detect lycopene.

It was also observed that all current lycopene quantification methods require a standard sample of this molecule to calibrate the equipment [1,52,56]. Additionally, several research teams have calibrated the colorimeter using calibrated HPLC to obtain a non-invasive, portable, and fast system for the measurement of lycopene in fruits [1,53,56].

It was also found that there is a limited amount of work on the estimation of lycopene using image processing in the ultraviolet, visible, and infrared spectra [10]. Although there are many computer vision works that have analyzed the quality, maturity, and defects of fruits, they have not focused on the lycopene content [2–4,58,64,65,88,89]. Raman spectroscopy has also been scarcely explored to solve this problem due to the cost, size, and controlled working conditions [84,85,89]. This could change in the future thanks to advances in technology, as the electronics industry offers us cameras with low cost and sensitivity in the ultraviolet, visible, and near infrared range, as Wu et al. [59] has claimed.

An important phenomenon in the study of the generation and degradation of lycopene in fruits is the effect of UV, VIS, and infrared radiation. Liu et al. [34] is the only study found on this topic.

Another interesting aspect to highlight is the correlation higher than $R^2 = 0.9$ reached by HPLC when the lycopene content in tomato is measured. However, for other fruits, this efficiency drops to $R^2 = 0.72$.

On the other hand, optical absorption is used in coloration quantification of the fruit epicarp. This measurement is related to the change in lycopene content and is a tool to estimate this important molecule. This descriptor is generally determined by using the a/b quantity in CIE L*a*b space, which it is the most sensitive factor in identifying measurements from 10.375 mg/100 g wet weight [56] and the a* component allows measurement of contents of 29.25 mg/100 g wet weight [52]. Between important relations, we note that there is a general increase in reflectance and transmittance in the region of 405 to 780 nm [64,73]. This optical property behavior coincides with the changes in the absorption spectrum reported by [90]. This fact is related to the maturity stage of the fruit and the change in lycopene content.

Although, HPLC continues to be the most reliable equipment (par excellence) to determine the amount of lycopene in fruits. Portable and easy-to-use equipment, such as the colorimeter and artificial vision systems, are beginning to increase in use because they help the agri-food industry to monitor the content of lycopene in fruits from its cultivation to its consumption without destroying the samples. Additionally, its reliability is increasing due to technological tools, such as artificial intelligence, and improvements in sensors, such as low power consumption and resolution.

6. Conclusions

The current review provides a description of the techniques used for quantification of lycopene contents in various fruits: High-performance liquid chromatography (HPLC), colorimetry, NIR spectroscopy, UV-Vis spectroscopy, Raman spectroscopy, multispectral images (MSI), and hyperspectral images (HSI). The HPLC and spectrophotometry methods can provide more efficient results, but their measurement procedures are lengthy and complicated. In contrast, both multispectral, hyperspectral, and colorimeter imaging techniques are fast and non-contact, and suitable for online applications but still face many challenges regarding their accuracy. NIR spectroscopy, UV-Vis spectroscopy, and Raman spectroscopy are moderately reliable with respect to HPLC. Likewise, current techniques have certain limitations that restrict their comprehensive applications in industrial food inspections. A solution that several research teams are implementing to measure lycopene content in the field is the calibration of colorimeters with respect to HPLC. Studies that considered this method of determining lycopene contents demonstrated that the use of the CIELAB model facilitates the calculations and provides high reliability, although the original color model of the sensors is the RGB model. A relevant finding is that the efficiency in the lycopene measurement in tomato turns out to be more efficient than in other fruits. This motivates the discovery of techniques to improve the estimation in various fruits that contain lycopene. In general, the introduction of artificial intelligence algorithms, the internet of things, parallel processing hardware, and the reduction of equipment costs are areas of future study that lead to the early translation of laboratory results to field applications.

Author Contributions: M.-J.V.-A., M.-G.B.-S. and A.-I.B.-G.; methodology, J.-A.P.-M. and J.-E.B.-Á.; validation, A.E.-C.; formal analysis, J.-E.B.-Á. and J.P.-O.; investigation, J.-A.P.-M.; writing—original draft preparation, M.-G.B.-S. and M.-J.V.-A.; writing—review and editing, A.-I.B.-G.; and supervision, A.-I.B.-G. All authors have read and agreed to the published version of the manuscript.

Funding: This research was funded by CONACyT and Tecnológico Nacional de México.

Conflicts of Interest: The authors declare no conflict of interest.

References

1. Vazquez-Cruz, M.A.; Jimenez-Garcia, S.N.; Luna-Rubio, R.; Contreras-Medina, L.M.; Vazquez-Barrios, E.; Mercado-Silva, E.; Torres-Pacheco, I.; Guevara-Gonzalez, R.G. Application of neural networks to estimate carotenoid content during ripening in tomato fruits (*Solanum lycopersicum*). *Sci. Hortic.* **2013**, *162*, 165–171. [CrossRef]

2. Villaseñor-Aguilar, M.J.; Bravo-Sánchez, M.G.; Padilla-Medina, J.A.; Vázquez-Vera, J.L.; Guevara-González, R.G.; García-Rodríguez, F.J.; Barranco-Gutiérrez, A.I. A maturity estimation of bell pepper (*Capsicum annuum* L.) by artificial vision system for quality control. *Appl. Sci.* **2020**, *10*, 5097. [CrossRef]

3. Villaseñor-Aguilar, M.-J.; Botello-Álvarez, J.E.; Pérez-Pinal, F.J.; Cano-Lara, M.; León-Galván, M.F.; Bravo-Sánchez, M.-G. Fuzzy-Classification of the maturity of the tomato using a vision system. *J. Sens.* **2019**, *2019*, 3175848. [CrossRef]

4. Santoyo-Mora, M.; Sancen-Plaza, A.; Espinosa-Calderon, A.; Barranco-Gutierrez, A.I.; Prado-Olivarez, J. Nondestructive Quantification of the Ripening Process in Banana (Musa AAB Simmonds) Using Multispectral Imaging. *J. Sens.* **2019**, *2019*, 6742896. [CrossRef]

5. Chen, P. Use of optical properties of food materials in quality evaluation and materials sorting. *J. Food Process Eng.* **1978**, *2*, 307–322. [CrossRef]

6. Birth, G.S. How light interacts with foods. In *Quality Detection in Foods*; Gaffney, J.J., Ed.; American Society of Agricultural and Biological Engineers: St. Joseph, MI, USA, 1976; Volume 6.

7. Polder, G.; Van Der Heijden, G.W.A.M.; Van der Voet, H.; Young, I.T. Measuring surface distribution of carotenes and chlorophyll in ripening tomatoes using imaging spectrometry. *Postharvest Biol. Technol.* **2004**, *34*, 117–129. [CrossRef]

8. Jahns, P.; Holzwarth, A.R. The role of the xanthophyll cycle and of lutein in photoprotection of photosystem II. *Biochim. Biophys. Acta BBA Bioenerg.* **2012**, *1817*, 182–193. [CrossRef] [PubMed]

9. Liu, C.; Liu, W.; Chen, W.; Yang, J.; Zheng, L. Feasibility in multispectral imaging for predicting the content of bioactive compounds in intact tomato fruit. *Food Chem.* **2015**, *173*, 482–488. [CrossRef]

10. Liu, H.; Mao, J.; Yan, S.; Yu, Y.; Xie, L.; Hu, J.G.; Li, T.; Abbasi, A.M.; Guo, X.; Liu, R.H. Evaluation of carotenoid biosynthesis, accumulation and antioxidant activities in sweetcorn (*Zea mays* L.) during kernel development. *Int. J. Food Sci. Technol.* **2010**, *53*, 381–388. [CrossRef]

11. Quinlan, R.F.; Shumskaya, M.; Bradbury, L.M.T.; Beltrán, J.; Ma, C.; Kennelly, E.J.; Wurtzel, E.T. Synergistic interactions between carotene ring hydroxylases drive lutein formation in plant carotenoid biosynthesis. *Plant Physiol.* **2012**, *160*, 204–214. [CrossRef] [PubMed]

12. Walter, M.H.; Floss, D.S.; Strack, D. Apocarotenoids: Hormones, mycorrhizal metabolites and aroma volatiles. *Planta* **2010**, *232*, 1–17. [CrossRef] [PubMed]

13. Bonet, M.L.; Canas, J.A.; Ribot, J.; Palou, A. Carotenoids and their conversion products in the control of adipocyte function, adiposity and obesity. *Arch. Biochem. Biophys.* **2015**, *572*, 112–125. [CrossRef] [PubMed]

14. Bohn, T.; Desmarchelier, C.; Dragsted, L.O.; Nielsen, C.S.; Stahl, W.; Rühl, R.; Keijer, J.; Borel, P. Host-related factors explaining interindividual variability of carotenoid bioavailability and tissue concentrations in humans. *Mol. Nutr. Food Res.* **2017**, *61*, 1600685. [CrossRef]

15. Pan, W.H.; Yeh, N.H.; Yang, R.Y.; Lin, W.H.; Wu, W.C.; Yeh, W.T.; Sung, M.-K.; Lee, H.-S.; Chang, S.-J.; Huang, C.-J.; et al. Vegetable, fruit, and phytonutrient consumption patterns in Taiwan. *J. Food Drug Anal.* **2016**, *25*, e1–e12. [CrossRef] [PubMed]

16. Rao, A.V.; Agarwal, S. Role of lycopene as antioxidant carotenoid in the prevention of chronic diseases: A review. *Nutr. Res.* **1999**, *19*, 305–323. [CrossRef]

17. Global Carotenoids Market—Premium Insight, Competitive News Feed Analysis, Company Usability Profiles, Market Sizing & Forecasts to 2025. Available online: https://www.marketresearch.com/360iResearch-v4164/Global-Carotenoids-Premium-Insight-Competitive-13036149/ (accessed on 15 June 2021).

18. Desmarchelier, C.; Bore, P. Overview of carotenoid bioavailability determinants: From dietary factors to host genetic variations. *Trends Food Sci. Technol.* **2017**, *69*, 270–280. [CrossRef]

19. Rao, A.V.; Shen, H. Effect of low dose lycopene intake on lycopene bioavailability and oxidative stress. *Nutr. Res.* **2002**, *22*, 1125–1131. [CrossRef]

20. Rao, A.V.; Ray, M.R.; Rao, L.G. Lycopene. *Adv. Food Nutr. Res.* **2006**, *51*, 99–164. [PubMed]

21. Rao, A.V.; Rao, L.G. Carotenoids and human health. *Pharmacol. Res.* **2007**, *55*, 207–216. [CrossRef] [PubMed]

22. Chauhan1, K.; Sharma, S.; Agarwal, N.; Chauhan, B. Lycopene of tomato fame: Its role in health and disease. *Int. J. Pharm. Sci. Rev. Res.* **2011**, *10*, 99–115.

23. Durairajanayagam, D.; Agarwal, A.; Ong, C.; Prashast, P. Lycopene and male infertility. *Asian J. Androl.* **2014**, *16*, 420–425. [PubMed]

24. Mezzomo, N.; Ferreira, S.R.S. Carotenoids Functionality, Sources, and Processing by Supercritical Technology: A Review. *J. Chem.* **2016**, *2016*, 3164312. [CrossRef]

25. Hashimoto, H.; Uragami, C.; Cogdell, R.J. Carotenoids and Photosynthesis. In *Carotenoids in Nature*; Stange, C., Ed.; Springer: Cham, Switzerland, 2016; pp. 111–139.

26. Ermakov, I.V.; Gellermann, W. Optical detection methods for carotenoids in human skin. *Arch. Biochem. Biophys.* **2015**, *572*, 101–111. [CrossRef] [PubMed]

27. Whigham, L.D.; Redelfs, A.H. Optical detection of carotenoids in living tissue as a measure of fruit and vegetable intake. In *2015 37th Annual International Conference of the IEEE Engineering in Medicine and Biology Society (EMBC)*; IEEE: Piscataway, NJ, USA, 2015; pp. 8197–8200.

28. Brecht, J.K.; Saltveit, M.E.; Talcott, S.T.; Schneider, K.R.; Felkey, K.; Bartz, J.A. Fresh-cut vegetables and fruits. *Hortic. Rev. Am. Soc. Hortic. Sci.* **2004**, *30*, 185–251.

29. Khudairi, A.K.; Arboleda, O.P. Phytochrome-mediated Carotenoid Biosynthesis and Its Influence by Plant Hormones. *Physiol. Plant.* **1971**, *24*, 18–22. [CrossRef]

30. Ronnie, L.T.; Jen, J.J. Phytochrome-mediated Carotenoids Biosynthesis in Ripening Tomatoes. *Plant Physiol.* **1975**, *56*, 452–453.

31. Barrett, D.M.; Anthon, G.E. Color Quality of Tomato Products. In *Color Quality of Fresh and Processed Foods*; Culver, C.A., Wrolstad, R.E., Eds.; American Chemical Society: Washington, DC, USA, 2008; pp. 131–139.

32. Meléndez-Martínez, A.J.; Britton, G.; Vicario, I.M.; Heredia, F.J. Relationship between the colour and the chemical structure of carotenoid pigments. *Food Chem.* **2007**, *101*, 1145–1150. [CrossRef]

33. Periago, M.J.; Martinez-Valverde, I.; Ros, G. Propiedades químicas, biológicas y valor nutritivo del licopeno. *An. Vet.* **2001**, *66*, 51–66.

34. Liu, L.H.; Zabaras, D.; Bennett, L.E.; Aguas, P.; Woonton, B.W. Effects of UV-C, red light and sun light on the carotenoid content and physical qualities of tomatoes during post-harvest storage. *Food Chem.* **2009**, *115*, 495–500. [CrossRef]

35. Zielinska, M.A.; Wesołowska, A.; Pawlus, B.; Hamułka, J. Health Effects of Carotenoids during Pregnancy and Lactation. *Nutrients* **2017**, *9*, 838. [CrossRef]

36. Bhat, R.; Ameran, S.B.; Voon, H.C.; Karim, A.A.; MinTze, L. Quality attributes of starfruit (*Averrhoa carambola* L.) juice treated with ultraviolet radiation. *Food Chem.* **2011**, *127*, 641–644. [CrossRef] [PubMed]

37. Agilent Technologies. Agilent 1200 Infinity Series Diode Array Detectors User Manual. 2015. Available online: https://www.agilent.com/cs/library/usermanuals/public/G4212-90013_DAD-A-B_USR_EN.pdf (accessed on 17 July 2021).

38. Dumas, Y.; Dadomo, M.; Di Lucca, G.; Grolier, P. Effects of environmental factors and agricultural techniques on antioxidant content of tomatoes. *J. Sci. Food Agric.* **2003**, *83*, 369–382. [CrossRef]

39. Ilić, Z.S.; Fallik, E. Light quality manipulation improves vegetable quality at harvest and postharvest: A review. *Environ. Exp. Bot.* **2017**, *139*, 79–90. [CrossRef]

40. Burns, J.; Fraser, P.D.; Bramley, P.M. Identification and quantification of carotenoids, tocopherols and chlorophylls in commonly consumed fruits and vegetables. *Phytochemistry* **2003**, *62*, 939–947. [CrossRef]

41. Pesek, A.; Warthesen, J.J. Photodegradation of carotenoids in a vegetable juice system. *J. Food Sci.* **1987**, *52*, 744–746. [CrossRef]

42. Lee, M.T.; Chen, B.H. Stability of lycopene during heating and illumination in a model system. *Food Chem.* **2002**, *78*, 425–432. [CrossRef]

43. Steffen, H.; Zumstein, P.; Rice, R.G. Fruit and Vegetables Disinfection at SAMRO, Ltd. Using Hygienic Packaging by Means of Ozone and UV Radiation. *Ozone Sci. Eng.* **2010**, *32*, 144–149. [CrossRef]

44. Stevens, C.; Khan, V.A.; Wilson, C.L.; Lua, J.Y.; Chalutz, E.; Drobyc, S. The effect of fruit orientation of postharvest commodities following low dose ultraviolet light-C treatment on host induced resistance to decay. *Crop Prot.* **2005**, *24*, 756–759. [CrossRef]

45. Bhat, R. Impact of ultraviolet radiation treatments on the quality of freshly prepared tomato (*Solanum lycopersicum*) juice. *Food Chem.* **2016**, *213*, 635–640. [CrossRef] [PubMed]

46. Noga, L.; Noga, G.; Fiebig, A.; Hunsche, M. Effects of continuous red light and short daily UV exposure during postharvest on carotenoid concentration and antioxidant capacity in stored tomatoes. *Sci. Hortic.* **2017**, *226*, 97–103.

47. Kumar, M.; Ahuja, S.; Dahuja, A.; Kumar, R.; Singh, B. Gamma radiation protects fruit quality in tomato by inhibiting the production of reactive oxygen species (ROS) and ethylene. *J. Radioanal. Nucl. Chem.* **2014**, *301*, 871–880. [CrossRef]

48. Silva-Sena, G.; de Santana, E.N.; Nascimento, R.G.E.; Neto, J.O.; de Oliveira, C.A.; de Figueiredo, S.G. Effect of Gamma Irradiation on Carotenoids and Vitamin C Contents of Papaya Fruit (*Carica papaya* L.) Cv. Golden. *Food Process. Technol.* **2014**, *5*, 337–339.

49. Artal, P.; Schwiegerling, J.; Schwiegerling, J. Handbook of Visual Optics. *Handb. Vis. Opt.* **2017**, *1*, 449–450.

50. Skoog, D.A.; Holler, F.J.; Nieman, T.A. *Principios de Análisis Instrumental*; McGraw-Hill Interamericana de España: Madrid, Spain, 2001.

51. Cámara, M.; Torrecilla, J.S.; Caceres, J.O.; Mata, M.C.S.; Fernández-Ruiz, V. Neural Network Analysis of Spectroscopic Data of Lycopene and β-Carotene Content in Food Samples Compared to HPLC-UV-Vis. *J. Agric. Food Chem.* **2009**, *58*, 72–75. [CrossRef]

52. Perkins-Veazie, P.; Collins, J.K.; Pair, S.D.; Roberts, W. Lycopene content differs among red-fleshed watermelon cultivar. *J. Sci. Food Agric.* **2001**, *81*, 983–987. [CrossRef]

53. Sadler, G.; Dezman, J.D.D. Rapid extraction of lycopene and β-carotene from reconstituted tomato paste and pink grapefruit homogenates. *J. Food Sci.* **1990**, *55*, 1460–1461. [CrossRef]
54. Barba, A.I.O.; Hurtado, M.C.; Mata, M.C.S.; Ruiz, V.F.; Tejada, M.L.S. De Application of a UV–vis detection-HPLC method for a rapid determination of lycopene and $β$-carotene in vegetables. *Food Chem.* **2006**, *95*, 328–336. [CrossRef]
55. Pedro, A.M.K.; Ferreira, M.M. Nondestructive determination of solids and carotenoids in tomato products by near-infrared spectroscopy and multivariate calibration. *ACS Publ.* **2005**, *77*, 2505–2511. [CrossRef]
56. Arias, R.; Lee, T.-C.; Logendra, L.; Janes, H. Correlation of lycopene measured by HPLC with the L*, a*, b* color readings of a hydroponic tomato and the relationship of maturity with color and lycopene content. *J. Agric. Food Chem.* **2000**, *48*, 1697–1702. [CrossRef]
57. Li, B.; Lecourt, J.; Bishop, G. Advances in Non-Destructive Early Assessment of Fruit Ripeness towards Defining Optimal Time of Harvest and Yield Prediction—A Review. *Plants* **2018**, *7*, 3. [CrossRef]
58. Mendoza, F.; Aguilera, J.M. Application of image analysis for classification of ripening bananas. *J. Food Sci.* **2004**, *69*. [CrossRef]
59. Wu, D.; Sun, D.-W. Colour measurements by computer vision for food quality control—A review. *Trends Food Sci. Technol.* **2013**, *29*, 5–20. [CrossRef]
60. Camelo, A.F.L.; Gómez, P.A. Comparison of color indexes for tomato ripening. *Hortic. Bras.* **2004**, *22*, 534–537. [CrossRef]
61. Hobson, G.E.; Adams, P.; Dixon, T.J. Assessing the colour of tomato fruit during ripening. *J. Sci. Food Agric.* **1983**, *34*, 286–292. [CrossRef]
62. Wyszecki, G.; Stiles, W.S. *Color Science*; Wiley: New York, NY, USA, 1982.
63. Bleys, J. *Language Strategies for the Domain of Colour*; Language Science Press, 2016.
64. Va Roy, J.; Keresztes, J.C.; Wouters, N.; De Ketelaere, B.; Saeys, W. Measuring colour of vine tomatoes using hyperspectral imaging. *Postharvest Biol. Technol.* **2017**, *129*, 79–89. [CrossRef]
65. Ye, X.; Izawa, T.; Zhang, S. Rapid determination of lycopene content and fruit grading in tomatoes using a smart device camera. *Cogent Eng.* **2018**, *5*, 1504499. [CrossRef]
66. Tilahun, S.; Park, D.S.; Seo, M.H.; Hwang, I.G.; Kim, S.H.; Choi, H.R.; Jeong, C.S. Prediction of lycopene and $β$-carotene in tomatoes by portable chroma-meter and VIS/NIR spectra. *Postharvest Biol. Technol.* **2018**, *136*, 50–56. [CrossRef]
67. Huang, Y.; Lu, L.R.; Kunjie, C. Prediction of firmness parameters of tomatoes by portable visible and near-infrared spectroscopy. *J. Food Eng.* **2018**, *222*, 185–1984. [CrossRef]
68. Reich, G. Near-infrared spectroscopy and imaging: Basic principles and pharmaceutical applications. *Adv. Drug Deliv. Rev.* **2005**, *57*, 1109–1143. [CrossRef]
69. Nicola, B.M.; Beullens, K.; Bobelyn, E.; Peirs, A.; Saeys, W.; Theron, K.I.; Lammertyn, J. Nondestructive measurement of fruit and vegetable quality by means of NIR spectroscopy: A review. *Postharvest Biol. Technol.* **2007**, *46*, 99–118. [CrossRef]
70. Pan, L.; Lu, R.; Zhu, Q.; McGrath, J.M.; Tu, K. Measurement of moisture, soluble solids, sucrose content and mechanical properties in sugar beet using portable visible and near-infrared spectroscopy. *Postharvest Biol. Technol.* **2015**, *102*, 42–50. [CrossRef]
71. Parker, R.S.; Swanson, J.E.; You, C.S.; Edwards, A.J.; Huang, T. Bioavailability of carotenoids in human subjects. *Proc. Nutr. Soc.* **1999**, *58*, 155–162. [CrossRef] [PubMed]
72. Xie, L.; Ying, Y. Use of near-infrared spectroscopy and least-squares support vector machine to determine quality change of tomato juice. *J. Zhejiang Univ. Sci. B* **2009**, *10*, 465–471. [CrossRef] [PubMed]
73. Tamburini, E.; Costa, S.; Rugiero, I.; Pedrini, P.; Marchetti, M.G. Quantification of Lycopene, Beta-Carotene, and Total Soluble Solids in Intact Red-Flesh Watermelon (Citrullus lanatus) Using On-Line Near-Infrared Spectroscopy. *Sensors* **2017**, *17*, 746. [CrossRef] [PubMed]
74. Eliasson, C.; Matousek, P. Noninvasive Authentication of Pharmaceutical Products through Packaging Using Spatially Offset Raman Spectroscopy. *Anal. Chem.* **2007**, *79*, 1696–1701. [CrossRef] [PubMed]
75. Ellis, D.I.; Brewster, V.L.; Dunn, W.B.; Allwood, J.W.; Golovanov, A.P.; Goodacre, R. Fingerprinting food: Current technologies for the detection of food adulteration and contamination. *Chem. Soc. Rev.* **2012**, *41*, 5706–5727. [CrossRef]
76. Tao, F.; Ngadi, M. Applications of spectroscopic techniques for fat and fatty acids analysis of dairy foods. *Curr. Opin. Food Sci.* **2017**, *17*, 100–112. [CrossRef]
77. Hoskins, L.C. Resonance Raman spectroscopy of beta-carotene and lycopene, a physical chemistry experiment. *J. Chem. Educ.* **1984**, *61*, 4608. [CrossRef]
78. Baranska, M.; Schulz, H.; Nothnagel, T. Tissue-specific accumulation of carotenoids in carrot roots. *Planta* **2006**, *224*, 1028–1037. [CrossRef]
79. da Silva, C.E.; Vandenabeele, P.; Edwards, H.G.M.; de Oliveira, L.F.C. Raman spectra of carotenoids in natural products. *Anal. Bioanal. Chem.* **2008**, *392*, 1489–1496. [CrossRef] [PubMed]
80. de Oliveira, V.E.; Castro, H.V.; Edwards, H.G.M.; de Oliveira, L.F.C. Carotenes and carotenoids in natural biological samples: A Raman spectroscopic analys. *J. Raman Spectrosc.* **2010**, *41*, 642–650. [CrossRef]
81. Merlin, J.C. Resonance Raman spectroscopy of carotenoids and carotenoid-containing systems. *Pure Appl. Chem.* **1985**, *57*, 785–792. [CrossRef]
82. Schulz, H.; Baranska, M. Identification and quantification of valuable plant substances by IR and Raman spectroscopy. *Vib. Spectrosc.* **2007**, *43*, 13–25. [CrossRef]

83. Withnall, R.; Chowdhry, B.Z.; Silver, J.; Edwards, H.G.M.; de Oliveira, L.F.C. Raman spectra of carotenoids in natural products. *Spectrochim. Acta Part A Mol. Biomol. Spectrosc.* **2003**, *59*, 2207–2212. [CrossRef]

84. Trebolazabala, J.; Maguregui, M.; Morillas, H.; de Diego, A.; Madariaga, J.M. Portable Raman spectroscopy for an in-situ monitoring the ripening of tomato (*Solanum lycopersicum*) fruits. *Spectrochim. Acta Part A Mol. Biomol. Spectrosc.* **2017**, *180*, 138–143. [CrossRef] [PubMed]

85. Qin, J.; Chao, K.; Kim, M.S. Evaluating carotenoid changes in tomatoes during postharvest ripening using Raman chemical imaging. In *Sensing for Agriculture and Food Quality and Safety III*; IEEE: Piscataway, NJ, USA, 2011; Volume 8027, p. 802703.

86. Lieber, C.A.; Mahadevan-Jansen, A. Automated Method for Subtraction of Fluorescence from Biological Raman Spectra. *Appl. Spectrosc.* **2003**, *57*, 1363–1367. [CrossRef]

87. Qin, J.; Kim, M.S.; Chao, K.; Chan, D.E.; Delwiche, S.R.; Cho, B.-K. Line-scan hyperspectral imaging techniques for food safety and quality applications. *Appl. Sci.* **2017**, *7*, 125. [CrossRef]

88. Qin, J.; Lu, R. Measurement of the optical properties of fruits and vegetables using spatially resolved hyperspectral diffuse reflectance imaging technique. *Postharvest Biol. Technol.* **2008**, *49*, 355–365. [CrossRef]

89. Gastélum-Barrios, A.; García-Trejo, J.F.; Macías-bobadilla, G.; Toledano-Ayala, M. Portable System to Estimate Ripeness and Lycopene Content in Fresh Tomatoes Based on Image Processing. In *2018 XIV International Engineering Congress (CONIIN)*; IEEE: Piscataway, NJ, USA, 2018.

90. Qin, J.; Chao, K.; Kim, M.S. Investigation of Raman chemical imaging for detection of lycopene changes in tomatoes during postharvest ripening. *J. Food Eng.* **2011**, *107*, 277–288. [CrossRef]

MDPI
St. Alban-Anlage 66
4052 Basel
Switzerland
Tel. +41 61 683 77 34
Fax +41 61 302 89 18
www.mdpi.com

Applied Sciences Editorial Office
E-mail: applsci@mdpi.com
www.mdpi.com/journal/applsci